METHODS IN MOLECULAR BIOLOGY

Series Editor
John M. Walker
School of Life and Medical Sciences
University of Hertfordshire
Hatfield, Hertfordshire, UK

For further volumes:
http://www.springer.com/series/7651

For over 35 years, biological scientists have come to rely on the research protocols and methodologies in the critically acclaimed *Methods in Molecular Biology* series. The series was the first to introduce the step-by-step protocols approach that has become the standard in all biomedical protocol publishing. Each protocol is provided in readily-reproducible step-by-step fashion, opening with an introductory overview, a list of the materials and reagents needed to complete the experiment, and followed by a detailed procedure that is supported with a helpful notes section offering tips and tricks of the trade as well as troubleshooting advice. These hallmark features were introduced by series editor Dr. John Walker and constitute the key ingredient in each and every volume of the *Methods in Molecular Biology* series. Tested and trusted, comprehensive and reliable, all protocols from the series are indexed in PubMed.

Xenotransplantation

Methods and Protocols

Second Edition

Edited by

Cristina Costa

Institut d'Investigació Biomèdica de Bellvitge (IDIBELL), Hospital Duran i Reynals, L'Hospitalet de LLobregat, Barcelona, Spain

Humana Press

Editor
Cristina Costa
Institut d'Investigació Biomèdica de
Bellvitge (IDIBELL)
Hospital Duran i Reynals
L'Hospitalet de LLobregat, Barcelona, Spain

ISSN 1064-3745 ISSN 1940-6029 (electronic)
Methods in Molecular Biology
ISBN 978-1-0716-0257-7 ISBN 978-1-0716-0255-3 (eBook)
https://doi.org/10.1007/978-1-0716-0255-3

This Humana imprint is published by the registered company Springer Science+Business Media, LLC, part of Springer Nature.
The registered company address is: 1 New York Plaza, New York, NY 10004, U.S.A.

Preface

Many major advances have been made in the last decade in the xenotransplantation field, many of them associated with technological progress. Progress continues thanks to the great benefit that xenotransplantation can provide to human health and the conviction of our community of the feasibility of success in multiple clinical applications. The long waiting lists for allogeneic cells, tissues, and organs and the impossibility of the current system to meet the demand clearly justify research in xenotransplantation. This new book focuses on describing technology that is now available to the field for both generating basic knowledge and setting the bases for the improvement and development of clinical applications based on xenotransplantation. It comprises some updated chapters of our previous book (Methods in Molecular Biology 885, 2012), as well as many new chapters describing complementary technology of high relevance to current research.

The complexity of the xenotransplantation field requires specialization of scientists and experts in a variety of aspects that need to be simultaneously studied and addressed to develop a successful clinical application. All potential xenogeneic therapies benefit from the basic work done to characterize the xenogeneic interactions at the cellular and molecular levels. The major contribution of innate immunity to xenograft rejection is well known thanks to multiple studies that have opened new ground to reveal the relevance of cells and molecules of the innate immune system in transplant rejection. Thus, not only proteins but also carbohydrates and possibly free radicals play critical roles in triggering the xenogeneic immune responses. Good tools and protocols are certainly needed for their study. Interestingly, an additional level of complexity is attained using 3D cell culture models that provide highly valuable information of cellular and physiological processes under multiple experimental conditions that mimic the in vivo setting. Importantly, efficient technologies are being developed to engineer the xenogeneic cells, organs, and source animals, particularly at the germ-line level, that should allow the modification of xenografts in order to prevent rejection and facilitate function. The attractive new tools for genome editing based on CRISPR/Cas9 technology are especially suited for this purpose. Lastly, a great amount of preclinical work is required to test the new developments in relevant in vivo models comprising both small animal models and major work in non-human primates such as baboon.

Recent progresses in xenotransplantation have renewed and attracted new interest as the expectations for therapeutic application are increasing for both solid organs and cellular xenografts. The substantial improvements in xenograft survival in the various preclinical studies, especially for solid organ xenotransplantation such as orthotopic heart and kidney, have been thoroughly reviewed in the introduction and pave the way for its clinical application. Each cell type, tissue, and organ of interest for xenotransplantation has its particularities. Thus, teams usually specialize and acquire a profound understanding of each system and the associated technology to help them progress efficiently. In this regard, the book includes an updated description of xenotransplantation models of pig kidney and lung in primate models, as well as a new chapter dedicated to liver xenotransplantation, all presented by experts in each subject. The detailed description of the currently used heart xenotransplantation models can still be found in our first book (Methods in Molecular Biology

885, 2012). Interestingly, a novel tolerogenic approach (based on intra-bone bone marrow transplantation) which could be applied to the different transplantation settings is also included.

The development of new therapies based on the use of scaffolds, cells, and engineered tissue of animal origin is certainly object of intense study. Acellular scaffolds and chemically treated tissues such as bioprosthetic heart valves (mainly of porcine and bovine origin) are broadly used in clinical practice. Xenogeneic cells and avascular tissues are simpler than solid organs and seem to pose less hurdles to attain long-term graft survival. Specific chapters are thus included that describe protocols of cell and tissue preparation and transplantation in relevant animal models. In this accordance, recent accomplishments at the level of clinical trials in xenotransplantation of pancreatic islets and anterior lamellar corneal grafts further encourage to continue with these efforts and lead us to believe that xenotransplantation will be for long an object of interest and research.

In the way toward the clinical application of xenotransplantation, it is also important to keep high safety standards and develop all the necessary tools to prevent potential zoonosis. A great advance has been made in this regard comprising detection and monitoring, as well as the application of genome editing technologies for eliminating porcine retroviruses. Finally, the regulatory, legal, ethical, and educational aspects of the xenotransplantation field are key to the success of xenogeneic applications and its acceptance by society in general.

With this second methodological book dedicated to xenotransplantation, we intend to continue covering many subjects related to this field that can be of special help to the scientific community. The xenotransplantation field will certainly benefit from a tool that helps them integrate the great amount of knowledge needed for progress in this technologically challenging area. We have high expectations for this second book as the first one was already a great success and contributed to the progress in the field. It can be helpful to all scientists, but specially so to the young researchers and students. This support can facilitate their work and encourage them to continue in this promising area. Nonetheless, other scientific communities such as basic researchers and clinicians working in allotransplantation and/or regenerative medicine can learn and obtain valuable information from this new volume.

Barcelona, Spain ***Cristina Costa***

Contents

Contributors

YUICHI ARIYOSHI • *Yamada Laboratory, Department of Surgery, Columbia Center for Translational Immunology, Columbia University Medical Center, New York, NY, USA*

AGNES M. AZIMZADEH • *Center for Transplantation Sciences and Division of Cardiac Surgery, Department of Surgery, Massachusetts General Hospital, Charlestown, MA, USA*

SHIRLEY BACHAR ABRAMOVITCH • *Department of Cell Research and Immunology, The George S. Wise Faculty of Life Sciences, Tel Aviv University, Tel Aviv, Israel*

JI BAO • *Laboratory of Pathology, Key Laboratory of Transplant Engineering and Immunology, NHFPC, West China Hospital, Sichuan University, Chengdu, Sichuan, China*

SUZANNE BERTERA • *Institute of Cellular Therapeutics, Allegheny Health Network Research Institute, Allegheny Health Network, Pittsburgh, PA, USA*

GILLES BLANCHO • *Centre de Recherche en Transplantation et Immunologie, Inserm, Université de Nantes, Nantes, France; Institut de Transplantation Urologie Néphrologie, CHU Nantes, Nantes, France*

RITA BOTTINO • *Institute of Cellular Therapeutics, Allegheny Health Network Research Institute, Allegheny Health Network, Pittsburgh, PA, USA*

HONG BU • *Laboratory of Pathology, Key Laboratory of Transplant Engineering and Immunology, NHFPC, West China Hospital, Sichuan University, Chengdu, Sichuan, China*

LARS BURDORF • *Center for Transplantation Sciences and Division of Cardiac Surgery, Department of Surgery, Massachusetts General Hospital, Charlestown, MA, USA*

KELLY CASÓS • *Infectious Diseases and Transplantation Division, Institut d'Investigació Biomèdica de Bellvitge (IDIBELL), L'Hospitalet de Llobregat, Barcelona, Spain; Department of Cardiac Surgery and Reparative Therapy of the Heart, Vall d'Hebron Research Institute (VHIR), University Hospital Vall d'Hebron, Universitat Autònoma de Barcelona, Barcelona, Spain*

DAVID K. C. COOPER • *Xenotransplantation Program, Department of Surgery, University of Alabama at Birmingham, Birmingham, AL, USA*

CRISTINA COSTA • *Infectious Diseases and Transplantation Division, Institut d'Investigació Biomèdica de Bellvitge (IDIBELL), L'Hospitalet de Llobregat, Barcelona, Spain*

NIEVES DOMÉNECH • *Instituto de Investigación Biomédica A Coruña (INIBIC)-Complejo Hospitalario Universitario A Coruña (CHUAC), Coruña, Spain; Centro de Investigación Biomédica en Red (CIBER) de Enfermedades Cardiovasculares (CIBERCV), Madrid, Spain*

A. FERNÁNDEZ • *National Centre for Biotechnology (CNB-CSIC) and Biomedical Research Networking Centre for Rare Diseases (CIBERER-ISCIII), Madrid, Spain*

MANUEL GALIÑANES • *Department of Cardiac Surgery and Reparative Therapy of the Heart, Vall d'Hebron Research Institute (VHIR), University Hospital Vall d'Hebron, Universitat Autònoma de Barcelona, Barcelona, Spain*

MARC GÜELL • *Translational Synthetic Biology Laboratory, Department of Experimental and Health Sciences, Pompeu Fabra University, PRBB, Barcelona, Spain*

BENT HONORÉ • *Department of Biomedicine, Aarhus University, Aarhus, Denmark*

María Jorqui-Azofra • *Department of Law, Public University of Navarra, Pamplona, Navarra, Spain*
Carmela A. Knoll • *Institute of Cellular Therapeutics, Allegheny Health Network Research Institute, Allegheny Health Network, Pittsburgh, PA, USA*
Michael F. Knoll • *Institute of Cellular Therapeutics, Allegheny Health Network Research Institute, Allegheny Health Network, Pittsburgh, PA, USA*
Sebastián G. Kuguel • *Infectious Diseases and Transplantation Division, Institut d'Investigació Biomèdica de Bellvitge (IDIBELL), L'Hospitalet de Llobregat, Barcelona, Spain*
Yi Li • *Laboratory of Pathology, Key Laboratory of Transplant Engineering and Immunology, NHFPC, West China Hospital, Sichuan University, Chengdu, Sichuan, China*
Jacqueline S. Manji • *Cardiac Sciences Program, I.H. Asper Clinical Research Institute, Winnipeg Regional Health Authority and St. Boniface Hospital, Winnipeg, MB, Canada*
Rizwan A. Manji • *Department of Surgery, University of Manitoba, Winnipeg, MB, Canada; Cardiac Sciences Program, I.H. Asper Clinical Research Institute, Winnipeg Regional Health Authority and St. Boniface Hospital, Winnipeg, MB, Canada*
Maribel Marquina • *Infectious Diseases and Transplantation Division, Institut d'Investigació Biomèdica de Bellvitge (IDIBELL), L'Hospitalet de Llobregat, Barcelona, Spain*
L. Montoliu • *National Centre for Biotechnology (CNB-CSIC) and Biomedical Research Networking Centre for Rare Diseases (CIBERER-ISCIII), Madrid, Spain*
D. Muñoz-Santos • *National Centre for Biotechnology (CNB-CSIC) and Biomedical Research Networking Centre for Rare Diseases (CIBERER-ISCIII), Madrid, Spain*
Nalú Navarro-Alvarez • *Center for Transplantation Sciences, Massachusetts General Hospital, Harvard Medical School, Boston, MA, USA; Department of Gastroenterology and Transplant, Instituto Nacional de Ciencias Médicas y Nutrición Salvador Zubirán, Mexico City, Mexico; Department of Molecular Biology, Universidad Panamericana, Medical School, Mexico City, Mexico*
Vered Padler-Karavani • *Department of Cell Research and Immunology, The George S. Wise Faculty of Life Sciences, Tel Aviv University, Tel Aviv, Israel*
Anu Paul • *Department of Cell Research and Immunology, The George S. Wise Faculty of Life Sciences, Tel Aviv University, Tel Aviv, Israel*
Magdiel Pérez-Cruz • *Infectious Diseases and Transplantation Division, Institut d'Investigació Biomèdica de Bellvitge (IDIBELL), L'Hospitalet de Llobregat, Barcelona, Spain; Division of Blood and Marrow Transplantation, Stanford University Medical Center, Stanford, CA, USA; Immunologie et Neurogénétique Expérimentales et Moléculaires (INEM), Le Studium-CNRS, Orleans, France*
Richard N. Pierson III • *Center for Transplantation Sciences and Division of Cardiac Surgery, Department of Surgery, Massachusetts General Hospital, Charlestown, MA, USA*
Thomas Pomposelli • *Yamada Laboratory, Department of Surgery, Columbia Center for Translational Immunology, Columbia University Medical Center, New York, NY, USA*
Robert Rieben • *Department for BioMedical Research (DBMR), University of Bern, Bern, Switzerland*
Pilar Sánchez-Corral • *Instituto de Investigación Hospital Universitario La Paz (IdiPAZ), Madrid, Spain; CIBER de Enfermedades Raras (CIBERER), Madrid, Spain*
Mitsuhiro Sekijima • *Yamada Laboratory, Department of Surgery, Columbia Center for Translational Immunology, Columbia University Medical Center, New York, NY, USA*

RICCARDO SFRISO • *Department for BioMedical Research (DBMR), University of Bern, Bern, Switzerland; Graduate School for Cellular and Biomedical Sciences, University of Bern, Bern, Switzerland*

ROBERTA SOMMAGGIO • *Infectious Diseases and Transplantation Division, Institut d'Investigació Biomèdica de Bellvitge (IDIBELL), L'Hospitalet de Llobregat, Barcelona, Spain*

KAZUHIRO TAKEUCHI • *Yamada Laboratory, Department of Surgery, Columbia Center for Translational Immunology, Columbia University Medical Center, New York, NY, USA*

MASSIMO TRUCCO • *Institute of Cellular Therapeutics, Allegheny Health Network Research Institute, Allegheny Health Network, Pittsburgh, PA, USA*

MIREIA URIBE-HERRANZ • *Infectious Diseases and Transplantation Division, Institut d'Investigació Biomèdica de Bellvitge (IDIBELL), L'Hospitalet de Llobregat, Barcelona, Spain*

BERTRAND VABRES • *Ophthalmology Department, CHU Nantes, Nantes, France*

PARSIA A. VAGEFI • *Department of Surgery, UT Southwestern Medical Center, Dallas, TX, USA*

BERNARD VANHOVE • *Centre de Recherche en Transplantation et Immunologie, Inserm, Université de Nantes, Nantes, France; Xenothera, Nantes, France*

YUJIA WANG • *Laboratory of Pathology, Key Laboratory of Transplant Engineering and Immunology, NHFPC, West China Hospital, Sichuan University, Chengdu, Sichuan, China*

QIONG WU • *Laboratory of Pathology, Key Laboratory of Transplant Engineering and Immunology, NHFPC, West China Hospital, Sichuan University, Chengdu, Sichuan, China*

KAZUHIKO YAMADA • *Yamada Laboratory, Department of Surgery, Columbia Center for Translational Immunology, Columbia University Medical Center, New York, NY, USA*

Chapter 1

Introduction: The Present Status of Xenotransplantation Research

David K. C. Cooper

Abstract

There is a well-known worldwide shortage of deceased human donor organs for clinical transplantation. The transplantation of organs from genetically engineered pigs may prove an alternative solution. In the past 5 years, there have been sequential advances that have significantly increased pig graft survival in nonhuman primates. This progress has been associated with (1) the availability of increasingly sophisticated genetically engineered pigs; (2) the introduction of novel immunosuppressive agents, particularly those that block the second T-cell signal (costimulation blockade); (3) a better understanding of the inflammatory response to pig xenografts; and (4) increasing experience in the management of nonhuman primates with pig organ or cell grafts. The range of investigations required in experimental studies has increased. The standard immunologic assays are still carried out, but increasingly investigations aimed toward other pathobiologic barriers (e.g., coagulation dysregulation and inflammation) have become more important in determining injury to the graft.

Now that prolonged graft survival, extending to months or even years, is increasingly being obtained, the function of the grafts can be more reliably assessed. If the source pigs are bred and housed under biosecure isolation conditions, and weaned early from the sow, most microorganisms can be eradicated from the herd. The potential risk of porcine endogenous retrovirus (PERV) infection remains unknown, but is probably small. Attention is being directed toward the selection of patients for the first clinical trials of xenotransplantation.

Key words Organs, Pancreatic islets, Pigs, genetically engineered, Xenotransplantation

Abbreviations

CIITA	Dominant-negative mutant major histocompatibility complex (MHC) class II transactivator gene
CMAH-KO	Cytidine monophosphate-*N*-acetylneuraminic acid hydroxylase gene knockout
EPCR	Endothelial cell protein C receptor
Gal	Galactose-α1,3-galactose
GTKO	α1,3-galacosyltransferase gene knockout
hCRP	Human complement regulatory protein
HLA	Human leukocyte antigen
HO-1	Hemeoxygenase-1
Neu5Gc	*N*-glycolylneuraminic acid

Cristina Costa (ed.), *Xenotransplantation: Methods and Protocols*, Methods in Molecular Biology, vol. 2110, https://doi.org/10.1007/978-1-0716-0255-3_1,

PERV	Porcine endogenous retrovirus
SLA	Swine leukocyte antigen
TKO	Triple gene knockout (pigs that do not express the three known pig xenoantigens, Gal, Neu5Gc, or Sda)

1 Introduction

There is a well-known worldwide shortage of deceased human donor organs for clinical transplantation (Fig. 1), and patients on the waiting list for an organ transplant continue to die on a daily basis. In the USA alone, more than 20 such patients die each day [1]. Xenotransplantation (i.e., cross-species transplantation, specifically the transplantation of genetically engineered pig organs into humans) could resolve this problem.

Surgeons, scientists, and physicians have been working toward overcoming the barriers to the transplantation of pig organs in primates for more than three decades [2]. Progress was initially slow, but research activity has been consistently expanding during the past 30 years (Fig. 2).

In the past 5 years, there have been sequential advances that have significantly increased pig graft survival in nonhuman primates. This progress has been associated with (1) the availability of increasingly sophisticated genetically engineered pigs [3, 4]; (2) the introduction of novel immunosuppressive agents, particularly those that block the second T-cell signal (costimulation blockade) [5]; (3) a better understanding of the inflammatory response

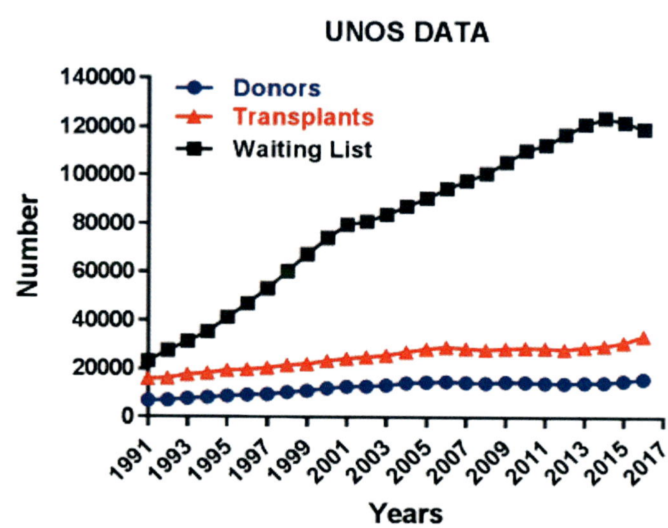

Fig. 1 The current United Network for Organ Sharing (UNOS) data (in the USA) [123] detailing the numbers of patients awaiting organ transplantation, deceased donors, and organ transplants performed over the last 30 years in the USA

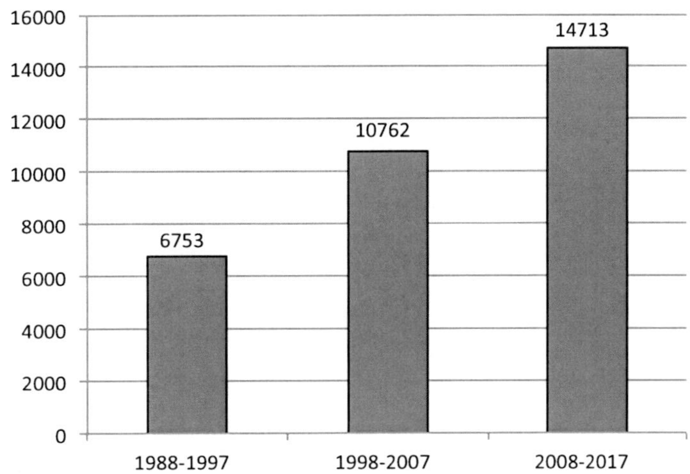

Fig. 2 Number of scientific publications in the field of xenotransplantation (1988–2017)

to a pig xenograft; and (4) increasing experience in the management of nonhuman primates with pig organ or cell grafts.

2 Genetic Engineering of Organ-Source Pigs

The transplantation of an organ from a wild-type (i.e., genetically unmodified) pig into a nonhuman primate initially resulted almost uniformly in hyperacute rejection, the graft generally undergoing antibody-mediated rejection within minutes or hours (Fig. 3a) [6–8]. Experience with the genetic engineering of pigs for the purposes of xenotransplantation began in the 1990s with the introduction of pigs expressing a human complement-regulatory protein (hCRP) [9–11]. Efforts were also made to generate pigs expressing H-transferase alone or in combination with an hCRP [12, 13]. Hearts and kidneys from hCRP-transgenic pigs demonstrated much more resistance to primate complement activation than organs from wild-type pigs, and graft survival increased to days, weeks, or even months, compared with minutes or hours [14–17].

Pigs in which the major carbohydrate xenoantigen, galactose-α 1,3-galactose (Gal) [18] had been knocked out (α1,3-galactosyltransferase gene-knockout [GTKO] pigs), first suggested in 1993 [19], followed [20, 21], and resulted in a similar prolongation of graft function as organs from hCRP-transgenic pigs [22–24]. However, coagulation dysregulation between pig and primate resulted in thrombotic microangiopathy in the grafts (Fig. 3b) [25, 26] and consumptive coagulopathy in the recipient, as had been previously demonstrated following the transplantation of organs from wild-

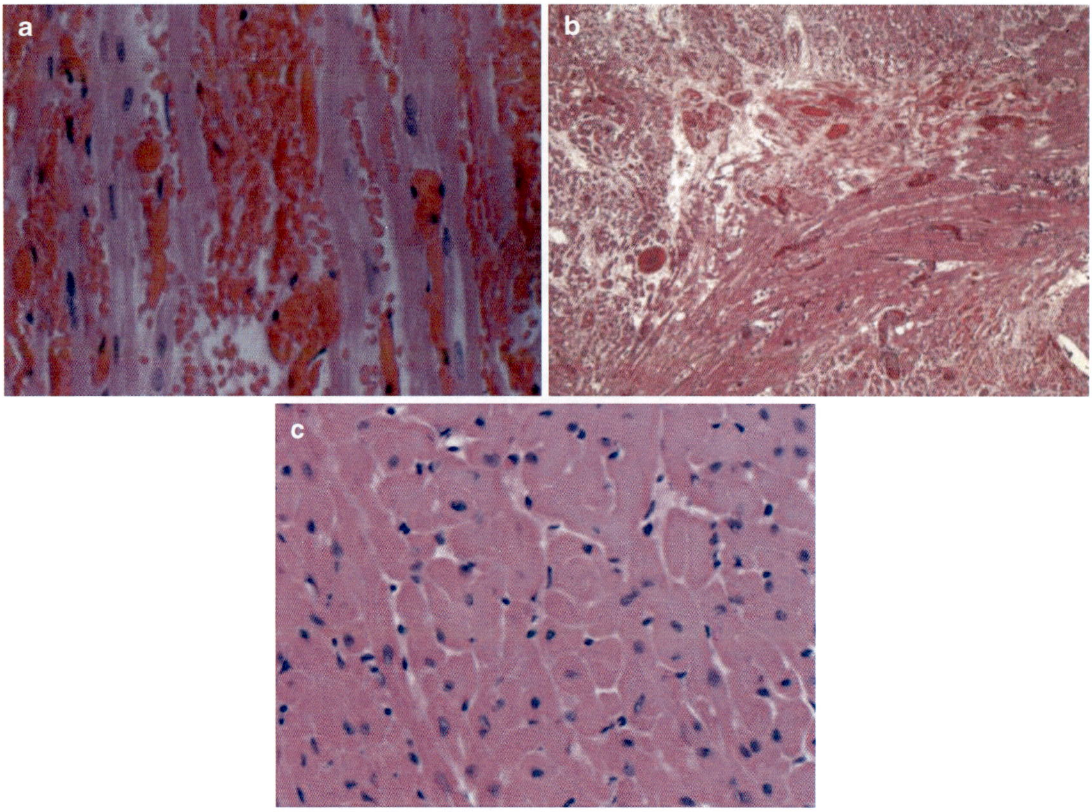

Fig. 3 (**a**) Histopathology of hyperacute rejection in a wild-type (i.e., genetically unmodified) pig heart graft. Complement-mediated injury associated with the binding of baboon natural preformed antibodies to antigens expressed on the vascular endothelium of the pig organ results in intravascular thrombosis, endothelial disruption, and interstitial hemorrhage (H&E, magnification ×400). (**b**) Histopathological features of severe thrombotic microangiopathy in a GTKO pig heart graft (H&E, x100). In thrombotic microangiopathy, platelet and fibrin accumulation occlude vessels and cause ischemic damage to the graft (reproduced with permission from Kuwaki et al. [22]). (**c**) Myocardial biopsy from an immunosuppressed baboon that received a GTKO/CD46/hTBM pig heart one year previously. The myocardium appears normal with no histopathologic changes (H&E ×200) (courtesy Dr. Muhammad Mohiuddin)

type pigs [27, 28] and organs expressing a human complement-regulatory protein [8, 29–31].

This complication was prevented by expression in the pig of one or more human coagulation-regulatory proteins, such as thrombomodulin and/or endothelial protein C receptor (EPCR) (Fig. 3c) ([32–35], reviewed in [36]). Attention was then paid to the documented systemic inflammatory response that developed in the presence of a pig graft [37–42], and human “anti-inflammatory” transgenes (e.g., hemeoxygenase-1 (hHO-1) or A20) were expressed in the pig [43–46].

Over the years, other xenoantigens were identified, and these (N-glycolylneuraminc acid [Neu5Gc; [47, 48]] and Sda [a product of the enzyme β-1,4 *N*-acetylgalactosaminyltransferase 2 gene

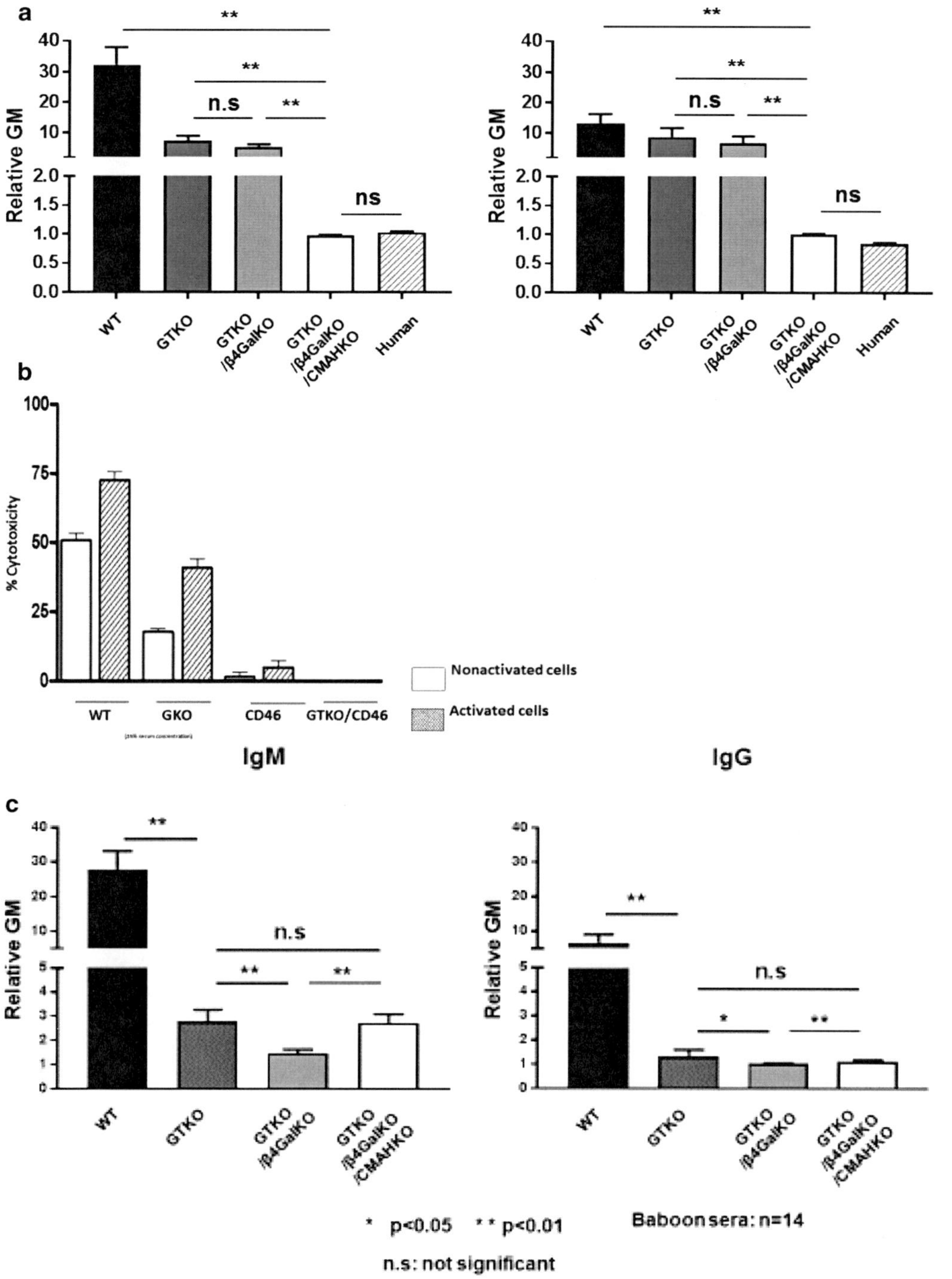

Fig. 4 (**a**) Human IgM (left) and IgG (right) antibody binding to wild-type (WT), GTKO, double-knockout (DKO, i.e., KO of Gal and Sda), and TKO pig red blood cells (RBCs). Binding to TKO pig RBCs was not significantly different from human IgM and IgG binding to human RBCs of blood type O. (**b**) Pooled human serum complement-dependent cytotoxicity (hemolysis) to WT, GTKO, DKO (KO of Gal and Sda), and TKO pig RBCs. Cytotoxicity of the same serum to autologous human O RBCs was tested as a control. (**c**) Baboon (an Old World monkey) IgM and IgG antibody binding to WT, GTKO, DKO (KO of Gal and Sda), and TKO pig RBCs. (**d**) Capuchin monkey (a New World monkey) IgM and IgG antibody binding to WT, GTKO, DKO (KO of Gal and Sda), and TKO pig RBCs

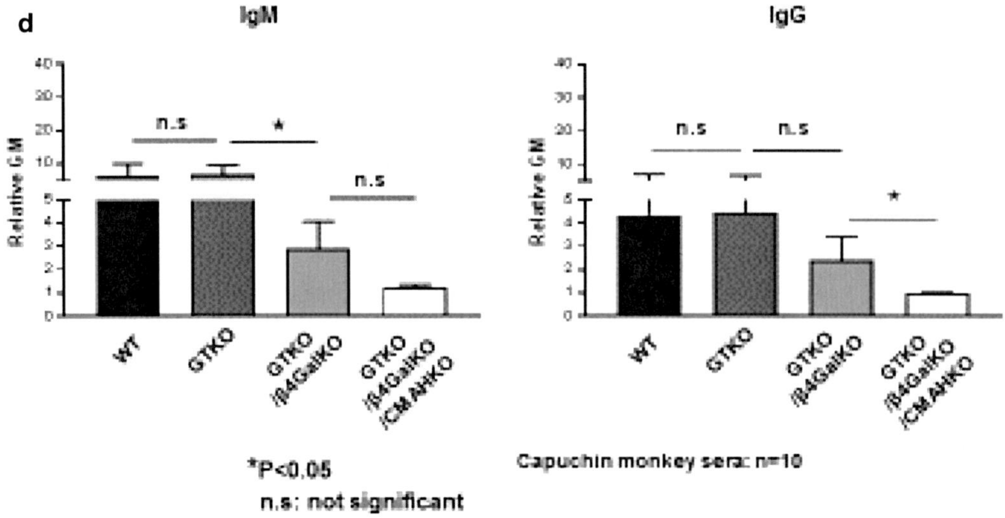

Fig. 4 (continued)

(βGalNT2) [49]] were knocked out, providing us recently with "triple gene-knockout" (TKO) pigs [50], with or without added human transgenes [46]. Human antibody binding to TKO pig cells was greatly reduced, with many human subjects documented to have no antibody (and no cytotoxicity) to these pig cells (Fig. 4a, b). My own group is currently testing the transplantation of organs from pigs with no fewer than nine genetic manipulations (TKO. CD46.CD55.TBM.EPCR.HO-1.CD47) in immunosuppressed baboons. There is considerable in vitro evidence that these pigs should be adequate as sources of organs for initial clinical trials of pig kidney or heart xenotransplantation.

However, there are factors that might make the testing of organs from TKO pigs in nonhuman primates less successful than in human recipients. Like pigs, Old World monkeys, including baboons, express Neu5Gc on their vascular endothelial cells and therefore do not make antibodies against this antigen (unlike humans, who do make anti-Neu5Gc antibodies). When Neu5Gc has been knocked out in the pig, its absence exposes the expression of a hitherto unknown antigen against which Old World monkeys develop natural antibodies (Fig. 4c) [51–53]. By contrast, New World monkeys do not express any natural antibodies against TKO pig cells (Fig. 4d) [51–53] and therefore would be a preferable surrogate for humans as recipients of TKO pig grafts. Unfortunately, the small size of the New World monkeys available for experimental studies (generally <5 kg) would make pig organ transplantation impractical, though they might prove ideal for studies involving the transplantation of TKO pig islets, skin, corneas, and cells, including red blood cells.

3 Controlling the Adaptive Immune Response

Fortunately, the adaptive immune response to a genetically engineered pig organ is weaker than to a wild-type pig organ [54, 55] (Fig. 5). Nevertheless, exogenous immunosuppressive therapy is still required.

The optimal immunosuppressive regimen remains to be determined. A regimen based on blockade of the CD40/CD154 costimulation pathway [5], initially by an anti-CD154 (40 L) monoclonal antibody (mAb) [56] and subsequently by an anti-CD40mAb, has been associated with very prolonged survival of both heterotopic (nonlife-supporting) heart [32, 33] and life-supporting kidney [34, 35, 52, 57] grafts in nonhuman primates. Graft and recipient survival have extended to months or even years. However, these agents are not yet approved for clinical use, and so various combinations of FDA-approved agents continue to be explored, but with less encouraging results [58, 59]. Blockade of the indirect pathway of T-cell activation provided by agents such as anti-CD40mAb may be particularly important to the success of xenotransplantation [5]. However, combinations of FDA-approved (conventional) immunosuppressive agents, including CTLA4-Ig, have not been fully tested in nonhuman primates transplanted with grafts from TKO pigs expressing numerous human transgenes.

The complex interactions of the immune, coagulation, and inflammatory responses to a xenograft are illustrated by the observation that, although inflammatory cytokines (e.g., interferon-

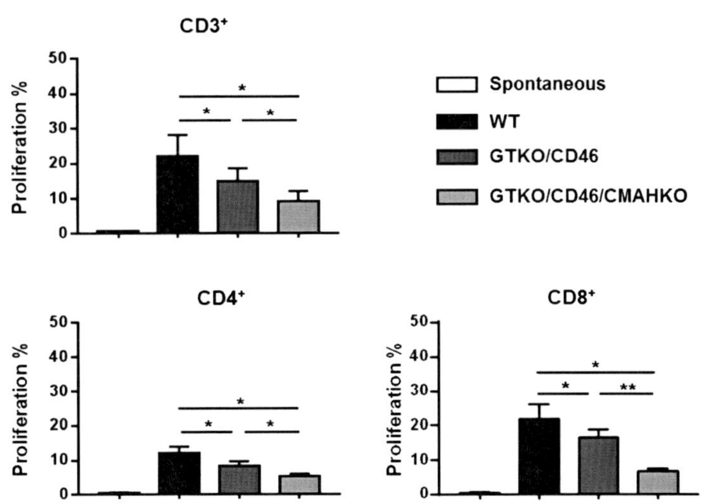

Fig. 5 Human T-cell proliferative response to WT, GTKO.hCD46, and GTKO.hCD46.CMAHKO (Neu5GcKO) pig peripheral blood mononuclear cells (PBMCs) in mixed lymphocyte reaction (MLR) (TKO pig PBMCs were not available to us at that time)

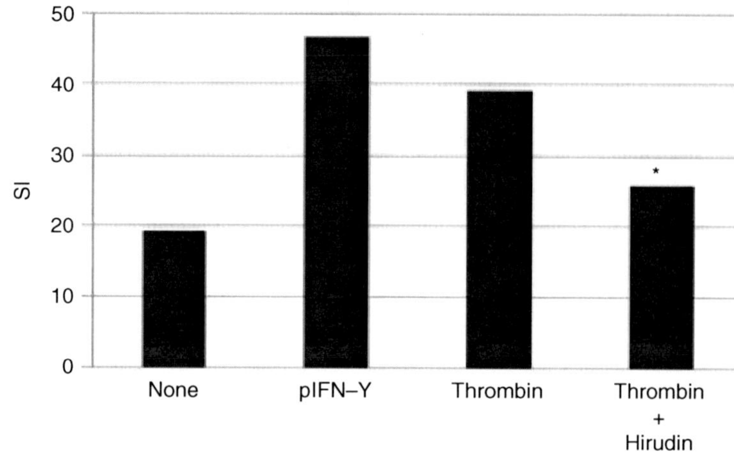

Fig. 6 Thrombin activates the human cellular response to GTKO pig peripheral blood mononuclear cells (PBMCs). The degree of activation of GTKO pig PBMCs by thrombin was comparable to that resulting from stimulation of the cells by porcine interferon-gamma (pIFN-γ). Thrombin-stimulated activation of the human cellular response was reduced by the addition of hirudin, confirming that thrombin was the stimulatory factor (modified from Ezzelarab et al. [60])

gamma (IFN-γ)) activate the T-cell response, thrombin activates this response to the same extent (Fig. 6) [60].

Although most genetic engineering of the pig has been directed to protecting the graft from the innate immune response, there are genetic manipulations that can protect from the adaptive immune response, if found necessary. These manipulations include swine leukocyte antigen (SLA) class I knockout [61] and SLA class II knockdown [62, 63] (as class II knockout is believed to be lethal). Both SLA I and II have been shown to be xenoantigens in some patients [64–66], and elicited antibodies to these targets may develop. In addition, expression of CTLA4-Ig in the graft has a protective immunosuppressive effect [67, 68].

4 The Systemic Inflammatory Response

There is increasing evidence for a systemic inflammatory response to the presence of a pig xenograft that can be prolonged [37–42], but its exact role remains to be investigated further. Prevention or suppression of this response would certainly be beneficial, as inflammation is known to augment the adaptive immune response [69].

Furthermore, inflammation downregulates expression of natural pig thrombomodulin, but not of transgenic human thrombomodulin [70]. The expression of natural pig thrombomodulin was downregulated after exposure to TNFα in vitro, whereas the expression of transgenic human thrombomodulin was upregulated.

Transgenically– expressed human thrombomodulin would appear to be resistant to downregulation by inflammation, and therefore the pig vascular endothelial cells continue to be protected against inflammation.

The beneficial effect of interleukin-6 blockade (e.g., by blockade of IL-6 receptors by tocilizumab) was thought to be important [35, 39], until it was demonstrated that this agent does not block *primate* IL-6 binding to *pig* IL-6 receptors ([71], Zhang G et al.,). Tocilizumab might therefore be detrimental to graft survival (even though very prolonged kidney graft survival has been documented in baboons receiving this agent) [34, 35]. Although we have monitored C-reactive protein as an indicator of the state of systemic inflammation, our recent data suggest that monitoring of serum amyloid A (SAA) may be preferable [42, 72].

Once again, genetic manipulation of the organ-source pig may be preferable to the systemic administration of a drug that suppresses the inflammatory response. There is some evidence that expression of human hemeoxygenase-1 (hHO-1) or A20 is beneficial in protecting the graft from the inflammatory response [43–45].

5 Pig Graft Survival

The above approaches have proved successful in extending life-supporting pig kidney graft survival for several months and even to >400 days (Fig. 7) [34, 35, 52, 57]. Pig heterotopic heart graft survival has extended to a remarkable >900 days (Fig. 7) [33], but these baboons have received continuous heparin infusions, which would be clinically impractical. Recently, using immunosuppressive and adjunctive regimens based on those of Mohiuddin, orthotopic (life-supporting) heart transplantation in the pig-to-baboon model has been reported to maintain survival for up to approximately 6 months (maximum 195 days) [73]. This significant advance on the previous studies in this difficult model was achieved largely by improved ex vivo graft preservation. Although pig liver and lung grafts, which appear to be susceptible to much more pathobiological injury than other organs, have functioned for much shorter periods (<1 month), nevertheless, some progress has been made [1, 74–78].

Pancreatic islet transplantation (in nonhuman primates with streptozotocin-induced diabetes) has maintained normoglycemia for months or years (Fig. 8) [79–81], and in some of the most successful cases, this state has even been achieved with islets from wild-type pigs under an anti-CD154mAb immunosuppressive regimen [81]. Monkeys with a Parkinson-like disorder have been clinically improved by the transplantation of dopamine-producing cells from genetically engineered pigs [82, 83]. Pig corneal

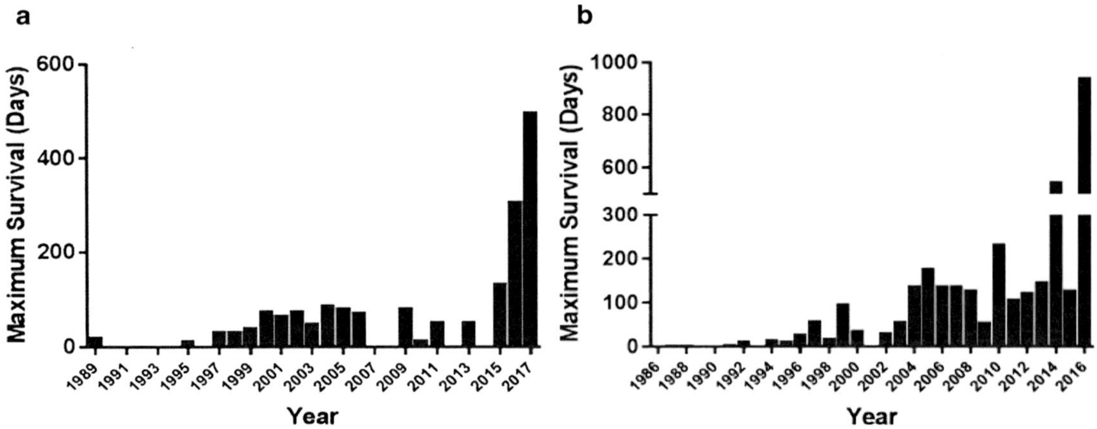

Fig. 7 Survival after pig organ transplantation in nonhuman primates (1986–2017). (**a**) After life-supporting pig kidney xenotransplantation, maximum survival has improved from 23 days in 1989 to >1 year. (**b**) After heterotopic (nonlife-supporting) pig heart xenotransplantation in nonhuman primates, maximum survival has improved from <8 h in 1986 to 945 days

transplants in nonhuman primates have also shown encouraging results [84], and some initial clinical experience has been gained in China [85]. The only other significant clinical trials have been related to the transplantation of encapsulated islets in non-immunosuppressed diabetic patients [86, 87]. Although there was relatively little improvement in the control of diabetes, there have been no safety concerns associated with these trials.

Other aspects of xenotransplantation of considerable potential clinical value relate to (1) pig skin transplantation in the acute treatment of patients with burns (reviewed in [88]); (2) the transfusion of genetically engineered pig red blood cells in patients with massive acute blood loss or in those with chronic hematological disorders, such as sickle cell disease, requiring frequent blood transfusions but in whom allosensitization has developed, making identifying suitable human red blood cell donors increasingly difficult (see below for a discussion on allosensitization in xenotransplantation) (reviewed in [89]); (3) improved survival of biological (bioprosthetic) heart valves in patients with stenotic or incompetent native heart valves requiring replacement (reviewed in [90]); and (4) implantation of porcine chondrocytes to repair cartilage defects (reviewed in [91]). In all of the above cases, TKO pig xenografts (with or without added protective transgenes) offer major improvements.

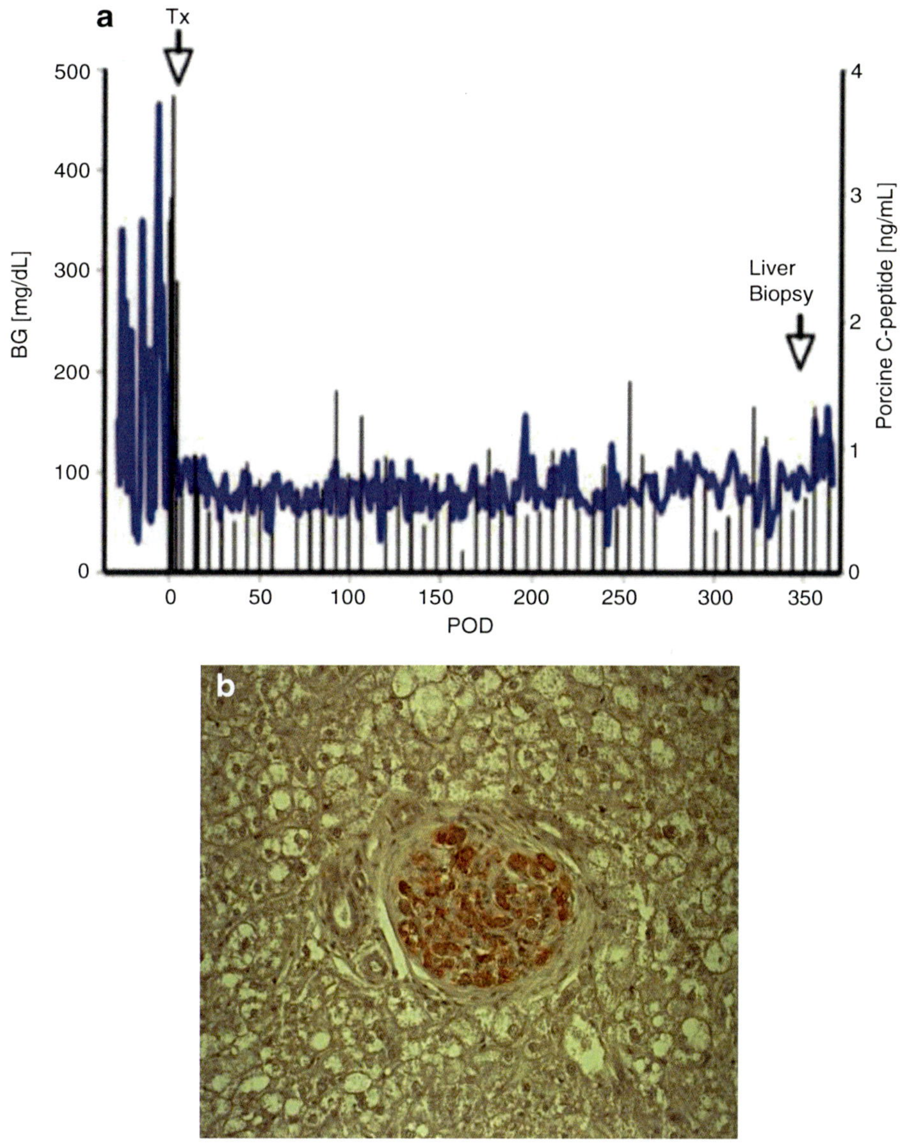

Fig. 8 (**a**) Blood glucose and pig C-peptide levels in a streptozotocin-induced diabetic cynomolgus monkey before and after intraportal transplantation of islets from a pig expressing the human complement-regulatory protein, CD46. No exogenous insulin was administered after the transplant. The normoglycemic monkey was electively euthanized after 12 months. Day 0 = day of islet transplantation (reproduced with permission from Bottino et al. [80]). (**b**) Insulin immunostaining (in red) of a liver section in a monkey recipient of islets from a pig transgenic for human CD46, showing a healthy pig islet 12 months after transplantation (magnification ×200) (reproduced with permission from van der Windt et al. [79])

6 Laboratory Assays

As one would expect, the range of investigations required in the above numerous experimental studies has increased. The standard immunologic assays (e.g., serum levels of anti-pig antibodies (by flow cytometry), T- and B-cell subset counts (by flow cytometry), serum cytotoxicity assays, and mixed lymphocyte reaction (MLR)) are still carried out, but increasingly investigations aimed toward other pathobiologic barriers (e.g., coagulation dysregulation (Fig. 9) and inflammation) have become more important in determining injury to the graft. For example, when antibody-mediated rejection is developing, there are negative correlations between platelet count and fibrinogen level and the extent of glomerular and interstitial thrombi and vasculitis in the graft [59].

As genetic engineering techniques become more sophisticated (Table 1), more effort is being directed to providing organ-source pigs with multiple genetic manipulations, all directed to protect the pig organ from the human immune, inflammatory, and coagulation responses (Table 2). In view of possible epigenetic effects, confirmation of expression of a specific human transgene in the actual organ to be transplanted is becoming increasingly important. An absence of the anticipated expression may result in early graft failure (Fig. 10).

Exploration of newer assays and markers/indicators of rejection or graft failure are underway in several laboratories, but factors such as (1) specificity, (2) sensitivity, (3) the rapidity with which a result can be obtained, and (4) cost require assessment in the evaluation. Areas of interest include (1) circulating pig-specific DNA as a biomarker of rejection [92, 93], (2) expression of mature microRNA [94], and (3) analyzing changes in molecular and genomic expression of the cardiac xenograft [95].

7 Physiological Aspects of Pig Organ Xenotransplantation

Until relatively recently, it was not possible to seriously study the function of transplanted pig organs because survival was impacted by the immune response. Now that prolonged graft survival is increasingly being obtained in recipients in which the immune response is being adequately controlled, the function of the graft can be more reliably assessed. To date, most data have been obtained from life-supporting pig kidney grafts (as the majority of heart grafts have been placed heterotopically [in the abdomen] and have not supported the circulation).

Two observations have been made that are worthy of comment. The first is the abnormally rapid growth of pig kidneys in nonhuman primate recipients during the first 3 months after

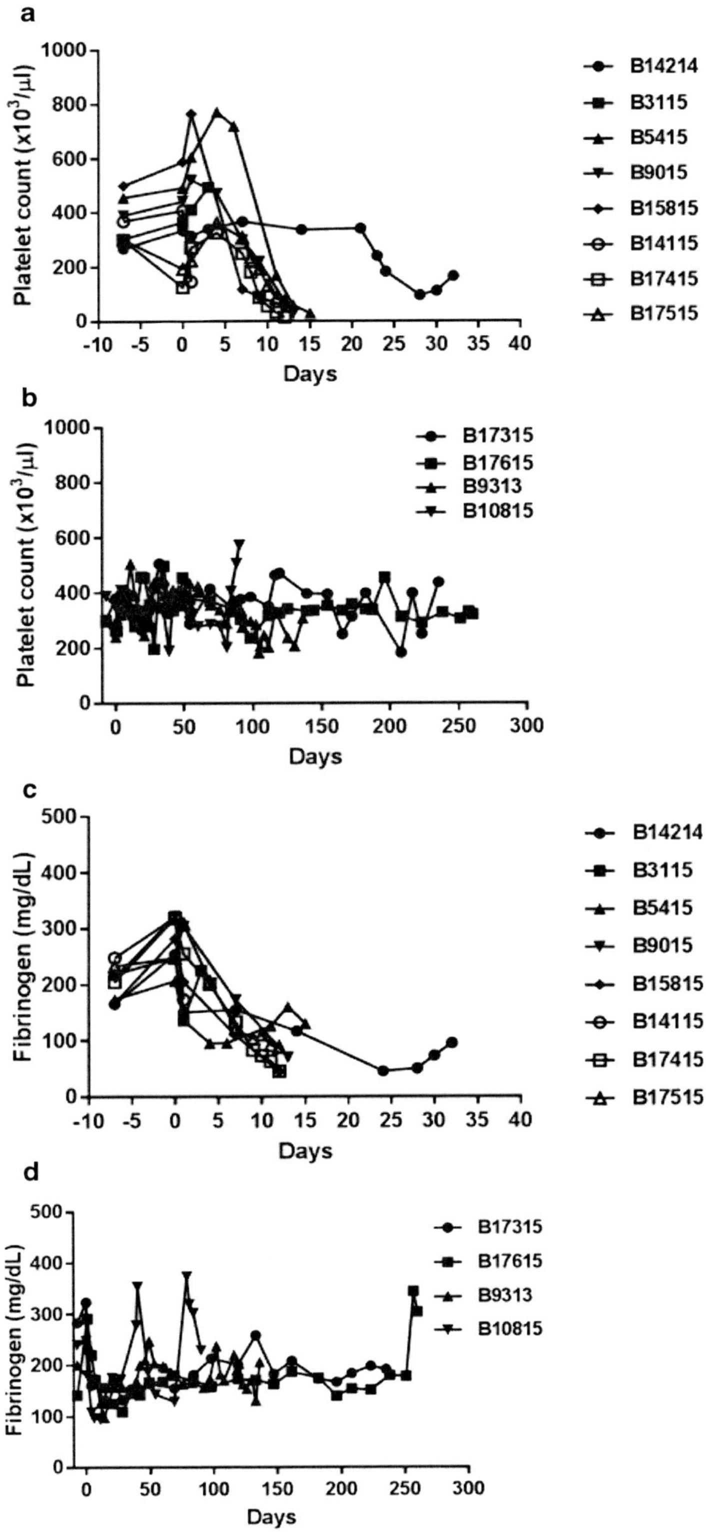

Fig. 9 Changes in key parameters in baboons developing genetically engineered pig kidney graft failure. When the kidney grafts were undergoing early antibody-mediated rejection, falls in (**a**) platelet count and (**c**) plasma fibrinogen were observed, but not in the absence of rejection when graft survival was prolonged (**b** and **d**) (note the differences in the time scale on the *x*-axis)

Table 1
Timeline for application of evolving techniques for genetic engineering of pigs employed in xenotransplantation

Year	Technique
1992	Microinjection of randomly integrating transgenes
2000	Somatic cell nuclear transfer (SCNT)
2002	Homologous recombination
2011	Zinc finger nucleases (ZFNs)
2013	Transcription activator-like effector nucleases (TALENs)
2014	CRISPR/Cas9[a]

[a]CRISPR/Cas9 clustered randomly interspaced short palindromic repeats and the associated protein 9

transplantation (Fig. 11) (reviewed in [96]). The grafts grow at the same rate as they would if still in the rapidly growing pig rather than in the much more slowly growing nonhuman primate. After 2–3 months, the rate of growth declines to the equivalent of that of the recipient primate. The factors associated with this intriguing observation (that has now been made by several groups [34, 35, 97, 98]) are uncertain, but the possibility that some pig growth hormone remains in the graft has been considered. The rapid growth of a pig kidney may not be problematic in the "elastic" confines of the abdomen, but rapid growth of an orthotopically placed pig heart in the chest could result in early graft or lung dysfunction [73]. The problem therefore needs to be resolved. One approach being investigated is to genetically engineer the pig to inactivate the growth hormone receptor gene [99]. Organs from miniature swine would obviously not grow so rapidly, but this case would require much more genetic engineering of miniature swine (to knockout the three xenoantigen genes and introduce multiple human transgenes) than making one further manipulation to the existing multi-genetically engineered pigs.

A second observation that may be of importance is that baboons with long-functioning pig kidneys appear to be unaware that they are becoming hypovolemic and dehydrated, resulting in reductions in central venous and arterial pressures and subsequently in inadequate perfusion of the transplanted kidney (leading to an increase in serum creatinine above normal levels) [100]. No histopathological or laboratory features of rejection are present, and the serum creatinine immediately returns to the normal range when intravenous fluid is administered to increase the venous and arterial pressures. It is known that pig renin does not function in primates (as it does not cleave angiotensinogen to angiotensin-1), and this case is being explored as a possible causative factor to explain this

Table 2
Selected genetically modified pigs currently available for xenotransplantation research

Complement regulation by human complement-regulatory gene expression
CD46 (membrane cofactor protein)
CD55 (decay-accelerating factor)
CD59 (protectin or membrane inhibitor of reactive lysis)
Gal or nonGal antigen "masking" or deletion
Human H-transferase gene expression (expression of blood type O antigen)
Endo-beta-galactosidase C (reduction of Gal antigen expression)
α1,3-galactosyltransferase gene-knockout (GTKO)
Cytidine monophosphate-*N*-acetylneuraminic acid hydroxylase (CMAH) gene knockout (NeuGcKO)
β4GalNT2 (β1,4 *N*-acetylgalactosaminyltransferase) gene knockout (β4GalNT2KO of Sd^{a}KO)
Suppression of cellular immune response by gene expression or downregulation
CIITA-DN (MHC class II transactivator knockdown, resulting in swine leukocyte antigen class II knockdown)
Class I MHC-knockout (MHC-IKO)
HLA-E/human β2-microglobulin (inhibits human natural killer cell cytotoxicity)
Human FAS ligand (CD95L)
Human GnT-III (*N*-acetylglucosaminyltransferase III) gene
Porcine CTLA4-Ig (cytotoxic T-lymphocyte antigen 4 or CD152)
Human TRAIL (tumor necrosis factor-alpha-related apoptosis-inducing ligand)
Anticoagulation and anti-inflammatory gene expression or deletion
Human von Willebrand factor (vWF)
Human tissue factor pathway inhibitor (TFPI)
Human thrombomodulin
Human endothelial protein C receptor (EPCR)
Human CD39 (ectonucleoside triphosphate diphosphohydrolase-1)
Anticoagulation, anti-inflammatory, and anti– apoptotic gene expression
Human A20 (tumor necrosis factor-alpha-induced protein 3)
Human heme oxygenase-1 (HO-1)
Human CD47 (species-specific interaction with SIRP-α inhibits phagocytosis)
Porcine asialoglycoprotein receptor 1 gene-knockout (ASGR1-KO) (decreases platelet phagocytosis)
Prevention of porcine endogenous retrovirus (PERV) activation or PERV-KO
PERV siRNA
PERV-KO

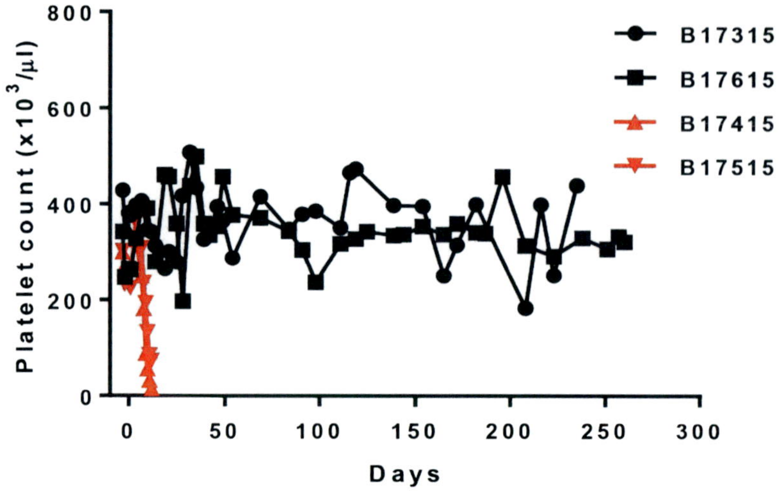

Fig. 10 Rapid development of thrombocytopenia (consumptive coagulopathy) in two baboons with life-supporting GTKO.hCD46 pig kidney grafts (indicated in red) and maintenance of normal platelet counts in two baboons (treated identically) with life-supporting GKTO.hCD46.hEPCR pig kidney grafts (indicated in black) (modified from Iwase et al. [35])

"clinical" observation. After a clinical pig kidney transplant, if the patient ensured his/her fluid intake was sufficient, perhaps the reduction in circulating volume would not occur. The ultimate solution, however, would be to genetically engineer the organ-source pig to produce human renin (in addition to pig renin).

8 Allosensitization and Xenosensitization

All of the pigs being produced for the purposes of xenotransplantation will be of blood group O (non-A), and so ABO-incompatibility will not be problematic. Whether a patient sensitized to human leukocyte antigens (HLA) will be able to undergo a pig kidney transplant without the risk of early antibody-mediated rejection is a topic that has been investigated for some years [101]. Patients who are awaiting an organ from a deceased human donor are at a disadvantage if they have been allosensitized by a previous kidney graft, blood transfusions, or pregnancies. Obtaining a compatible organ may take many years.

For several years, the relatively limited evidence available was mixed in that some studies indicated that these patients would be at no greater risk than non-sensitized patients [101–104], but others suggested they might or would be [101]. Recent data suggest that some patients have anti-HLA antibodies that cross-react with swine leukocyte antigens (SLA) [64–66]. However, it has also been demonstrated that this problem may be solved by genetic

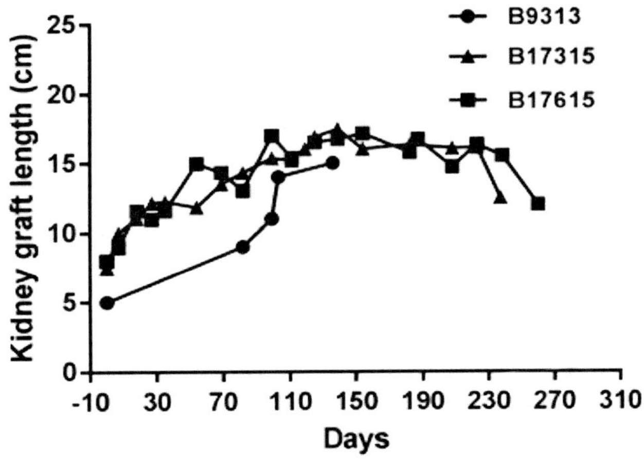

Fig. 11 Increases in the length of the pig kidneys in three baboons with genetically engineered pig kidney grafts that functioned for >136, >237, and >260 days, respectively (all of the experiments being terminated for infectious complications) (reproduced with permission from Iwase et al. [96])

engineering of the pig by replacing specific "incompatible" SLA antigens by "compatible" SLA antigens (or possibly by replacing them with compatible HLA antigens) (Martens G and Ladowski J, personal communications). Thus, ultimately, it should be possible for all allosensitized patients to receive a pig graft with no greater risk than non-sensitized patients.

If a pig xenograft is transplanted into a patient as a bridge to allotransplantation, it needs to be ascertained whether, if the patient becomes sensitized to SLA, this would delay obtaining a compatible allograft or be detrimental to the outcome of allotransplantation. The current, albeit limited, evidence is that allosensitization will *not* result from xenosensitization (reviewed in [105]). If, indeed, some anti-HLA antibodies recognize SLA, then the question needs to be asked as to why no groups have reported a detrimental effect of initial exposure to SLA on the outcome of subsequent allotransplantation. Both HLA and SLA genes encode proteins on the cell surface (antigens), but the HLA genes are at least 100 kbp longer than SLA and are greater than SLA in number [106]. As the HLA system is more complex than the SLA system, the greater number and complexity of anti-HLA antibodies might result in their recognition of SLA. By contrast, anti-SLA antibodies may be too restricted in variety and number to recognize HLA. This hypothesis requires investigation.

9 Safety Aspects of Clinical Xenotransplantation

Xenotransplantation presents a risk that a pig microorganism may be transferred with the organ to the human recipient, resulting in an infectious complication in the recipient. What has raised more concern is the possibility that the patient may transfer that microorganism to his or her close contacts (e.g., family, friends, or even the nursing and medical staff). If the source pigs are bred and housed under biosecure isolation conditions, and weaned early from the sow, then most microorganisms can be eradicated from the herd [107]. Nevertheless, concern has been raised that the porcine endogenous retroviruses (PERVs, that are present within the nucleus of every pig cell) may be pathologic to the human recipient.

There appears to be little or no evidence that PERVs are detrimental to the pig (e.g., they do not appear to be associated with any known infectious or neoplastic condition). Furthermore, humans have equivalent retroviruses in every cell, known as HERVs, which also have not been judged to be factors in any disease process. Nevertheless, concern has been raised that PERVs may be pathologic in humans or may recombine with HERVs to form new pathologic viruses.

Although PERVs have been demonstrated to infect human cells in vitro, this state has only been achieved under specific laboratory conditions, and there has been no evidence of PERV infection of humans after various clinical xenotransplantation procedures (e.g., pig skin or spleen transplantation). Although genetic engineering techniques have been used to prevent PERV activation [108–110] or knockout PERV [111], it is generally considered that the risk of PERV infection is so small that these procedures will not be necessary when organ xenotransplantation enters clinical trials. However, the final decision will be made by the national regulatory authorities (e.g., the Food and Drug Administration (FDA) in the USA) [112]. If genetic manipulation of PERV is required, then this may set back the clinical introduction of xenotransplantation for some considerable time. Furthermore, it is not known if the knockout of 62 PERV genes will damage the pig cell so much that no further genetic manipulation can be made.

The risk of neoplasia in the transplantation of a pig organ/tissues seems very low [113].

10 Selection of Patients for the First Clinical Trials of Xenotransplantation

Increasing thought is being given to this matter [114–116]. Although trials of pig corneal transplantation and encapsulated pig islet transplantation have already been undertaken, to my knowledge, these trials have not involved grafts from genetically engineered pigs or prolonged intensive immunosuppressive therapy.

Our own group believes that pig kidney transplantation may form the basis of the initial clinical trial [117]. There is (1) a great need for kidney transplantation (and thus for "donor" kidneys), and many patients wait for >5 years for a kidney from a deceased human donor to become available (during which period patient mortality is not insignificant), (2) the quality of life of a patient supported by dialysis is poor, (3) the results of *life-supporting* pig kidney transplantation in nonhuman primates are better than of any other life-supporting organ and steadily improving, and (4) if the pig graft fails, then this situation may not be fatal for the patient, who could be "rescued" by dialysis.

We have suggested that the pig kidney transplant might be "preemptive" (i.e., carried out before the patient needs chronic dialysis support (as the results of kidney allotransplantation are better under these conditions)), while the patient would be waiting for a suitable allograft [117]. If the pig kidney graft functions for even a modest period of time (e.g., 12 months or longer), then this development would at least reduce the length of time the patient would require dialysis until an allograft becomes available.

An alternative scenario that we have considered is heart xenotransplantation in neonates with complex, life-threatening congenital heart conditions, such as hypoplastic left heart syndrome. The results of cardiac allotransplantation in this age group are excellent, with an approximate 60% 25-year allograft survival (reviewed in [118]), but it is frequently not possible to obtain a suitable allograft before the baby dies. A pig xenograft would be readily available whenever required. Furthermore, there is evidence that the neonate's immune system is relatively "immature" and thus more easily "manipulated," and therefore possibly less likely to reject a pig graft. Evidence for this scenario is provided by the groundbreaking studies on clinical ABO-incompatible heart transplantation in infants and young children carried out by West and her colleagues [119, 120]. Indeed, if the transplant is combined with thymectomy (which is almost always carried out in infants undergoing heart transplantation) and possibly donor-specific thymus transplantation, then immunological tolerance might be achieved, thus potentially allowing all pharmacologic immunosuppressive therapy to be discontinued [121].

11 Comment

Other topics that require consideration include (1) regulation of xenotransplantation by national regulatory authorities (e.g., the Food and Drug Administration in the USA), and (2) the impact that xenotransplantation will make on healthcare costs.

National regulatory authorities will likely be most concerned with regard to the safety of xenotransplantation (i.e., the risk of an infectious complication), not specifically for the patient, who may well be prepared to take a risk if he/she is close to death, but to the public at large. The benefit of xenotransplantation to a relatively large number of patients (which, if patients with diabetes are included, will run into millions) will be judged against any perceived potential risk to the public. Monitoring the patients for complications that might be associated with infection or neoplasia will be foremost in the considerations of the regulatory authorities.

The impact of xenotransplantation on healthcare costs is relatively unknown as there are many factors to be taken into consideration (e.g., (1) the very considerable reduction in the costs associated with chronic dialysis [117], and/or the maintenance of a very sick patient in an intensive care unit for days or weeks, (2) the negation of the costs associated with the retrieval of deceased human organs [122], and (3) the hitherto unknown acquisition costs of obtaining a genetically engineered pig organ).

Personally, I believe xenotransplantation will prove to be a very major advance in medicine and will ultimately bring curative therapy to millions of patients with potentially life-threatening disease or conditions that seriously impact the quality of life. Indeed, I have frequently labeled it as "the next great medical revolution."

Acknowledgments

Disclosure: The author declares no conflicts of interest.

Funding: Work on xenotransplantation at the University of Alabama at Birmingham is supported in part by NIH grant #U19 AI090959/08.

References

1. Ekser B, Cooper DKC, Tector AJ (2015) The need for xenotransplantation as a source of organs and cells for clinical transplantation. Int J Surg 23(Pt B):199–204
2. Cooper DKC (ed) (2018) Recollections of pioneers in xenotransplantation research. Nova Science Publishers, New York
3. Butler JR, Ladowski JM, Martens GR, Tector M, Tector AJ (2015) Recent advances in genome editing and creation of genetically modified pigs. Int J Surg 23(Pt B):217–222
4. Cooper DKC, Ekser B, Ramsoondar J, Phelps C, Ayares D (2016) The role of genetically-engineered pigs in xenotransplantation research. J Pathol 238:288–299
5. Samy KP, Butler JR, Li P, Cooper DKC, Ekser B (2017) The role of costimulation blockade in solid organ and islet xenotransplantation. J Immunol Res 2017:8415205. https://doi.org/10.1155/2017/8415205. Corrigendum J Immunol Res 2018 doi.org/10.1155/2018/6343608
6. Lexer G, Cooper DKC, Rose AG, Wicomb WN, Rees J, Keraan M, Du Toit E (1986) Hyperacute rejection in a discordant (pig to baboon) cardiac xenograft model. J Heart Transplant 5:411–418
7. Cooper DKC, Human PA, Lexer G et al (1988) Effects of cyclosporine and antibody adsorption on pig cardiac xenograft survival in the baboon. J Heart Transplant 7:238–246
8. Cooper DKC, Ezzelarab MB, Hara H et al (2016) The pathobiology of pig-to-primate xenotransplantation: a historical review. Xenotransplantation 23:83–105
9. Fodor WL, Williams BL, Matis LA et al (1994) Expression of a functional human complement inhibitor in a transgenic pig as a model for the prevention of xenogeneic hyperacute organ rejection. Proc Natl Acad Sci U S A 91:11153–11157
10. White DJG, Langford GA, Cozzi E et al (1995) Production of pigs trangenic for

human DAF: a strategy for xenotransplantation. Xenotransplantation 2:213–217

11. Cozzi E, White DJ (1995) The generation of transgenic pigs as potential organ donors for humans. Nat Med 1:964–966
12. Costa C, Zhao L, Burton WV et al (1999) Expression of the human alpha1,2-fucosyltransferase in transgenic pigs modifies the cell surface carbohydrate phenotype and confers resistance to human serum-mediated cytolysis. FASEB J 13:1762–1773
13. Costa C, Zhao L, Burton WV et al (2002) Transgenic pigs designed to express human CD59 and H-transferase to avoid humoral xenograft rejection. Xenotransplantation 9:45–57
14. Cozzi E, Vial C, Ostlie D et al (2003) Maintenance triple immunosuppression with cyclosporin A, mycophenolate sodium and steroids allows prolonged survival of primate recipients of hDAF porcine renal xenografts. Xenotransplantation 10:300–310
15. Baldan N, Rigotti P, Calabrese F et al (2004) Ureteral stenosis in HDAF pig-to-primate renal xenotransplantation: a phenomenon related to immunological events? Am J Transplant 4:475–481
16. Lambrigts D, Sachs DH, Cooper DK (1998) Discordant organ xenotransplantation in primates - world experience and current status. Transplantation 66:547–561
17. Cooper DK, Satyananda V, Ekser B et al (2014) Progress in pig-to-nonhuman primate transplantation models (1998-2013): a comprehensive review of the literature. Xenotransplantation 21:397–419
18. Good AH, Cooper DKC, Malcolm AJ et al (1992) Identification of carbohydrate structures that bind human antiporcine antibodies: implications for discordant xenografting in man. Transplant Proc 24:559–562
19. Cooper DKC, Koren E, Oriol R (1993) Genetically engineered pigs. Lancet 342:682–683
20. Phelps CJ, Koike C, Vaught TD et al (2003) Production of alpha 1,3-galactosyltransferase-deficient pigs. Science 299:411–414
21. Kolber-Simonds D, Lai L, Watt SR et al (2004) Production of alpha-1,3-galactosyltransferase null pigs by means of nuclear transfer with fibroblasts bearing loss of heterozygosity mutations. Proc Natl Acad Sci U S A 101:7335–7340
22. Kuwaki K, Tseng YL, Dor FJ et al (2005) Heart transplantation in baboons using α1,3-galactosyltransferase gene-knockout pigs as donors: initial experience. Nat Med 11:29–31
23. Tseng YL, Kuwaki K, Dor FJ et al (2005) α 1,3-galactosyltransferase gene-knockout pig heart transplantation in baboons with survival approaching six months. Transplantation 80:1493–1500
24. Yamada K, Yazawa K, Shimizu A et al (2005) Marked prolongation of porcine renal xenograft survival in baboons through the use of α1,3- galactosyltransferase gene-knockout donors and the cotransplantation of vascularized thymic tissue. Nat Med 11:32–34
25. Shimizu A, Hisashi Y, Kuwaki K et al (2008) Thrombotic microangiopathy associated with humoral rejection of cardiac xenografts from alpha1,3-galactosyltransferase gene-knockout pigs in baboons. Am J Pathol 172:1471–1481
26. Shimizu A, Yamada K, Robson SC, Sachs DH, Colvin RB (2012) Pathologic characteristics of transplanted kidney xenografts. J Am Soc Nephrol 23:225–235
27. Ierino FL, Kozlowski T, Siegel JB et al (1998) Disseminated intravascular coagulation in association with the delayed rejection of pig-to-baboon renal xenografts. Transplantation 66:1439–1450
28. Kozlowski T, Shimizu A, Lambrigts D et al (1999) Porcine kidney and heart transplantation in baboons undergoing a tolerance induction regimen and antibody adsorption. Transplantation 67:18–30
29. Houser SL, Kuwaki K, Knosalla C et al (2004) Thrombotic microangiopathy and graft arteriopathy in pig hearts following transplantation into baboons. Xenotransplantation 11:416–425
30. Kuwaki K, Knosalla C, Dor FJ et al (2004) Suppression of natural and elicited antibodies in pig-to-baboon heart transplantation using a human anti-human CD154 mAb-based regimen. Am J Transplant 4:363–372
31. Shimizu A, Yamada K, Yamamoto S et al (2005) Thrombotic microangiopathic glomerulopathy in human decay accelerating factor-transgenic swine-to-baboon kidney xenografts. J Am Soc Nephrol 16:2732–2745
32. Mohiuddin MM, Singh AK, Corcoran PC et al (2014) One-year heterotopic cardiac xenograft survival in a pig to baboon model. Am J Transplant 14:488–489
33. Mohiuddin MM, Singh AK, Corcoran PC et al (2016) Chimeric 2C10R4 anti-CD40 antibody therapy is critical for long-term survival of GTKO.hCD46.hTBM pig-to-primate cardiac xenograft. Nat Commun 7:11138

34. Iwase H, Liu H, Wijkstrom M et al (2015) Pig kidney graft survival in a baboon for 136 days: longest life-supporting organ graft survival to date. Xenotransplantation 22:302–309
35. Iwase H, Hara H, Ezzelarab M et al (2017) Immunological and physiological observations in baboons with life-supporting genetically engineered pig kidney grafts. Xenotransplantation 24. https://doi.org/10.1111/xen.12293
36. Wang L, Cooper DKC, Burdorf L, Wang Y, Iwase H (2018) Overcoming coagulation dysregulation in pig solid organ transplantation in nonhuman primates: recent progress. Transplantation 10:1050–1058
37. Ezzelarab MB, Ekser B, Azimzadeh A et al (2015) Systemic inflammation in xenograft recipients precedes activation of coagulation. Xenotransplantation 22:32–47
38. Ezzelarab MB, Cooper DKC (2015) Systemic inflammation in xenograft recipients (SIXR): a new paradigm in pig-to-primate xenotransplantation? Int J Surg 23(Pt B):301–305
39. Iwase H, Ekser B, Zhou H et al (2015) Further evidence for sustained systemic inflammation in xenograft recipients (SIXR). Xenotransplantation 22:399–405
40. Iwase H, Liu H, Li T et al (2017) Therapeutic regulation of systemic inflammation in xenograft recipients. Xenotransplantation 24(2). https://doi.org/10.1111/xen.12296
41. Li T, Lee W, Hara H et al (2017) An investigation of extracellular histones in pig-to-baboon organ xenotransplantation. Transplantation 101:2330–2339. https://doi.org/10.1097/TP.0000000000001676
42. Li T, Iwase H, Hara H, Long C, Ezzelarab M, Wang Y, Cooper DK (2018) Serum amyloid A as a marker of inflammation in xenotransplantation. Eur J Inflamm. https://doi.org/10.1177/2058739218780046
43. Oropeza M, Petersen B, Carnwath JW et al (2009) Transgenic expression of the human A20 gene in cloned pigs provides protection against apoptotic and inflammatory stimuli. Xenotransplantation 16:522–534
44. Petersen B, Lucas-Hahn A, Lemme E et al (2010) Generation and characterization of pigs transgenic for human hemeoxygenase-1 (hHO-1). Xenotransplantation 17:102–103
45. Petersen B, Ramackers W, Lucas-Hahn A et al (2011) Transgenic expression of human heme oxygenase-1 in pigs confers resistance against xenograft rejection during ex vivo perfusion of porcine kidneys. Xenotransplantation 18:355–368
46. Cooper DKC, Ezzelarab MB, Iwase H, Hara H (2018) Perspectives on the optimal genetically-engineered pig in 2018 for initial clinical trials of kidney or heart xenotransplantation. Transplantation 102 (12):1974–1982. https://doi.org/10.1097/TP.0000000000002443
47. Bouhours D, Pourcel C, Bouhours JE (1996) Simultaneous expression by porcine aorta endothelial cells of glycosphingolipids bearing the major epitope for human xenoreactive antibodies (gal alpha1-3Gal), blood group H determinant and N-glycolylneuraminic acid. Glycoconj J 13:947–953
48. Zhu A, Hurst R (2002) Anti-N-glycolylneuraminic acid antibodies identified in healthy human serum. Xenotransplantation 9:376–381
49. Byrne GW, Du Z, Stalboerger P, Kogelberg H, McGregor CG (2014) Cloning and expression of porcine beta1,4 N-acetylgalactosaminyl transferase encoding a new xenoreactive antigen. Xenotransplantation 21:543–554
50. Estrada JL, Martens G, Li P, Adams A et al (2015) Evaluation of human and non-human primate antibody binding to pig cells lacking GTA1/CMAH/beta4GALNT2 genes. Xenotransplantation 22:194–202
51. Lutz AJ, Li P, Estrada JL et al (2013) Double knockout pigs deficient in N-glycolylneuraminic acid and galactose alpha-1,3-galactose reduce the humoral barrier to xenotransplantation. Xenotransplantation 20:27–35
52. Adams AB, Kim SC, Martens GR et al (2018) Xenoantigen deletion and chemical immunosuppression can prolong renal xenograft survival. Ann Surg 268:564–573. https://doi.org/10.1097/SLA.0000000000002977
53. Li Q, Shaikh S, Iwase H et al (2019) An investigation of carbohydrate antigen expression and anti-pig antibodies in New World capuchin monkeys. Xenotransplantation 26 (3):e12498
54. Wilhite T, Ezzelarab C, Hara H et al (2012) The effect of Gal expression on pig cells on the human T-cell xenoresponse. Xenotransplantation 19:56–63
55. Ezzelarab MB, Ayares D, Cooper DKC (2015) Transgenic expression of human CD46: does it reduce the primate T-cell response to pig endothelial cells? Xenotransplantation 22:487–489
56. Bühler L, Awwad M, Basker M et al (2000) High-dose porcine hematopoietic cell transplantation combined with CD40 ligand

blockade in baboons prevents an induced anti-pig humoral response. Transplantation 69:2296–2304

57. Higginbotham L, Mathews D, Breeden C et al (2015) Pre-transplant antibody screening and anti-CD154 costimulation blockade promote long-term xenograft survival in a pig-to-primate kidney transplant model. Xenotransplantation 22:221–230
58. Yamamoto T, Li Q, Hara H et al (2018) B cell phenotypes in baboons with pig artery patch grafts receiving conventional immunosuppressive therapy. Transpl Immunol 51:12–20. https://doi.org/10.1016/j.trim.2018.08.005
59. Yamamoto T, Hara H, Wang L, et al (2018) Life-supporting kidney xenotransplantation from genetically-engineered pigs in baboons: a comparison of two immunosuppressive regimens. Transplantation. 2019 Jun 24. doi:10.1097/TP2796 [Epub ahead of print]
60. Ezzelarab C, Ayares D, Cooper DK et al (2012) Human T-cell proliferation in response to thrombin-activated GTKO pig endothelial cells. Xenotransplantation 19:311–316
61. Reyes LM, Estrada JL, Wang ZY et al (2014) Creating class I MHC-null pigs using guide RNA and the Cas9 endonuclease. J Immunol 193:5751–5757
62. Hara H, Witt W, Crossley T et al (2013) Human dominant-negative class II transactivator transgenic pigs – effect on the human anti-pig T cell immune response and immune status. Immunology 140:39–46
63. Iwase H, Ekser B, Satyananda V et al (2015) Initial in vivo experience of pig artery patch transplantation in baboons using mutant MHC (CIITA-DN) pigs. Transpl Immunol 32:99–108
64. Martens GR, Reyes LM, Butler JR et al (2017) Humoral reactivity of renal transplant-waitlisted patients to cells from GGTA1/CMAH/B4GalNT2, and SLA class I knockout pigs. Transplantation 101: e86–e92
65. Ladowski JM, Reyes LM, Martens GR et al (2018) Swine leukocyte antigen class II is a xenoantigen. Transplantation 102:248–254
66. Ladowski JM, Martens GR, Reyes LM et al (2018) Examining the biosynthesis and xenoantigenicity of class II swine leukocyte antigen protein. J Immunol 200:2957–2964
67. Phelps CJ, Ball SF, Vaught TD et al (2009) Production and characterization of transgenic pigs expressing porcine CTLA4-Ig. Xenotransplantation 16:477–485
68. Klymiuk N, van Buerck L, Bähr A et al (2012) Xenografted islet cell clusters from INSLEA29Y transgenic pigs rescue diabetes and prevent immune rejection in humanized mice. Diabetes 61:1527–1532
69. Solhjou Z, Athar H, Xu Q et al (2015) Emerging therapies targeting intra-organ inflammation in transplantation. Am J Transplant 15:305–311
70. Hara H, Iwase H, Miyagawa Y et al (2017) Reducing the inflammatory response by expressing human thrombomodulin in pigs. Xenotransplantation 24:19–20. (Abstract) (PL.3.11)
71. Gao H, Lu L, Zhao Y et al (2017) Human IL-6, IL-17, IL-1β and TNF-α differently regulate the expression of pro-inflammatory related genes, tissue factor, and swine leukocyte antigen class I in porcine aortic endothelial cells. Xenotransplantation 24(2). https://doi.org/10.1111/xen.12291
72. Zhang G, Hara H, Yamamoto T et al (2018) Serum amyloid A as an indicator of impending xenograft failure. Int J Surg 60:283–290
73. Langin M, Mayr T, Reichart B et al (2018) Consistent success in life-supporting porcine cardiac xenotransplantation. Nature 564:430–433. https://doi.org/10.1038/s41586-018-0765-z
74. Ekser B, Markmann JF, Tector AJ (2015) Current status of liver xenotransplantation. Int J Surg 23(Pt B):240–246
75. Kubicki N, Laird C, Burdorf L, Pierson RN III, Azimzadeh AM (2015) Current status of pig lung xenotransplantation. Int J Surg 23 (Pt B):247–254. https://doi.org/10.1016/j.ijsu.2015.08.019
76. Cooper DKC, Dou K-F, Tao K-S, Yang Z-X, Tector AJ, Ekser B (2016) Pig liver xenotransplantation: a review of progress towards the clinic. Transplantation 100:2039–2047
77. Laird C, Burdorf L, Pierson RN (2016) 3rd. Lung xenotransplantation: a review. Curr Opin Organ Transplant 21:272–278
78. Shah JA, Patel MS, Elias N et al (2017) Prolonged survival following pig-to-primate liver xenotransplantation utilizing exogenous coagulation factors and costimulation blockade. Am J Transplant 17:2178–2185
79. van der Windt DJ, Bottino R, Casu A et al (2009) Long-term controlled normoglycemia in diabetic non-human primates after transplantation with hCD46 transgenic porcine islets. Am J Transplant 9:2716–2726
80. Bottino R, Wijkstrom M, Van der Windt DJ et al (2014) Pig-to-monkey islet

xenotransplantation using multi-transgenic pigs. Am J Transplant 14:2275–2287

81. Shin JS, Kim JM, Kim JS et al (2015) Long-term control of diabetes in immunosuppressed nonhuman primates (NHP) by the transplantation of adult porcine islets. Am J Transplant 15:2837–2850
82. Martin C, Plat M, Nerriere-Daguin V et al (2005) Transgenic expression of CTLA4-Ig by fetal pig neurons for xenotransplantation. Transgenic Res 14:373–384
83. Leveque X, Cozzi E, Naveilhan P, Neveu I (2011) Intracerebral xenotransplantation: recent findings and perspectives for local immunosuppression. Curr Opin Org Transplantation 16:190–194
84. Choi HJ, Kim MK, Lee HJ et al (2011) Efficacy of pig-to-rhesus lamellar corneal xenotransplantation. Invest Ophthalmol Vis Sci 52:6643–6650
85. Zhang MC, Liu X, Jin Y, Jiang DL, Wei XS, Xie HT (2015) Lamellar keratoplasty treatment of fungal corneal ulcers with acellular porcine corneal stroma. Am J Transplant 15:1068–1075
86. Matsumoto S, Tan P, Baker J et al (2014) Clinical porcine islet xenotransplantation under comprehensive regulation. Transplant Proc 46:1992–1995
87. Cooper DKC, Matsumoto S, Abalovich A et al (2016) Progress in clinical encapsulated islet xenotransplantation. Transplantation 100:2301–2308
88. Yamamoto T, Iwase H, King TW, Hara H, Cooper DKC (2018) Skin xenotransplantation: historical review and clinical potential. Burns 44:1738–1749. https://doi.org/10.1016/j.burns.2018.02.029
89. Smood B, Hara H, Wang Z-Y, Schoel LJ, Tector M, Cooper DKC (2019) Genetically-engineered pigs as sources for clinical red blood cell transfusion: what pathobiological barriers need to be overcome? Blood Rev 35:7–17
90. Smood B, Hara H, Cleveland D, Cooper DKC (2019) In search of the ideal valve prosthesis: optimizing genetic modifications to prevent bioprosthetic valve degeneration. Ann Thorac Surg 108:624–635
91. Sommaggio R, Uribe-Herranz M, Marquina M, Costa C (2016) Xenotransplantation of pig chondrocytes: therapeutic potential and barriers for cartilage repair. Eur Cell Mater 32:24–39
92. Zhou M, Lu Y, Zhao C et al (2019) Circulating pig-specific DNA as a novel biomarker for monitoring xenograft rejection. Xenotransplantation 26(4):e12522
93. Agbor-Enoh S, Chan JL, Singh A et al (2018) Circulating cell-free DNA as a biomarker of tissue injury: assessment in a cardiac xenotransplantation model. J Heart Lung Transplant 37:967–975
94. Song Z, Cooper DKC, Cai Z, Mou L (2018) Expression and regulation profile of mature microRNA in the pig: relevance to xenotransplantation. Biomed Res Int 2018:2983908. https://doi.org/10.1155/2018/2983908
95. Byrne GW, Du Z, Sun Z, Asmann YW, McGregor CG (2011) Changes in cardiac gene expression after pig-to-primate orthotopic xenotransplantation. Xenotransplantation 18:14–27
96. Iwase H, Klein E, Cooper DK (2018) Physiological aspects of pig kidney transplantation in primates. Comp Med 68:332–340
97. Soin B, Ostlie D, Cozzi E et al (2000) Growth of porcine kidneys in their native and xenograft environment. Xenotransplantation 7:96–100
98. Tanabe T, Watanabe H, Shah JA et al (2017) Role of intrinsic (graft) versus extrinsic (host) factors in the growth of transplanted organs following allogeneic and xenogeneic transplantation. Am J Transplant 17:1778–1790
99. Hinrichs A, Klymiuk N, Reichart B et al (2017) Inactivation of the GHR gene – a strategy to overcome excess growth of orthotopic pig-to-baboon cardiac xenografts? Xenotransplantation 24:39–40. Abstract (#O6.2)
100. Iwase H, Yamamoto T, Cooper DKC (2019) Episodes of hypovolemia/dehydration in baboons with pig kidney transplants: a new syndrome of clinical importance? Xenotransplantation 28:e12472. https://doi.org/10.1111/xen.12472
101. Cooper DKC, Tseng YL, Saidman SL (2004) Allo- and xeno-antibody cross-reactivity in transplantation. Transplantation 77:1–5
102. Hara H, Ezzelarab M, Rood PP et al (2006) Allosensitized humans are at no greater risk of humoral rejection of GT-KO pig organs than other humans. Xenotransplantation 13:357–365
103. Wong BS, Yamada K, Okumi M et al (2006) Allosensitization does not increase the risk of xenoreactivity to α1,3-Galactosyltransferase gene-knockout miniature swine in patients on transplantation waiting lists. Transplantation 82:314–319
104. Zhang Z, Hara Z, Long C et al (2018) Immune responses of HLA-highly-sensitized

and non-sensitized patients to genetically-engineered pig cells. Transplantation 102: e195–e204. https://doi.org/10.1097/TP.0000000000002060

105. Li Q, Hara H, Breimer ME, Wang Y, Cooper DKC (2018) Is sensitization to pig antigens detrimental to subsequent allotransplantation? Xenotransplantation 25:e12393. https://doi.org/10.1111/xen.12393
106. Lunney JK (1994) Current status of the swine leukocyte antigen complex. Vet Immunol Immunopathol 43:19–28
107. Fishman JA (2018) Infectious disease risks in xenotransplantation. Am J Transplant 18:1857–1864
108. Dieckhoff B, Karlas A, Hofmann A et al (2007) Inhibition of porcine endogenous retroviruses (PERVs) in primary porcine cells by RNA interference using lentiviral vectors. Arch Virol 152:629–634
109. Dieckhoff B, Petersen B, Kues WA, Kurth R, Niemann H, Denner J (2008) Knockdown of porcine endogenous retrovirus (PERV) expression by PERV-specific shRNA in transgenic pigs. Xenotransplantation 15:36–45
110. Ramsoondar J, Vaught T, Ball S et al (2009) Production of transgenic pigs that express porcine endogenous retrovirus small interfering RNAs. Xenotransplantation 16:164–180
111. Niu D, Wei HJ, Lin L et al (2017) Inactivation of porcine endogenous retrovirus in pigs using CRISPR-Cas9. Science 357:1303–1307
112. Cooper DKC, Pierson RN III, Hering BJ et al (2017) Regulation of clinical xenotransplantation – time for a reappraisal? Transplantation 102:1766–1769
113. Jagdale A, Iwase H, Klein E, Cooper DKC (2018) Will donor-derived neoplasia be problematic after clinical pig organ or cell xenotransplantation. Xenotransplantation 9: e12469. https://doi.org/10.1111/xen.12469
114. Ekser B, Gridelli B, Tector AJ, Cooper DKC (2009) Pig liver xenotransplantation as a bridge to allotransplantation in acute liver failure: which patients might benefit? Transplantation 88:1041–1049
115. Cooper DKC, Wijkstrom M, Hariharan S et al (2017) Selection of patients for initial clinical trials of solid organ xenotransplantation. Transplantation 101:1551–1558
116. Chan JL, Miller JG, Singh AK, Horvath KA, Corcoran PC, Mohiuddin MM (2018) Consideration of appropriate clinical applications for cardiac xenotransplantation. Clin Transpl 32:e13330. https://doi.org/10.1111/ctr.13330
117. Jagdale A, Cooper DKC, Iwase H, Gaston RS (2018) Chronic dialysis in patients with end-stage renal disease: relevance to kidney xenotransplantation. Xenotransplantation 20:e12471. https://doi.org/10.1111/xen.12471
118. Cleveland D, Banks CA, Hara H et al (2019) The case for cardiac xenotransplantation in neonates: is now the time to reconsider xenotransplantation for hypoplastic left heart syndrome? Pediatr Cardiol 40:437–444. https://doi.org/10.1007/s00246-018-1998-1
119. West LJ, Pollock-Barziv SM et al (2001) ABO-incompatible heart transplantation in infants. N Engl J Med 344:793–800
120. West LJ (2005) Targeting antibody-mediated rejection in the setting of ABO-incompatible infant heart transplantation: graft accommodation vs. B cell tolerance. Curr Drug Targets Cardiovasc Haematol Disord 5:223–232
121. Cooper DKC, Hara H, Iwase H, Banks CA, Cleveland D (2018) An approach to induction of tolerance to pig cardiac xenografts in neonates. Xenotransplantation 20:e12454. https://doi.org/10.1111/xen.12454
122. Saari R, Cooper DKC (2017) Financial aspects of organ procurement from deceased donors in the USA – relevance to xenotransplantation. Xenotransplantation 24(4). https://doi.org/10.1111/xen.12322
123. UNOS United Network for Organ Sharing. http://www.unos.org. Last accessed on September 30, 2018

Chapter 2

Tools for Molecular Studies in Xenotransplantation

Sebastián G. Kuguel, Mireia Uribe-Herranz, and Cristina Costa

Abstract

Understanding the molecular bases of xenograft rejection is one of the highest priorities in the xenotransplantation field. This information is needed for the successful development of genetic modifications of the animal source of xenogeneic cells and organs that prevent rejection. Furthermore, the identification of physiological incompatibilities in the xenogeneic setting is also necessary for developing the appropriate strategies that allow long-term xenograft function. As the pig is the species of choice for the development of the majority of xenogeneic applications, the cloning of pig genes or cDNA is a key step to elucidate the interactions of pig and human molecules. In addition, there are currently multiple bioinformatic tools which facilitate the study in silico of protein structures, molecular interactions, and docking sites. Thus, we describe a basic cloning method that comprises total RNA extraction, reverse transcription (RT), and polymerase chain reaction (PCR) for cDNA amplification and include some links for databases and bioinformatics tools available on the Internet for the subsequent analyses and predictions. Finally, some procedures of protein expression and analysis of protein interactions by surface plasmon resonance (SPR) are described for elucidating the molecular mechanisms of xenograft function and rejection.

Key words Pig, Human, cDNA cloning, RT-PCR, Bioinformatic tools, Protein interactions

1 Introduction

Genetic and biochemical analyses are key for understanding the molecular bases of any pathological process and allow the development of biomedical therapies. In the field of xenotransplantation, it is critical to understand the molecular bases of xenograft rejection and identify physiological incompatibilities [1–3]. This knowledge is necessary for developing the appropriate strategies to overcome rejection and promote long-term function of the xenograft. In this context, the cloning of pig genes is the first step toward elucidating the role of pig and human molecules in these processes. Subsequent analyses may include structural predictions followed by experimental studies of molecular interactions across species, studies of expression of the native proteins, and functional assays. To briefly compile the methodology of this work flow, this chapter describes procedures for isolating and analyzing a cDNA of interest and its

Cristina Costa (ed.), *Xenotransplantation: Methods and Protocols*, Methods in Molecular Biology, vol. 2110,
https://doi.org/10.1007/978-1-0716-0255-3_2,

corresponding protein. The isolation of a gene or a portion of it can be key for some specific purposes such as for knockout or knock-in technologies [4, 5]. Genomic sequencing can also be used for characterization of the genome of genetically modified pigs [6].

The sequence of the whole pig genome has been obtained by the Swine Genome Sequencing Consortium and made public [7]. Their initial approaches included hierarchical shotgun sequencing of BAC clones and whole-genome shotgun sequencing. The current annotated high-coverage genome assembly, named Sscrofa11.1, can be reached through NCBI (National Center for Biotechnology Information, USA) and was published in February 2017. It comprises a median total length of 2457.91 Mb. Even now, when the pig genome sequence is fully annotated, cDNA cloning remains useful to detect the native mRNA variants that encode different isoforms, assess polymorphisms and work with the more manageable cDNA.

Several strategies can be used to clone a specific pig cDNA such as hybridization of cDNA libraries, standard reverse transcription (RT) followed by polymerase chain reaction (PCR), or rapid amplification of cDNA ends (RACE) [8–10]. The cDNA libraries are especially useful when the whole genome is known (i.e., interpretation of results, high specificity of probes). Unfortunately, there are not many libraries available of pig cDNA, it involves a labor-intensive work, and your success still depends on the level of expression of your gene of interest in the tissue and condition used for creating the cDNA library. In our experience, the standard RT-PCR using various tissues as a source of cDNA is a very useful approach for obtaining a pig cDNA. Thus, we describe a basic cloning method that comprises total RNA extraction, reverse transcription, and PCR amplification. Note that following the RT-PCR, there are the necessary steps of sub-cloning and sequencing multiple clones to obtain a consensus sequence. We recommend using a topoisomerase-based method for sub-cloning. Regarding sequencing, we are not going to describe the procedure as it is now commonly done in specialized facilities at very reasonable cost. Finally, the RACE can also be performed to obtain the full cDNA [8–10]. It can provide further confirmation of the sequence obtained and lead to the discovery of new isoforms. Many genes lead to the generation of multiple mRNA variants resulting from alternative splicing [11], and a combination of techniques can be more effective in detecting such variants. In the case of a gene that produces multiple mRNA variants, it is unquestionable that identification of most or all variants is important to elucidate their function [8, 11].

Nowadays, with the huge amount of information generated by high-throughput techniques such as large-scale expression profiling or genome annotation, bioinformatic tools have become essential for any laboratory. Over the past years, numerous tools have emerged for storing and organizing information, analysis, and prediction of biological processes [12]. There are various major databases such as those from EMBL-EBI (European Molecular Biology Laboratory-European Bioinformatic Institute), NCBI, and DDBJ (DNA Databank of Japan). The nucleotide databases GenBank (NCBI), EMBL DataLibrary, and DDBJ share and store every nucleotide sequence already published or made public. Protein databases such as Swiss-Prot store primary amino acidic sequences, whereas the protein database from NCBI collects information from multiple sources. Moreover, there are plenty of servers which provide the scientist with high-quality resources of protein sequence, functional information, and tools for analysis (e.g., EBI, ExPASy). They provide tools that are especially useful for the analysis of newly cloned genes and cDNA such as comparison with homologues, detection of motifs, and prediction of function of the encoded proteins. In this regard, there are some programs such as SWISS-MODEL or PyMOL designed for modeling protein structure and interactions departing from the amino acid sequence and/or information stored in registries of crystallographic analyses. The structural analyses may involve determination of folding patterns and presence of different chains, grooves, and pockets. Furthermore, a dynamic simulation can be run in silico to predict the behavior of the studied protein when exposed to other proteins, peptides, carbohydrates, or small molecules. Notably, scarce information is available at this level that is of interest to the xenotransplantation field. The characterization of porcine SLA-II is one of such few works [13]. Thus, this area is new and can be used to generate relevant information. Here, we provide the links to some bioinformatic tools that can be used for this type of studies.

In the end, two major factors will impact the function of a pig protein in a xenogeneic setting: the level of expression and the characteristics of the interaction with the human molecules. The methods previously described by our group to assess expression of pig genes and proteins are still up to date and will not be covered in this chapter [14, 15]. Nevertheless, we have included the bases and general protocol for surface plasmon resonance (SPR) as this is a very powerful tool not sufficiently used for xenotransplantation studies. SPR can be used to determine the binding kinetics of pig and human proteins with their xenogeneic ligands. Molecular interactions across species are not always well preserved, even between pig and human. The physiological and immunological incompatibilities should be well characterized for the progress of the field as many of the deficient interactions may have a negative impact on xenograft survival. An example in the pig-to-human combination is that SLA I fails to provide inhibitory signaling through the human killer inhibitory receptors on NK cells [16]. For this reason, techniques such as SPR and calorimetry can provide useful information on kinetics, affinity, stoichiometry, and even thermodynamics of the

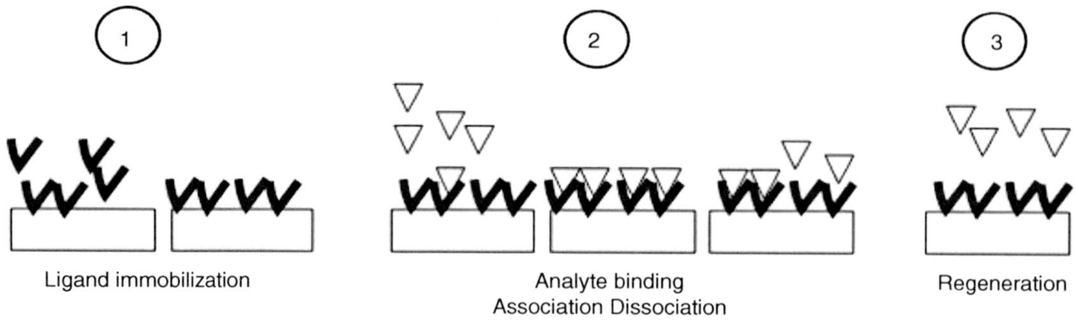

Fig. 1 Scheme of SPR process. (1) Ligand immobilization. (2) Analyte binding which involves association and dissociation to ligand. (3) Regeneration of the chip surface

molecular interactions [8, 17]. The basic procedure of SPR involves immobilization of the ligand on the chip, binding of the analyte under flow, and regeneration of the chip surface (Fig. 1). As a read out, the equipment captures changes in refractive index that result from increases in mass at the sensor surface and calculates the affinity and kinetics of the interaction. All this information can be subsequently used for understanding the xenografting process and design new therapeutic strategies.

2 Materials

Prepare and store all reagents at room temperature unless otherwise indicated. Use gloves to prepare and handle the reagents for RNA and DNA analysis for both protection and to prevent sample contamination. Unless stated specifically, use purified water such as Elix-deionized water (Millipore Corporation, Billerica, MA, USA) to prepare all solutions.

2.1 Cloning of Pig cDNA

2.1.1 Total RNA Extraction

1. Porcine cells or tissue (*see* **Note 1**). Use fresh or frozen.
2. Optional: 0.05% trypsin-EDTA.
3. RNA purification kit (*see* **Note 2**) including buffer RLT, buffer RW1 and buffer RPE, RNase-free water, spin columns, and collection tubes. Buffer RPE should be diluted with four volumes of 96%–100% ethanol the first time it is opened for use to prepare the working solution.
4. Optional: Diethyl pyrocarbonate (DEPC)-treated water: 0.1% v/v DEPC in water (*see* **Note 3**).
5. Optional: Reagent for RNA stabilization (*see* **Note 1**) such as RNAprotect Cell Reagent (QIAGEN), Allprotect Tissue Reagent (QIAGEN), or RNAlater (Merck).
6. β-mercaptoethanol (β-ME), to be added to buffer RLT before use (10 μL per mL of RLT).
7. 70% ethanol (in DEPC-treated water).

8. Phosphate-buffered saline (PBS).
9. Proteinase K (for RNA purification from the heart, muscle, and skin).
10. Sterile RNAse-free tips and tubes.
11. Gloves for handling all reagents and materials for RNA isolation.
12. Equipment: micropipettes, pipettors, microcentrifuge, vortexer, fume hood, mortar and pestle baked in oven to prevent ribonuclease contamination, blunt-ended needle, and syringe spectrophotometer (a micro-volume spectrophotometer such as NanoDrop One is especially convenient (Thermo Fisher Scientific)).

2.1.2 Standard RT-PCR

1. Kit for RT (*see* **Note 4**) containing various reagents: reverse transcriptase, amplification buffer, dithiothreitol (DTT), dNTPmix, RNase, oligo(dT)$_{20}$ or random hexamers, DEPC-treated water.
2. Kit for PCR (*see* **Note 5**) containing various reagents: thermostable DNA polymerase, dNTP mix, amplification buffer 10×, DEPC-treated water.
3. Specific degenerate primers (*see* **Note 6**).
4. cDNA template (*see* **Note** 7).
5. Sterile tips (preferably with filter) and PCR tubes.
6. Gloves for handling all reagents and materials.
7. DNA gel stain such as GelRed Nucleic Acid Gel Stain (Biotium, Fremont, CA, USA) or Midori Green Advance DNA Stain (Nippon Genetics Europe, Düren, Germany).
8. DNA size-marker such as 1Kb Plus DNA Ladder (Invitrogen™, Thermo Fisher Scientific).
9. Electrophoresis buffer TAE 50×: 121 g of Tris base, 28.6 mL of glacial acetic acid, and 50 mL of 0.5 M EDTA. Bring solution to a final volume of 500 mL.
10. Electrophoresis running buffer: dilute the 50× TAE 1/50 in water. Prepare in sufficient quantity for the electrophoresis.
11. Erlenmeyer flask made of heat-resistant glass (such as Pyrex) for preparation of agarose gel.
12. Agarose gel (*see* **Note 8**): For a 100-mL 1% agarose gel use 1 g of agarose in 1× electrophoresis buffer in the Erlenmeyer flask, bring to boil, and mix until the solution is clear. Cool down, add 5 μL of the GelRed stock solution, mix well, and pour in gel box. We conduct this last step and leave the gel to solidify.
13. Loading buffer 6×: 30% (w/v) sucrose and 0.35% (w/v) orange G sodium salt.
14. Equipment: micropipettes, pipettors, thermal cycler, horizontal electrophoresis unit with power supply and imaging system with ultraviolet illumination.

2.2 Bioinformatic Analyses

These procedures can be all conducted at a satisfactory level using a personal computer with Internet connection. For studies of higher complexity, there are specialized laboratories and companies that can provide service.

2.2.1 Sequence Analyses

1. Databases: GenBank (NIH genetic sequence database), EMBL-EBI (European Molecular Biology Laboratory-European Bioinformatic Institute), or DDBJ (DNA Databank of Japan) (the three form the International Nucleotide Sequence Database).
2. See the pig genome database at https://www.animalgenome.org/pig/genome/db/.
3. Multiple software tools such as the nucleotide and protein Basic Local Alignment Search Tool (BLAST) from the National Center for Biotechnology Information (NCBI) (*see* Subheading 3.2.1 for a more detailed list).

2.2.2 Structural Protein Analysis

1. Several software tools can be used such as PyMOL (DeLano Scientific LLC, Palo Alto, CA, USA) and UCSF Chimera (*see* **Note 9**).
2. Source of crystallographies of proteins: Protein Data Bank, which can be reached through various routes (RCSB, EMBL-EBI, NCBI, PDBj-BMRB), and Swiss-Model (from Biozentrum, Basel, Switzerland). Both contain repositories.

2.2.3 Molecular Dynamic and Docking

1. Simulated molecules to study in silico their interaction with a protein of interest can be loaded from an online library/database or created with software such as SWISS-MODEL or PyMOL (*see* **Note 9**).
2. Source of drugs to do the docking: PubChem (NCBI) and DrugBank.
3. There are some tools to run a protein dynamic. A free option is GROMACS, which provide a tutorial following this link: http://www.bevanlab.biochem.vt.edu/Pages/Personal/justin/gmx-tutorials/lysozyme/05_EM.html. Another option is AutoDock Vina, also with tutorials in the main website.

2.3 Recombinant Protein Expression and Purification

2.3.1 Transfection

1. Cell line, the choice depends on the final objective (*see* **Note 10**).
2. Culture media appropriate for the chosen cell line such as Dulbecco's modified Eagle's medium (DMEM) containing high glucose supplemented with 10% heat-inactivated fetal bovine serum (FBS), 200 U/mL penicillin, and 200 μg/mL streptomycin (*see* **Note 10**).

3. Purified DNA of expression vector alone and containing the gene of interest. There are many expression vectors, most available commercially or at molecular biology laboratories. Minimally, they contain a promoter such as the cytomegalovirus (CMV) and a polyadenylation signal.
4. 1 mg/mL polyethylenimine (PEI).
5. Tissue culture plastic: T25 flasks, pipettes, sterile tips.
6. Equipment: pipettors, tissue culture cabin, aspiration pump, or source of vacuum, 5% CO_2 incubator, 37 °C water bath, inverted microscope.

2.3.2 Production and Purification of Fusion Proteins in Mammalian Cells

1. All materials described in Subheading 2.3.1, including an expression vector containing the gene of interest fused to a GST tag (*see* **Note11**).
2. Appropriate matrix to purify the fusion proteins by affinity (*see* **Note 11**). For capturing the GST tag, the Bulk glutathione Sepharose 4B (GE Healthcare, Waukesha, WI, USA) can be used directly in tubes or packed in columns.
3. Elution buffer for GST: 10 mM glutathione. This mild buffer should suffice for successful elution.
4. PBS.
5. Disposable sterile tubes to be used for bulk purification. As mentioned, there is the alternative to use columns if preferred.
6. Buffer exchange devices such as protein desalting spin columns (Thermo Fisher Scientific). These materials are needed for all procedures that require that the protein of interest is resuspended in a specific buffer such as injection into animals or SPR.

2.4 Surface Plasmon Resonance

1. Ligand (protein) (*see* **Note 12**).
2. Analyte (protein) (*see* **Note 12**).
3. HBS-N buffer: 10 mM HEPES, pH 7.4, 150 mM NaCl (*see* **Note 13**).
4. HBS-EP buffer: 10 mM HEPES, pH 7.4, 150 mM NaCl, 3 mM EDTA, and 0.005% vol/vol polysorbate 20 (Surfactant P20, GE Healthcare Life Sciences) (*see* **Note 13**).
5. 100 mM *N*-ethyl-*N′*-(dimethyl-aminopropyl)-carbodiimide-hydrochloride (EDC).
6. 400 mM *N*-hydroxysuccinimide (NHS). Store at −20 °C.
7. 1 M ethanolamine. Store at 4 °C.
8. Filters of 0.22 μm.
9. Sterile Erlenmeyer flasks or glass bottles with connection for degassing with vacuum system.

10. Sensor surface Biacore chip (e.g., CM5) and Biacore tubes (GE Healthcare).
11. Equipment: Biacore system (e.g., T200) (GE Healthcare).

3 Methods

3.1 Cloning of Pig cDNA

3.1.1 Total RNA Extraction

1. For cells: add 600 μL of buffer RLT (lysis buffer) to disrupt sample if cellular sample is between 5×10^6 and 1×10^7 or 350 μL when processing less than 5×10^6 cells. Cells in monolayer can be lysed directly in the cell-culture dish (up to 10 cm diameter) or trypsinized and collected prior to lysis. Prior to RLT addition, make sure to completely aspirate the cell-culture media in the first case or loosening the pellet if the cells are harvested by standard trypsinization and centrifugation. For tissues, these can be grinded with a mortar and pestle (or equivalent commercial equipment) while kept frozen before addition of 600 μL of buffer RLT. For small portions of tissue ($\leq$15 mg) not subjected to stabilization, use 350 μL of buffer RLT.
2. In order to homogenize the lysate, pass it at least five times through a blunt 20-gauge needle (0.9 mm diameter) fitted to an RNase-free syringe. For cell lysates, proceed directly to **Step 3**. For tissue lysates, centrifuge for 3 min at full speed in the microfuge to remove debris and carefully transfer the lysate to a clean tube in preparation for **Step 3**.
3. Add one volume of ethanol 70% to the lysate (equal volume to that of the lysate after homogenization). Do not centrifuge.
4. Transfer up to 700 μL of the sample (including precipitate) to a spin column, close the lid gently, and centrifuge for 15 s ($\geq$8000 × *g*). Discard the flow-through.
5. Add 700 μL of buffer RW1 to the spin column. Close the lid and centrifuge for 15 s at $\geq$8000 × *g*. Discard the flow-through.
6. Add 500 μL of RPE working solution to the spin column, close the lid, and centrifuge for 15 s ($\geq$8000 × *g*). Discard the flow-through.
7. Add 500 μL of RPE working solution to the spin column, close the lid, and centrifuge for 2 min at $\geq$8000 × *g*. After centrifugation, transfer the column into a new 1.5 mL collection tube. Make sure not to carry ethanol from the flow through.
8. Add 30–50 μL of RNase-free (or DEPC-treated) water directly to the spin column membrane and centrifuge for 1 min at $\geq$8000 × *g* to elute RNA. Keep the flow-through as your RNA preparation.

9. If the expected RNA yield is >30 μg, repeat Step 8 using an additional 30–50 μL of RNase-free water. The eluate can be mixed with the previous collection or kept separately. Keep the RNA on ice and store at −80 °C for long-term storage.

3.1.2 Standard RT-PCR

1. In a sterile 0.5-mL tube, mix in the following order: 0.5 μL of primer Oligo(dT)$_{20}$ (50 μM) or random hexamers (50 ng/μL) (*see* **Note 14**), X μL of RNA (5–10 μg) (maximum 10.5 μL), 1 μL of 10 mM dNTP mix, and up to 12 μL of DEPC-treated water. Mix thoroughly by pipetting.
2. In a sterile 0.5-mL tube, mix in the following order: 0.5 μL of primer Oligo(dT)$_{20}$ (50 μM) or random hexamers (50 ng/μL) (*see* **Note 14**), X μL of RNA (5-10 μg) (maximum 10.5 μL), 1 μL of 10 mM dNTP mix, and up to 12 μL of DEPC-treated water. Mix thoroughly by pipetting.
3. In a different tube, prepare an additional mix in the following order: 4 μL of amplification buffer 5×, 2 μL of 0.1 M DTT, 20–40 units (1 μL) of RNase Inhibitor (40 units/μL), and DEPC-treated water (up to 7 μL final volume). This mixture can be prepared as a master mix by increasing the amounts proportionally to the number of samples to be processed (and a little more due to pipetting error).
4. For each sample, mix both solutions thoroughly by pipetting in the 0.5-mL tube containing the RNA (12 + 7 = total volume of 19 μL).
5. Preheat the tube in a thermal cycler for 2 min at 37 °C.
6. Add 1 μL of Moloney murine leukemia virus reverse transcriptase (M-MLV) for a final volume of 20 μL.
7. Put the tube in the thermal cycler under a program set for 50 min at 37 °C (*see* **Note 14**) followed by 15 min at 70 °C.
8. In order to amplify the target gene once the cDNA is obtained, mix in a sterile 0.5-mL tube the indicated reagents in the following order: 5 μL of amplification buffer 10×, 1.5 μL of 0.3 mM dNTP mix, 1.5 μL of 0.3 μM of each primer, five units of thermostable DNA polymerase, X μL of DEPC-treated water (up to 50 μL final volume), and 1–5 μL of cDNA (10–1000 ng)
9. Place the tubes into a preheated thermal cycler and follow the indicated protocol of temperatures and incubation times. First, apply an initial denaturalization cycle at 94 °C for 2 min. Then, repeat between 25 and 35 cycles the following program: one cycle of denaturalization at 94 °C for 15 s, one cycle of annealing at X °C (*see* **Note 15**) for 30 s, and one cycle of extension at 72 °C for 1 min. The last step is a final elongation cycle at 72 °C for 5 min.

10. Mix half of the reaction with loading buffer and load into the agarose gel slots for electrophoresis for approximately 30–45 min at a constant voltage (up to 10 V/cm) (*see* **Note 16**). Use a size marker for comparison in one or two of the lanes.
11. Visualize with ultraviolet illumination and take a picture. Determine the size of the fragments amplified. If a PCR product with the expected size is amplified, then the procedure should be followed by subcloning into a plasmid for sequencing and expression (*see* **Notes 5** and **17**).

3.2 Bioinformatic Analyses

3.2.1 Sequence Analyses

We provide here the Internet address and a brief explanation of some of the most common tools for nucleotide and amino-acid sequence analysis in a molecular laboratory. Many can be accessed through various routes as are of free access. We simply recommend locating them and follow their instructions and help links.

1. **The Basic Local Alignment Search Tool (BLAST)** (http://blast.ncbi.nlm.nih.gov/Blast.cgi) allows finding regions of local similarity between your sequence of interest and sequences available at databases (homologues, etc.).
2. **T-Coffee** (https://www.ebi.ac.uk/Tools/msa/tcoffee/) is a multiple nucleotide or amino acidic sequence alignment program.
3. **ExPASy** (**Ex**pert **P**rotein **A**nalysis **Sy**stem) (https://www.expasy.org/).

   This server from the SIB Swiss Institute of Bioinformatics contains a long list of proteomic tools such as the following:

   - **Translate**: Translates a nucleotide sequence into a protein sequence.
   - **Compute pI/Mw:** Calculates the theoretical isoelectric point (*pI*) and molecular weight (*Mw*) of a protein from a given sequence.
   - **SignalP:** Predicts the cleavage sites of the signal peptide of a given sequence.
   - **SMART:** Allows the identification and annotation of genetically mobile domains and the analysis of domain architectures in a given sequence (from EMBL).
   - **NetNGlyc:** Predicts the *N*-glycosylation sites in human proteins.
   - **NetOGlyc:** Predicts the *O*-glycosylation sites in mammalian proteins.
4. **DTU Bioinformatics** (http://www.bioinformatics.dtu.dk/).

This unit from the department of Bio and Health informatics from the Technical University of Denmark also provides access to multiple tools for analysis and predictions of proteins including **TargetP 1.1**, which predicts the subcellular location based on the sequence information.

3.2.2 Structural Protein Analysis

1. Search and download the protein of interest from a protein database in a Protein Data Bank (PDB) file format to be opened by PyMOL or UCSF Chimera (*see* **Note 9**). As there are few pig proteins crystallized and even less of those that are of interest to study the immune response, we recommend to look for available data from homologous proteins (mostly human), as an initial approach. Bovine proteins are certainly of interest as they show high homology with pig proteins (Fig. 2).
2. Load and investigate different features of the protein of interest by comparing between species, chains, and selected regions. It is highly advised to check some tutorials before using this kind of software. Specific amino acids can be localized in the protein and displayed using a color code. Also, surfaces can be visualized according to charges. Among the tools available, there is one which allows the configuration of new molecules or variants by modifying the protein object of study. It could be useful to the xenotransplantation field. Figure 2 displays an example of what can be done.

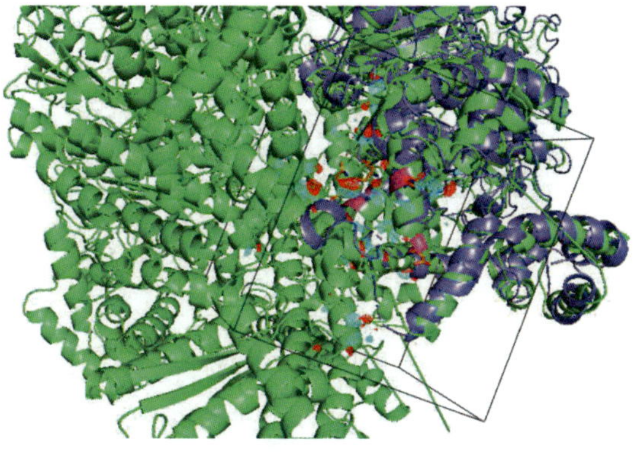

Fig. 2 Example of structural protein study. Comparison generated with Pymol between bovine ATPase α subunit (green) and that of E.Coli (blue). The scheme of one protein is superimposed to the other for the comparison. Residues that differ between the two species are shown in pink. The box drawn on top delimits a selected region for conducting the docking studies. In this particular case, the docking is simulated with ethanol as solvent, and the results show hydrophobic link forces (depicted as a red mesh) and polar link forces (as a pale blue mesh). The porcine homologue of this protein shows very high homology as assessed with BLAST (data not shown)

3.2.3 Molecular Dynamic and Docking

In brief, a simulated protein is delimited in a box to establish the area of study and is surrounded at that level with different molecules to assess the potential interactions. Before initiating any study, it is necessary to follow one of the tutorials specified in the methods section to understand these procedures (*see* **Note 9**). The software allows movement of the protein which can be visualized from all angles to observe the various pockets. Generally, proteins are first surrounded with ethanol molecules to calculate how much time each ethanol molecule remains in proximity to the protein depending on the exposed charges. Then, the protein can be cut and divided into smaller zones to speed up the process. A summary is presented as follows:

1. Generate topology: All the generated documents require the tail ".gmx" in their name in order to apply the various software tools. To get "help" information about any GROMACS module, you can invoke either of the following commands: "gmx help (module)" or "gmx (module) –h."
2. Define box and solvent: Generate a box, configure the box dimensions, and choose the solvent which will be surrounding the molecule of study.
3. Add ions: In order to change the charge of the system (solvent), water molecules can be replaced with other ions.
4. Energy minimization: To ensure that the system has no steric clashes or inappropriate geometry, it is recommended to apply an energy minimization (EM) procedure.
5. Equilibration: After geometry and solvent orientation correction, the solvent and the ions around the protein need to be equilibrated to prevent the collapse of the system.
6. Production of molecular dynamics: In this point, dynamics are ready to run and generate data for collection.
7. Data analysis depending on objectives.

3.3 Recombinant Protein Expression and Purification

3.3.1 Transfection

1. For transfection of adherent cells, these should be between 50% and 90% confluence (depending on the cell line) and maintained at 37 °C in a 5% CO_2 atmosphere.
2. In a sterile 0.5-mL tube, put 48 μL of PEI (1 mg/mL) followed by 100 μL of media without antibiotic and mix (*see* **Note 18**). Incubate the solution for 1 min at room temperature.
3. Add 8 μg of DNA and mix gently. Incubate for another 5 min at room temperature.
4. Add this mixture to the flask containing the cells and 3 mL of medium without antibiotic. Incubate it at 37 °C in 5% CO_2 during 6 h.

5. Remove transfection media and add fresh media (5 mL) for subsequent culture.
6. Cells should be ready for analyses and studies 48 h after the transfection. To obtain stably transfected cells, add selection media at this time point (48 h) (*see* **Note 19**).

3.3.2 Production and Purification of Fusion Proteins in Mammalian Cells

1. Transfect the chosen cell line with the appropriate expression vector containing the cDNA of the fusion protein following the method described in Subheading 3.3.1.
2. Harvest the culture supernatant if the recombinant fusion protein is secreted (contains signal peptide) or lyse the cells following standard methods if it is intracellular.
3. Prepare the affinity matrix for bulk method following the instructions of the manufacturer. The procedure depends on the matrix chosen.
4. Add the culture supernatant or cell lysate to the tube containing the prepared resin and incubate with agitation for 30 min at room temperature.
5. Centrifuge at 500 × *g* for 5 min and wash the matrix three times with PBS by aspirating the supernatant and resuspending in PBS after centrifugation (*see* **Note 20**).
6. Incubate the matrix with the appropriate elution buffer (glutathione 10 mM for GST fusion proteins) at room temperature with agitation during 10 min.
7. Centrifuge and keep the supernatant.
8. Analyze the protein yield following standard chromogenic methods of quantification such as Bradford or by measuring the absorbance at 280 nm. Purity can be assessed by SDS-PAGE, gel staining, and western blotting.
9. Use an exchange buffer device such as protein desalting spin columns following the instructions of the manufacturer when it is necessary to change the buffer (e.g., into PBS for injection in animals or specific buffer for SPR).

3.4 Surface Plasmon Resonance

1. First, it is important to conduct a pH scouting to find experimentally the best pH to bind the ligand to the sensor surface (*see* **Note 12**). To this end, use one channel of the sensor chip to test 3–4 conditions for binding (the surface can be regenerated between tests) and follow the resonance units (RU) as a guide. As the pH appropriate for binding is commonly close to the ligand isoelectric point (pI), it is recommended to assess at least one unit above and one below the pI (e.g., use 6, 7, and 8 for a pI of 7).
2. Activate the sensor chip with NHS and EDC following the instructions of the manufacturer.

3. Immobilize the ligand (protein) on the sensor chip using HBS-EP (*see* **Note 13**) as the running buffer. A ligand concentration 0.5 μM should be sufficient for a successful immobilization. Remember to dilute the ligand with the appropriate buffer (chosen at the pH scouting step).
4. Leave one channel of the chip empty or loaded with the tag protein alone (e.g., GST) in order to subtract the background.
5. Once you have immobilized the ligand, pass the ethanolamine solution to block the surface of the sensor chip.
6. Dilute the analyte sample with the running buffer (usually HBS-N, but HBS-EP may work better for some proteins) (*see* **Note 13**). Set up minimally five different concentrations and include a control with only buffer. Pass them randomly to ensure there is no additive effect.
7. Data analysis is performed using the Biacore software provided by the manufacturer. To obtain conclusive results, repeat the experiment two more times.

4 Notes

1. Isolated tissues are wrapped with aluminum foil labeled directly with marker for identification, snap frozen in liquid nitrogen, and stored at −80 °C until use. The use of RNA stabilization reagents is highly recommended to prevent early degradation and should be applied following carefully the recommendations of the manufacturers.
2. There are many kits now available for RNA purification. We get good results with the kits from QIAGEN such as the RNeasy Mini Kit or the RNeasy Midi Kit. However, you need to pay attention to the expected yield considering the amount of starting material, cells or tissue, as there are thresholds above and below that can be critical. Following the recommendation of the manufacturer, no more than 30 mg of animal tissue or 1×10^7 cells should be used as starting material for the RNeasy Mini Kit. Moreover, for small amounts of expected RNA, one should use the RNeasy Micro Kit and preferably more than 250,000 cells as starting material.
3. DEPC-treated water is now commercially available, and it is often provided with kits. Nevertheless, it is good to have the recipe in case more is needed. Water is treated with 0.1% DEPC to inactivate the RNase enzymes in the water. For preparation, add 1 mL of DEPC into 1 L of deionized water and stir overnight under the fume hood. Autoclave the solution afterward to remove any traces of DEPC. Buy DEPC in small aliquots as it can be used only for 1 month after opening.

4. There are various options for carrying out RT. The reverse transcriptase can be from Moloney murine leukemia virus (MMLV) or avian myeloblastosis virus (AMV). Both can be obtained from various suppliers. There are also genetically modified enzymes such as the Invitrogen™ SuperScript III RT, which show higher performance overall and can be purchased when difficulties arise, and the routine procedure fails to produce the desired amplified cDNA. In addition, many commercially kits are available that facilitate the task as they include all the reagents needed for the RT.
5. There are also multiple options for conducting a PCR. The choice of the thermostable DNA polymerase is especially important as it can be a Taq polymerase or a high-fidelity DNA polymerase, which possesses 3′ to 5′ exonuclease proofreading activity. A high-fidelity polymerase is recommended such as the Platinum™ SuperFi™ DNA polymerase (Invitrogen™) when the RT-PCR is intended for cDNA cloning. However, sometimes the Taq polymerase is more effective in producing amplification. We like the GoTaq® Green Master Mix (Promega, Madison, WI, USA), which comes as a premixed ready-to-use solution with the appropriate reaction buffers, dNTPmix and $MgCl_2$. Now they also have a second-generation DNA polymerase called the GoTaq® G2. Note that the subsequent subcloning system will vary depending on the type of polymerase. We use the pcDNA3.1 directional TOPO expression Kit (Invitrogen™) when using the high-fidelity polymerase and a TOPO TA Cloning Kit (Invitrogen™) when using a Taq polymerase.
6. The expected sequence of a pig gene can now be found in the pig genome database and be used for the design of the primers. When this information is lacking, degenerate primers for the amplification of a cDNA of interest are designed based on known sequences of other species. Thus, the sequences of mammalian homologues are aligned, and the primers are designed to be located at the ends of the coding sequence and contain all the conserved nucleotides as well as the combinations that cover all the variations observed (Table 1). For cloning a porcine cDNA, use for instance bovine, human, and murine sequences. An appropriate primer size would be 20–22 bp. We usually try to have a conserved nucleotide at the primer end and preferably a C or a G to add stability.
7. Note that the source of cDNA is critical for successful amplification of the cDNA of interest. Thus, it is highly recommended that the specific mRNA is well represented in the chosen cell/tissue. This can be estimated from information available/published for human genes. It is a good idea to start with a battery of tissues as some may work better than others.

Table 1
Letter code used for the generation of degenerate primers

Letter code	Mixed bases
B	T + C + G
H	A + T + C
W	A + T
S	C + G
K	T + G
M	A + C
Y	C + T
R	A + G

Each letter corresponds to a combination of two or three different bases

8. The percentage of agarose that is appropriate varies depending on the size of the expected PCR product/s, but is commonly used between 1% and 2%. A 1% agarose gel is appropriate for fragments between 1–2 kb, whereas a higher agarose percentage such as 2% will separate better the smaller fragments (e.g., 300–500 bp).
9. Depending on the desired level of complexity of the in silico analyses, it may be a good option to seek support from experts to guarantee the generation of useful results.
10. Multiple established cell lines can be used for expression of pig and human proteins. It may be interesting to test a couple of them simultaneously as it is common to find major differences in expression levels between cell lines. We use extensively Chinese hamster ovary (CHO) and human embryonic kidney 293-EBNA cells. The CHO are good for recombinant expression and have normal culture requirements (DMEM/10% FBS). The 293-EBNA cells are cultured as well with DMEM/10% FBS, but it is necessary to add 250 μg/mL G418 to maintain the EBNA-1 system that increases protein expression. This system will only be advantageous when using plasmids with OriP such as pCEP4 (Invitrogen™).
11. There are several systems available for purification of tagged proteins by affinity chromatography that involve, for instance, prepacked ready-to-use columns. We explain the procedure for the GST tag because we find it especially efficient compared with others. Nevertheless, other tags such as the His tail combined with nickel columns or other purification systems can be used. The position of the tag in the fusion protein can be important depending on the objectives and the type of protein

studied. Most plasmids available place the GST at the amino-terminus of the protein, but this case can be a problem for type I proteins (extracellular amino terminus) if the tag is kept for detection. Thus, the fusion protein has to be designed carefully and always keep the sequence in frame for proper translation.

12. SPR is a complex technique, and we strongly recommend seeking advice from experts and technical support when designing and conducting the first SPR assay. It is also convenient to become familiar with the terms used in SPR technology. In particular, the protein (or other type of molecule) immobilized on the sensor surface is called the ligand (independently of its native biological function). The ligand can be immobilized directly or indirectly via an immobilized capturing molecule such as an anti-GST antibody for GST-fusion proteins. The molecule that is used free in solution to determine binding to the ligand is called analyte. It is recommended that the analyte has a high molecular weight to better detect changes in the refractive index.
13. All buffers for the Biacore should be filtered through a 0.22 μm filter and degassed just before the assay to prevent the formation of aggregates and bubbles. Ready-to-use buffers can be also obtained from the manufacturer.
14. Note that the MMLV reverse transcriptase works at 37 °C, but when using the AMV reverse transcriptase, the RT reaction should be conducted at 50 °C. In addition, we regularly use oligo$(dT)_{20}$, not the random hexamers, and have good success for the amplification of pig cDNA.
15. The annealing temperature depends on the primers sequence. We estimate the annealing temperature with the following formula: $T(°C) = (4 \times (G + C)) + (2 \times (A + T))$. In the case of degenerate primers, we determine the range of potential temperatures. We usually apply an annealing temperature between 50 °C and 65 °C that corresponds to 5°–10° below the calculated annealing temperature. If it fails, then reduce the annealing temperature as it will increase the chances of amplification. However, it also increases the chances to obtain undesired DNA products.
16. Today there are different DNA markers which are not as toxic as ethidium bromide, such as GelRed® and MidoriGreen®. However, these reagents have not being used long enough to fully discard any potential harmful effects in the long term. Therefore, it is recommended to use them carefully. Also, it is important not losing the staining in the area where the PCR product is expected to migrate. This scenario is especially relevant for small fragments or poor amplifications that produce weak signal. Visualize the gel immediately after electrophoresis

and excise the desired fragment if present, as the DNA diffuses and the signal can be lost. For this reason, we do not recommend to stain the gels after electrophoresis.

17. We recommend using a topoisomerase-based system for subcloning the PCR product as mentioned in Note 5. To this end, we usually purify the PCR fragment directly from the electrophoresis gel using a kit such as GENECLEAN® II (MP Biomedicals, Santa Ana, CA, USA). Minimally, the vector chosen for subcloning should allow sequencing of the insert with common primers such as M13 forward and/or reverse. In addition, some of the vectors will be adequate to be used for recombinant expression of the insert.
18. It is important to perform the transfection without antibiotic; otherwise, PEI will become toxic to the cells.
19. Note that neither transfection or transduction ensure that 100% of the cells express the recombinant protein unless the cells are subsequently selected. If stable and/or long-term recombinant expression is needed, then use an expression vector that contains a gene that confers resistance to a selection antibiotic such as G418 or puromycin. Take into account that different cell lines display different levels of resistance to such antibiotics and that each antibiotic kills in a particular time frame. Thus, it is advisable to check the literature, choose the appropriate antibiotic and conditions, and if necessary, do some testing. Always treat simultaneously unmodified cells in a small flask or dish to confirm they all die.
20. Keep aliquots form each step to analyze the outcome of the procedure and to ensure you do not lose your protein of interest in any intermediate step.

Acknowledgments

This work was supported by the European Union Seventh Framework Programme (FP7/2007–2013) under Grant Agreement no. 603049 (TRANSLINK), the SLT002/16/00445 grant funded by the Department of Health of the Generalitat de Catalunya, Ministerio de Economía y Competitividad-ISCiii (PI15/00181), and Fundació La Marató de TV3 (2007-082430), all granted to CC, and cofunded by FEDER funds, a way to build Europe. IDIBELL benefits from the support of CERCA. S.G.K. was additionally supported by an IDIBELL summer internship and M.U.-H. by an IDIBELL fellowship.

References

1. Ekser B, Li P, Cooper DKC (2017) Xenotransplantation: past, present, and future. Curr Opin Organ Transplant 22:513–521. https://doi.org/10.1097/MOT.0000000000000463
2. Yamada K, Sykes M, Sachs DH (2017) Tolerance in xenotransplantation. Curr Opin Organ Transplant 22:522–528. https://doi.org/10.1097/MOT.0000000000000466
3. Cooper DK, Ezzelarab MB, Hara H et al (2016) The pathobiology of pig-to-primate xenotransplantation: a historical review. Xenotransplantation 23:83–105. https://doi.org/10.1111/xen.12219
4. Butler JR, Martens GR, Estrada JL et al (2016) Silencing porcine genes significantly reduces human-anti-pig cytotoxicity profiles: an alternative to direct complement regulation. Transgenic Res 25:751–759. https://doi.org/10.1007/s11248-016-9958-0
5. Cooper DK, Ekser B, Ramsoondar J, Phelps C, Ayares D (2016) The role of genetically engineered pigs in xenotransplantation research. J Pathol 238:288–299. https://doi.org/10.1002/path.4635
6. Wang Q, Shi N, Shang Y et al (2017) Comprehensive molecular characterization of a transgenic pig expressing hCD46 gene. Gene 626:376–385. https://doi.org/10.1016/j.gene.2017.05.057
7. Groenen MA, Archibald AL, Uenishi H et al (2012) Analyses of pig genomes provide insight into porcine demography and evolution. Nature 491:393–398. https://doi.org/10.1038/nature11622
8. Uribe-Herranz M, Casinghino SR, Bosch-Presegué L, Fodor WL, Costa C (2011) Identification of soluble and membrane-bound isoforms of porcine tumor necrosis factor receptor 2. Xenotransplantation 18:131–146. https://doi.org/10.1111/j.1399-3089.2011.00634.x
9. Bruel T, Guibon R, Melo S et al (2010) Epithelial induction of porcine suppressor of cytokine signaling 2 (SOCS2) gene expression in response to Entamoeba histolytica. Dev Comp Immunol 34:562–571. https://doi.org/10.1016/j.dci.2009.12.017
10. Zismann V, Nourbakhsh M (2014) Rapid mapping of RNA 3′ and 5′ ends. Methods Mol Biol 1182:19–25. https://doi.org/10.1007/978-1-4939-1062-5_2
11. Nygard A-B, Jørgensen CB, Cirera S, Fredholm M (2010) Investigation of tissue-specific human orthologous alternative splice events in pig. Anim Biotechnol 21:203–216. https://doi.org/10.1080/10495398.2010.497729
12. Doolittle RF (2010) The roots of bioinformatics in protein evolution. PLoS Comput Biol 6: e1000875
13. Gao FS, Xu CB, Long YH, Xia C (2009) Secondary structure and 3D homology modeling of swine leukocyte antigen class 2 (SLA-2) molecules. Immunobiology 214:475–482. https://doi.org/10.1016/j.imbio.2008.11.001
14. Uribe-Herranz M, Costa C (2012) Cloning and expression analyses of pig genes. Methods Mol Biol 885:17–31. https://doi.org/10.1007/978-1-61779-845-0_2
15. Uribe-Herranz M, Costa C (2012) Basic analyses of proteins of interest for xenotransplantation. Methods Mol Biol 885:33–45. https://doi.org/10.1007/978-1-61779-845-0_3
16. Seebach JD, Comrack C, Germana S et al (1997) HLA-Cw3 expression on porcine endothelial cells protects against xenogeneic cytotoxicity mediated by a subset of human NK cells. J Immunol 159:3655–3661
17. Otamiri M, Nilsson KG (1999) Analysis of human serum antibody-carbohydrate interaction using biosensor based on surface plasmon resonance. Int J Biol Macromol 26:263–268

Chapter 3

Proteomic Protocols for Differential Protein Expression Analyses

Bent Honoré

Abstract

Protocols are described for preparation of various types of samples for differential protein expression proteomics using nano liquid chromatography—tandem mass spectrometry (nLCMS/MS). Fresh frozen tissue, formalin-fixed paraffin-embedded (FFPE) tissue, cells, pull-downs, immunoprecipitations, and plasma or serum can be processed. For transplantation studies, it should be noted that samples must be comparable; thus, similar cells or organ systems from the same species can be compared when only a minority of the proteins change expression levels. For xenotransplantation studies, caution should be taken, as different species possess proteins with differences in primary structure. The sample preparation technique is based on spin filter devices using sodium deoxycholate (SDC) as detergent for extraction of proteins. The detergent is removed using phase extraction with ethyl acetate, and samples can be used for label-free quantification (LFQ) analysis or peptides can be labeled with tandem mass tags (TMT).

Key words Proteomics, Differential protein expression analysis, Spin filter device-based preparation, FFPE, Tissue, Cells, Pull-downs, Plasma, Serum, Label-free quantification, TMT labeling, nLC-MS/MS

1 Introduction

Proteomic methods possess a huge potential for identifying proteins, their expression levels, and posttranslational modifications in various samples occurring from, for example, cells, tissues, serum, or plasma. Importantly, samples to be compared should be comparable (i.e., the major part of the proteins in the sample should have similar expression levels and only a minority of the proteins showing abundance changes) [1]. This case is crucial in transplantation studies. Similar cells or organ systems can be compared between the same species. However, in xenotransplantation, the primary structure of proteins in one species will vary considerably from another species even though the same cells or organ systems are studied. An important barrier in a pig-to-man xenotransplantation is the expression of the Galα1-3Galβ1-4GlcNAc-R

Cristina Costa (ed.), *Xenotransplantation: Methods and Protocols*, Methods in Molecular Biology, vol. 2110, https://doi.org/10.1007/978-1-0716-0255-3_3,

carbohydrate (α-Gal epitope) on pig endothelial cells. This epitope will react with antibodies present in the recipient. The synthesis of the αGal epitope may be prevented by the deletion of the α1-3-galactosyltransferase gene. Because wild-type mice also express the αGal epitope, mice have been used as knockout models for xeno-transplantation studies. Thus, we have performed proteomics to compare two organ systems in α1-3-galactosyltransferase knockout mice versus wild-type mice [2]. This scenario is possible because only a minority of proteins change expression level upon knockout.

A number of years ago, proteomic analysis was performed by using two-dimensional polyacrylamide gel electrophoresis (2D-PAGE) for protein separation and quantification, and mass spectrometry (MS) was used for identification of the proteins. In recent years, with the development of MS, this technique has now become the predominant tool for the analysis of proteins and their expression levels [3]. Of paramount importance to obtain valuable results is the methodology used to prepare samples for analysis. One consideration in this regard is using sufficient detergents to conduct efficient extraction of the proteins from cells or tissues and at the same time obtain samples that are compatible with down-stream analysis. Several methods have been described. Here we present a method based on the digestion of proteins on spin filter devices originally introduced by Manza et al. [4], subsequently developed by Wisniewski et al. [5], and further combined with preparation of formalin-fixed paraffin-embedded (FFPE) tissue [6]. The latter protocols use the efficient detergent sodium dodecyl sulphate (SDS) to lyse the samples and then urea to remove SDS again as this detergent, at least in high concentrations, is not suitable for subsequent MS analysis. A later systematic investigation of several detergents, digestion conditions, and detergent removal methods have revealed that using the natural detergent present in bile, sodium deoxycholate (SDC), with subsequent removal by phase transfer (PT) using ethyl acetate [7], outperforms the use of SDS [8].

We here present a modified preparation technique based on the spin filters that involves exchanging SDS with SDC. Samples may be used for label-free quantification (LFQ), or labeling may be performed using tandem mass tags (TMT). It is possible to analyze various types of samples, including fresh-frozen tissue, FFPE tissue, cells, pull-downs, immunoprecipitations, and plasma or serum.

2 Materials

All solvents for LC-MS analyses are made using chemicals of the highest purity available (e.g., Optima™ LC-MS grade).

1. A kit for TMT labeling can be obtained from Thermo Fisher Scientific: TMT10plex Mass Tag Labeling Kits and Reagents. The kit contains some of the solutions mentioned below.
2. Pierce High pH Reverse-Phase Peptide Fractionation Kit (Thermo Fisher Scientific). It is a convenient kit for fractionation of samples, although there are other possibilities.
3. Microcon Centrifugal Filter Devices 30 K (Millipore). Prepare filters as described by the manufacturer.
4. 100 mM Triethylammonium bicarbonate (TEAB) buffer. 1 M TEAB stock solution is diluted with water for lower concentrations. The 100 mM is made by tenfold dilution with water.
5. 200 mM Tris(2-carboxyethyl)phosphine hydrochloride (TCEP), 20 mM TEAB solution: Use two volumes of 0.5 M TCEP and add two volumes of water and one volume of 100 mM TEAB.
6. 375 mM iodoacetamide. Immediately before use, dissolve one tube of iodoacetamide (9.3 mg) with 132 μL of 100 mM TEAB. Protect solution from light.
7. 10% trifluoroacetic acid (TFA). This solution is made by diluting one volume of TFA with nine volumes of water.
8. 5% hydroxylamine, 90 mM TEAB: 50 μL of a 50% hydroxylamine solution is added to 450 μL of 100 mM TEAB.
9. Additional solvents needed are ethanol, ethyl acetate, and xylene.
10. Make a 1 μg/μL Pierce trypsin protease (MS Grade) solution by dissolving 20 μg in 20 μL storage solution or use 50 mM acetic acid (dilute 2.862 μL acetic acid by adding water to 1 mL). Store at −20 °C.
11. Lysis buffer: 5% SDC, 20 mM TEAB. Weigh 0.5 g SDC and dissolve in 0.2 mL 1 M TEAB and 9.8 mL water.
12. Digestion buffer: 0.5% SDC, 20 mM TEAB. Weigh 0.05 g SDC and dissolve in 0.2 mL 1 M TEAB and 9.8 mL water.
13. Urea buffer (for protein determinations): 8 M urea, 100 mM Tris–HCl, pH 8.5. Weigh 1.2114 g Tris-base and dissolve in 80 mL water. Adjust pH to 8.5 with HCl. Add 48.048 g urea and dissolve with stirring. Finally, the volume is adjusted to 100 mL with water.
14. Bovine serum albumin (BSA) standard. Various concentrations of standard solutions can be made by diluting the commercially available bovine serum albumin standard (7% solution NIST®, Sigma-Aldrich, Merck KGaA), a 7% BSA solution (7 g/ 100 mL). Dilute it in a convenient concentration range (e.g., 0.2–5 μg/μL).

15. Tryptophan standard. A standard curve of tryptophan is constructed by dissolving 3.2 mg tryptophan in 50 mL of the appropriate buffer, giving a concentration of 64 μg/mL. This may be diluted with buffer in twofold steps giving concentrations of 64, 32, 16, 8, 4, 2, 1, 0.5, 0.25, 0.125, 0.0625, 0.03125 μg/mL.

16. Infrared spectrometer such as Direct Detect Spectrometer (Merck KGaA, Darmstadt). This tool is convenient for measuring the protein concentration in samples before further processing (*see* **Note 4**).

17. Microplate reader such as Enspire (Perkin Elmer). Convenient for measuring peptide concentration in digested samples by fluorescence (*see* **Note 8**).

18. Nano liquid chromatography–tandem mass spectrometry (nLC-MS/MS) system. It can be any high-resolution mass spectrometer connected to a liquid chromatography system (e.g., Orbitrap Fusion (MS) connected to a Dionex UltiMate™ 3000 RSLC nano system (LC), Fig. 1).

19. A number of buffers should be made for the LC system and changed regularly, every week or every second week. Use an ultrasonic bath to degas the solvents. Buffer A: water/formic acid, 99.9/0.1 (vol/vol); Buffer B: acetonitrile/formic acid, 99.9/0.1 (vol/vol); loading solvent: water/acetonitrile/formic acid, 97.9/2/0.1 (vol/vol/vol); wash solvent WPS: water/acetonitrile/formic acid, 19.9/80/0.1 (vol/vol/vol); rear seal wash solvent: water/methanol: 90/10 (vol/vol).

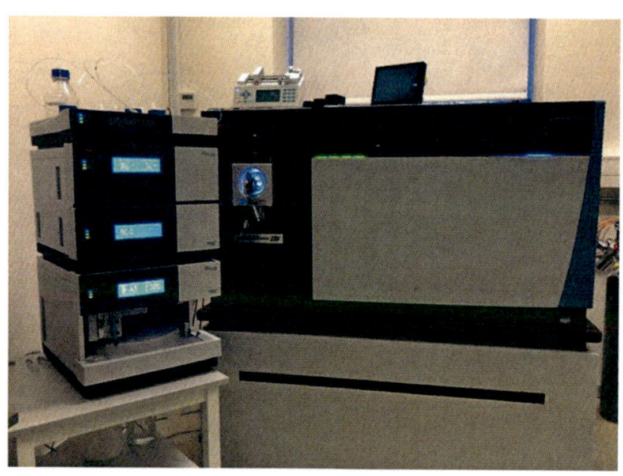

Fig. 1 nano liquid chromatography—tandem mass spectrometry (nLC-MS/MS) used for differential expression proteomics. Any high-resolution mass spectrometer can be used to analyze samples. Here we display a Dionex UltiMate™ 3000 RSLC nano system used to separate peptides by reverse phase chromatography before analysis using an Orbitrap Fusion Tribrid MS from Thermo Fisher Scientific Instruments

3 Methods

The techniques used to prepare samples depend on the type of samples to be analyzed. FFPE tissue needs to be deparaffinized and rehydrated before digestion. For fresh-frozen tissue or cells, this point is skipped as described below. Pull-downs, immunoprecipitations, plasma, or serum samples can also be analyzed.

3.1 Preparation of FFPE Tissue and Extraction of Proteins

1. FFPE sections (e.g., 10-μm-thick slices) are incubated in 1 mL of xylene in 2-mL microfuge tubes with gentle agitation at room temperature for 5 min after which, the solution is removed. Repeat this step (*see* **Note 1**).
2. Incubate FFPE sections with 1 mL of absolute ethanol and agitate for 5 min. Repeat this step one time.
3. Rehydrate slices with 1-mL volumes of ethanol/water (vol/vol) starting with 97/3, then 80/20, 70/30, and finally 50/50, each for 5 min with gentle agitation. Remove the ethanol/water and vacuum dry the sample.
4. Mix the dried FFPE tissue with lysis buffer (5% SDC, 20 mM TEAB) in a tissue to buffer ratio of approximately 1:20 (approx. 300 μL).
5. Incubate in a heating block with agitation (600 rpm) at 99 °C for 1 h and let it slowly chill to room temperature. This step presumably cleaves the inter- and intramolecular cross-links in the proteins in FFPE tissue [9].
6. Homogenize on ice (*see* **Note 2**).
7. Sonicate the suspension on ice (*see* **Note 3**).
8. Centrifuge the extract at 16,000 × *g* for 10 min. Take supernatant to a new 2-mL tube.
9. Measure the protein concentration by infrared spectrometry or fluorescence (*see* **Note 4**). Typically use 100 μg in 100 μL and continue with steps described in Subheading 3.4.

3.2 Preparation of Frozen Tissue or Cells and Extraction of Protein

1. Mix the frozen tissue or pelleted cells with lysis buffer (5% SDC, 20 mM TEAB) in a tissue to buffer ratio of approximately 1:20 (approx. 300 μL) and homogenize on ice if necessary (*see* **Note 2**).
2. Sonicate the suspension on ice (*see* **Note 3**).
3. Incubate in a heating block with agitation (600 rpm) at 99 °C for 5 min and let it slowly chill to room temperature.
4. Centrifuge the extract at 16,000 × *g* for 10 min. Take the supernatant to a new 2-mL tube.

5. Measure the protein concentration by infrared spectrometry or fluorescence (*see* **Note 4**). Typically use 100 μg in 100 μL and continue with Subheading 3.4.

3.3 Preparation of Proteins from Pull-Downs, Immunoprecipitations, Plasma, or Serum

1. Mix the protein solution (for plasma or serum, *see* **Note 5**) with digestion buffer (0.5% SDC, 20 mM TEAB). Optionally, wash a number of times using filter units (Microcon Centrifugal Filter Devices 30 K from Milipore) by adding 100 μL digestion buffer with subsequent centrifugation (*see* **Note 6**), and finally add 100 μL digestion buffer and proceed with steps in Subheading 3.4 (*see* **Note 5**).

3.4 Reduction and Alkylation

1. Add 5 μL of 200 mM TCEP, 20 mM TEAB solution to 100 μL of sample and incubate at 55 °C for 1 h.
2. Add 5 μL of freshly prepared 375 mM iodoacetamide to the sample and incubate for 30 min protected from light at room temperature.
3. Add the lysate to the filter unit, if not already done under Subheading 3.3 (Microcon Centrifugal Filter Devices 30 K from Millipore) and centrifuge at 14,000 × *g* for 15 min (*see* **Note 6**).
4. Add 100 μL of digestion buffer in the filter unit and centrifuge at 14,000 × *g* for 15 min. Repeat this step.

3.5 Protein Digestion

1. Add 100 μL of digestion buffer (0.5% SDC, 20 mM TEAB) to the filter unit (*see* **Note 7**).
2. Add trypsin (2.5 μg to 100 μg protein) and mix at 600 rpm in a thermomixer for 1 min.
3. Incubate the units in a wet chamber at 37 °C overnight.
4. Transfer the filter unit to new collection tubes.
5. Centrifuge the filter units at 14,000 × *g* for 10 min. Repeat with 100 μL of digestion buffer one or two times.

3.6 Phase Transfer

1. Add an equal volume of ethyl acetate.
2. Add 10% TFA to a final concentration of 0.5% (vol/vol) of the final volume obtained in **Step 5** of Subheading 3.5 and shake for 1 min.
3. Centrifuge at max. speed for 2 min.
4. Save the lower phase by removal of the upper phase.
5. Add an equal volume of ethyl acetate and shake for 1 min.
6. Centrifuge at max. speed for 2 min. Save the lower phase by removal of the upper phase.
7. Repeat **Steps 5** and **6**.
8. Dry in a vacuum centrifuge.

9. Resuspend in 100 mM TEAB, measure the peptide concentration (e.g., by fluorescence) (*see* **Note 8**), and adjust to a concentration of 1 μg/μL.

3.7 Peptide Labeling with Tandem Mass Tags (TMT)

1. Add 41 μL anhydrous acetonitrile to each tube with TMT reagent (10 plex or other; *see* **Note 9**) and add the label to 100 μL peptide (100 μg). Incubate for 1 h at room temperature. Add 8 μL of 5% hydroxylamine (in 90 mM TEAB) as described in the kit. Mix equal amounts of each TMT-labeled samples. Store at −80 °C. Less amounts of peptide may be labeled if the volumes are adjusted accordingly. Fractionation of samples is recommended to increase number of identified proteins (*see* **Note 10**). If more than 10 samples should be compared, then it is necessary to measure a standard sample together with each series of 10 plex measurements in order to take instrument and measurement variations into account (*see* **Note 9**). The standard can be made by combining equal amounts from each sample to be measured. One or two channels can then be reserved for the standard (e.g., if two channels are used, then there will be eight channels left for samples to be analyzed for each 10 plex series).

3.8 Quantitative Mass Spectrometry Using nLC-MS/MS

1. Unlabeled peptide samples from Subheading 3.6 may be analyzed by label-free quantitative nano liquid chromatography—tandem mass spectrometry LFQ nLC-MS/MS. We use an Orbitrap Fusion Tribrid instrument from Thermo Fisher Scientific Instruments connected to a Dionex UltiMate™ 3000 RSLC nano system. The gradient is constructed by mixing Buffer A with increasing volumes of Buffer B. For LFQ analysis, the MS is operated with the universal method for achieving raw data. Load a similar amount of 1 or 2 μg of digested peptide per analysis. Several parameters can be modified in order to increase the number of identified protein groups (*see* **Note 11**). For unlabeled samples, it is important that the combined LC-MS system works properly (*see* **Note 12**).
2. TMT-labeled peptides from 3.6 is analyzed by MS using (e.g., TMT SPS MS3).
3. Raw data are then imported into software in order to analyze the data (*see* **Note 13**).

4 Notes

1. If necessary, this step may be repeated more than two times in order to remove the paraffin completely, and no centrifugation should be performed between the steps.

2. Homogenization: Homogenizer VDI 12 from VWR used at maximum speed for 30 s.
3. Sonication: Sonicator Q125 from QSonica LLC. Settings: amplitude, 75%, three sonications for 59 s with 59 s rest in between.
4. Measurement of protein concentration can be performed conveniently by infrared spectrometry (Direct Detect Spectrometer, Merck KGaA, Darmstadt) using 2 μL volumes with bovine serum albumin as a standard (7% solution, NIST®, Standard Reference Material from Sigma-Aldrich, Merck). The concentration of this standard is around 70 μg/μL and may be diluted with appropriate buffer to concentrations between approximately 0.2 and 5 μg/μL. Alternatively, concentrations can be measured by fluorescence emission at 350 nm and excitation at 295 nm by using a microplate reader (e.g., EnSpire (PerkinElmer)). A standard curve of tryptophan is constructed from 3.2 mg tryptophan dissolved in 50 mL of the appropriate buffer giving a concentration of 64 μg/mL. This substance may be diluted with buffer in twofold steps giving concentrations of 64, 32, 16, 8, 4, 2, 1, 0.5, 0.25, 0.125, 0.0625, 0.03125 μg/mL. Measurements of protein and tryptophan are performed in a urea buffer (8 M urea, 100 mM Tris–HCl, pH 8.5) in order to denature the proteins. Similar final volumes are used for samples and standards. A final 100 μL solution is made by using 3 μL of standard to 97 μL urea which may be made from 3.3 μL diluted with 106.7 μL and finally adding 100 μL to the microplate. For optimal precision, determinations can be performed in triplicate. For human and mouse proteins, it can be anticipated that 0.0117 g of tryptophan corresponds to 1 g of protein [10]. Thus, 1 μg of tryptophan corresponds to 85.47 μg of protein.
5. Plasma and serum contain proteins in a very high dynamic range with more than ten orders of magnitude difference between the highest and the lowest abundant proteins [11]. If plasma or serum are analyzed directly, then it is a problem that the highly abundant proteins mask lower abundant proteins. In order to perform proteomic analysis, it is necessary to work on circumventing this issue. One possibility is to deplete samples for some of the most abundant proteins. This process may be performed with the Proteome Purify™ 12 kit (R&D Systems) that uses an antibody containing resin that immunodepletes for the 12 highest abundant proteins. Samples end up in PBS buffer with 0.02% sodium azide. Another principle is to compress the dynamic range of proteins by using the ProteoMiner™ protein enrichment kit (Bio-Rad Laboratories, Inc.) that uses hexapeptide libraries attached to beads in order to enrich for low and medium abundant

proteins and deplete samples for high abundant proteins [12]. These samples end up in 5% acetic acid and 8 M urea with 2% CHAPS. It is important to efficiently rinse the sample for detergents that may interfere with subsequent trypsin digestion and MS analysis. We have successfully analyzed protein pull-downs with a buffer composition including 10% glycerol and 1% Triton X-100. The samples were rinsed four times with digestion buffer using filter units (Microcon Centrifugal Filter Devices 30 K from Milipore), (*see* **Note 6**). Repeated centrifugations may remove some detergents. Otherwise, subsequent centrifugation steps with urea may clean up samples with SDS [5]. Another possibility to remove several detergents is to use phase transfer with ethyl acetate [13] essentially as described in Subheading 3.6.

6. The filter devices should be prepared as given by the manufacturer. Although the filter units may have an unpredictable loss of sample, we find them very convenient, because it is possible to rinse and change solvent of the sample by repeated centrifugations. We estimate the recovery generally to be well above 50%. Even though the molecular cut-off of the filters is 30 kDa, they are useful for preparation of samples with proteins way below 30 kDa probably because the proteins are denatured [14].
7. Trypsin digestion activity is higher by the addition of 0.01%–1% of SDC than in the absence of SDC. Thus, 0.5% is used for the digestion step as previously recommended [7].
8. Measurement of peptide concentration after digestion is conveniently performed by fluorescence. The tryptophan standard and samples should be measured in the appropriate buffer (e.g., TEAB, 0.1% formic acid, or others). Also, here, the peptides are dissolved in a volume a little larger than 100 μL (e.g., 120 μL), and measurements are then performed using 100 μL. The solution is saved after measurement. Protein concentrations may be estimated by the tryptophan measurements (*see* **Note 4**). If measurements are performed in a volatile buffer such as TEAB, then the sample can be vacuum dried and then redissolved at, for example, 1 μg/μL.
9. TMT isobaric label reagent set may be obtained in various kits for multiplex analysis (duplex, sixplex, 10-plex, and 11-plex, from Thermo Fisher Scientific) applicable for high-resolution mass spectrometers. If the number of samples included in the analysis exceeds the capacity of the kits, then it is still possible to use them by including a standard pooled from all the samples to be measured. For example, using a 10-plex kit and saving one or two channels for the standard leaves eight or nine samples that can be measured in each run [15]. A 10-plex kit contains

the labels (126, 127N, 127C, 128N, 128C, 129N, 129C, 130N, 130C, 131). Equal amounts of each labeling are mixed. For each 10-plex series, the total measurement in each channel is adjusted to the same level. The signals in the standard channels may be averaged (geometric mean).

10. Fractionation of samples is recommended to increase the number of proteins identified. A number of possibilities are available. We use a High pH Reverse-Phase Peptide Fractionation Kit from Pierce. The sample is dissolved in 300 μL of 0.1% TFA and separated in fractions using increasing concentrations of acetonitrile.

11. A number of parameters can be changed in order to increase the number of identified proteins. Chromatography column lengths can be increased, gradient times can be increased, samples may be analyzed a number of times (replicates, triplicates, etc.), and samples can be fractionated. A backside, however, is that the instrument time also increases. Column lengths for the Easy Spray unit may vary from, for example, 15 to 25, 50, or 75 cm. Generally, a longer length of the column gives a better separation with an increase in the number of peptides that can be identified. Longer column lengths also require longer gradient times. Gradient times may vary between, for example, 1 h (for shorter columns) and up to around 4–5 h (for longer columns). Our rough experience is that the approximate increase in number of identified proteins with each hour of increase in gradient using the same type of column is around one-third from 1 h to 2 h, about one half from 1 h to 3 h, and about two-thirds from 1 h to 4 h. The use of technical replicates is another way to increase the number of proteins analyzed. Using replicates instead of a single determination increases identifications about 15%, triplicates increases about 20%, and quadruplicates increases about 25%, as compared with a single measurement. Fractionation of samples is very efficient to increase the number, although it also increases the analysis time considerably. One possibility is to fractionate with the Pierce High pH Reversed-Phase Peptide Fractionation Kit giving eight fractions from each sample. This scenario increases identifications about twofold.

12. For LFQ analysis to work, it is necessary to have all parts of the system working properly with regular preventive maintenance. The LC system should be in optimal condition, change liquids regularly, for example, every 1 or 2 weeks. It is necessary to have good stability of the spray for several days with the same column for the whole series. It is a good practice to run samples in technical replicates with several hours or days between the replicates. After analysis, a principal component analysis may indicate whether the technical replicates are reproducible. Also,

limit the time the samples are placed in the autosampler at 5 °C before analysis. In our experience, they may work properly, at least for up to 2 days. However, several days is not recommended.

13. MaxQuant [16] is a freely available software which may be downloaded upon registration from http://www.biochem.mpg.de/5111795/maxquant. The software can analyze LFQ data and labeled peptides. The output should be further processed in, for example, Perseus [17], where data may be filtered, analyzed statistically, and reported. Another option is the commercially available Proteome Discoverer program, which also may be used to analyze LFQ and labeled data.

Acknowledgment

The skillful technical help of Mona Britt Hansen is highly appreciated. Financial support for MS instruments was obtained from A. P. Møller og Hustru Chastine Mc-Kinney Møllers Fond til almene Formaal.

References

1. Cox J, Hein MY, Luber CA, Paron I, Nagaraj N, Mann M (2014) Accurate proteome-wide label-free quantification by delayed normalization and maximal peptide ratio extraction, termed MaxLFQ. Mol Cell Proteomics 13(9):2513–2526. https://doi.org/10.1074/mcp.M113.031591
2. Thorlacius-Ussing L, Ludvigsen M, Kirkeby S, Vorum H, Honoré B (2013) Proteomic analysis of tissue from alpha1,3-galactosyltransferase knockout mice reveals that a wide variety of proteins and protein fragments change expression level. PLoS One 8(11):e80600. https://doi.org/10.1371/journal.pone.0080600
3. Ludvigsen M, Honoré B (2018) Trancriptomics and proteomics: integration? In: eLS. John Wiley & Sons, Ltd., Chichester. https://doi.org/10.1002/9780470015902.a0006188.pub2
4. Manza LL, Stamer SL, Ham AJ, Codreanu SG, Liebler DC (2005) Sample preparation and digestion for proteomic analyses using spin filters. Proteomics 5(7):1742–1745. https://doi.org/10.1002/pmic.200401063
5. Wisniewski JR, Zougman A, Nagaraj N, Mann M (2009) Universal sample preparation method for proteome analysis. Nat Methods 6 (5):359–362. https://doi.org/10.1038/nmeth.1322
6. Ostasiewicz P, Zielinska DF, Mann M, Wisniewski JR (2010) Proteome, phosphoproteome, and N-glycoproteome are quantitatively preserved in formalin-fixed paraffin-embedded tissue and analyzable by high-resolution mass spectrometry. J Proteome Res 9(7):3688–3700. https://doi.org/10.1021/pr100234w
7. Masuda T, Tomita M, Ishihama Y (2008) Phase transfer surfactant-aided trypsin digestion for membrane proteome analysis. J Proteome Res 7(2):731–740. https://doi.org/10.1021/pr700658q
8. Leon IR, Schwammle V, Jensen ON, Sprenger RR (2013) Quantitative assessment of in-solution digestion efficiency identifies optimal protocols for unbiased protein analysis. Mol Cell Proteomics 12(10):2992–3005. https://doi.org/10.1074/mcp.M112.025585
9. Yamashita S (2007) Heat-induced antigen retrieval: mechanisms and application to histochemistry. Prog Histochem Cytochem 41 (3):141–200. https://doi.org/10.1016/j.proghi.2006.09.001
10. Wisniewski JR, Gaugaz FZ (2015) Fast and sensitive total protein and peptide assays for proteomic analysis. Anal Chem 87

(8):4110–4116. https://doi.org/10.1021/ac504689z

11. Anderson NL, Anderson NG (2002) The human plasma proteome: history, character, and diagnostic prospects. Mol Cell Proteomics 1(11):845–867
12. Guerrier L, Righetti PG, Boschetti E (2008) Reduction of dynamic protein concentration range of biological extracts for the discovery of low-abundance proteins by means of hexapeptide ligand library. Nat Protoc 3 (5):883–890. https://doi.org/10.1038/nprot.2008.59
13. Yeung YG, Nieves E, Angeletti RH, Stanley ER (2008) Removal of detergents from protein digests for mass spectrometry analysis. Anal Biochem 382(2):135–137. https://doi.org/10.1016/j.ab.2008.07.034
14. Wisniewski JR, Zielinska DF, Mann M (2011) Comparison of ultrafiltration units for proteomic and N-glycoproteomic analysis by the filter-aided sample preparation method. Anal Biochem 410(2):307–309. https://doi.org/10.1016/j.ab.2010.12.004
15. Plubell DL, Wilmarth PA, Zhao Y, Fenton AM, Minnier J, Reddy AP, Klimek J, Yang X, David LL, Pamir N (2017) Extended multiplexing of tandem mass tags (TMT) labeling reveals age and high fat diet specific proteome changes in mouse epididymal adipose tissue. Mol Cell Proteomics 16(5):873–890. https://doi.org/10.1074/mcp.M116.065524
16. Tyanova S, Temu T, Cox J (2016) The MaxQuant computational platform for mass spectrometry-based shotgun proteomics. Nat Protoc 11(12):2301–2319. https://doi.org/10.1038/nprot.2016.136
17. Tyanova S, Temu T, Sinitcyn P, Carlson A, Hein MY, Geiger T, Mann M, Cox J (2016) The Perseus computational platform for comprehensive analysis of (prote)omics data. Nat Methods 13(9):731–740. https://doi.org/10.1038/nmeth.3901

Chapter 4

Specific Detection of Neu5Gc in Animal Tissues by Immunohistochemistry

Anu Paul, Shirley Bachar Abramovitch, and Vered Padler-Karavani

Abstract

One of the major obstacles in xenotransplantation is tissue expression of the non-human mammalian carbohydrate *N*-glycolylneuraminic acid (Neu5Gc). This 9-carbon backbone acidic sugar is the hydroxylated form of *N*-acetylneuraminic acid (Neu5Ac), and both constitute the two most common sialic acid types in mammals. Loss of CMP-Neu5Ac hydroxylase encoding gene in humans dictates the immunogenic nature of Neu5Gc-containing xenografts. Here we describe an immunohistochemistry method for the detection of Neu5Gc in mammalian-derived tissues using affinity-purified chicken anti-Neu5Gc IgY. Specificity is further demonstrated by competitive inhibition with free Neu5Gc or Neu5Gc-glycoproteins, but not with Neu5Ac or Neu5Ac-glycoproteins. This method can be used to evaluate potential Neu5Gc-immunogenicity of xenografts.

Key words Sialic acid, *N*-acetylneuraminic acid (Neu5Ac), *N*-glycolylneuraminic acid (Neu5Gc), Anti-carbohydrate antibodies, Xenotransplantation, Immunohistochemistry

1 Introduction

Cross-species transplantation (e.g., pig to humans) had been long envisioned as a potential solution for the limited supply of transplantable cells, tissues, and organs for humans, yet still hampered by adverse immune reactions in the recipient host [1–3]. One of the major obstacles for successful xenotransplantation is the humoral immune response developed against the xenografts [2]. While at first protein antigens seemed to be the main immunological targets, carbohydrate antigens have also been noticed to play a major role in the rejection of xenografts [4]. One of the first targets to be identified is the galactose-α1,3-galactose (αGal) [4–6]; however, the non-human acidic carbohydrate *N*-glycolylneuraminic acid (Neu5Gc) was then discovered as another culprit [7–10]. It differs from the human carbohydrate *N*-acetylneuraminic acid (Neu5Ac) by a single oxygen atom [7]. Humans cannot synthesize Neu5Gc [11], and all have serum anti-Neu5Gc antibodies [12–16] that

Cristina Costa (ed.), *Xenotransplantation: Methods and Protocols*, Methods in Molecular Biology, vol. 2110, https://doi.org/10.1007/978-1-0716-0255-3_4, © Springer Science+Business Media, LLC, part of Springer Nature 2020

recognize multiple presentations of diverse Neu5Gc-expressing glycans, glycoproteins, and glycolipids [16, 17]. These antibodies have been proposed to mediate the potential deleterious outcomes to the host upon exposure to Neu5Gc-expressing xenograft [4, 9, 10, 17], and recent progress in genetic engineering suggests that elimination of such carbohydrate antigens is a promising approach for reduced immunogenicity [3, 6, 18, 19].

Here we describe a method to assess the expression of Neu5Gc in tissues using a well-characterized anti-Neu5Gc antibody [16, 20, 21]. This mono-specific polyclonal anti-Neu5Gc antibody was produced in chickens, as they do not produce intrinsic Neu5Gc, similar to humans [20]. Developed antibodies were affinity-purified from Neu5Gc-immunized chicken serum to high purity and specificity [20]. While this antibody can be used for a variety of applications (e.g., immunohistochemistry, enzyme-linked immunosorbent assay (ELISA), Western blot, and flow cytometry), here we focus on the immunohistochemistry method. For specific detection of Neu5Gc in mammalian-derived tissues using this affinity-purified chicken anti-Neu5Gc IgY, we incorporated parallel competitive inhibition with the specific inhibitors Neu5Gc or Neu5Gc-glycoproteins and with the non-specific inhibitors Neu5Ac or Neu5Ac-glycoproteins. This method can be used to evaluate potential Neu5Gc-immunogenicity of xenografts.

2 Materials

2.1 Animals

1. C57BL/6 wild-type mice. Mice are bred, maintained, and used according to Animal Care and Use Committee protocols approved by the corresponding ethical committee/s (*see* **Note 1**).

2.2 Reagents

2. Affinity-purified polyclonal chicken anti-Neu5Gc IgY antibody (Cat. No.: 146903, BioLegend, San Diego, CA, USA).
3. Peroxidase-conjugated AffiniPure donkey-anti-chicken IgY (Cat. No.: 703-035-55, Jackson ImmunoResearch Laboratories, Inc., West Grove, PA, USA).
4. Purified chicken IgY isotype control (Cat. No.: 402101, BioLegend, San Diego, CA, USA).
5. *N*-glycolylneuraminic acid, NGNA (Cat. No.: 1758–9850, Inalco S.P.A., Milano, Italy).
6. *N*-acetylneuraminic acid (NANA, Sialic acid) (Cat. No.: 08371–81, Nacalai Tesque, Kyoto, Japan).
7. Chimpanzee serum–Chimpanzee sera (*see* **Note 2**).
8. Chicken serum (Cat. No.: C5405, Sigma-Aldrich, St. Louis, MO, USA).

9. DAB substrate kit (Cell Marque; Cat. No.: 957D-20, Sigma-Aldrich, St. Louis, MO, USA).
10. Paraformaldehyde, 96% extra pure, ACROS Organics (Cat. No.: 416780010, Fisher Scientific, MA, USA).
11. Sucrose (Cat. No.: S0389, Sigma-Aldrich, St. Louis, MO, USA).
12. Ethylene glycol (Cat. No.: 000518020100, Bio-Lab LTD, Jerusalem, Israel).
13. Tissue Plus liquid Optimal Cutting Temperature Compound (OCT). (Cat. No.: 4586, Scigen, Gardena, CA, USA).
14. Cold-water fish gelatin (Cat. No.: G7765, Sigma, St. Louis MO, USA).
15. Fluoromount-G® (Cat. No.: 0100-01, SouthernBiotech, Birmingham, AL USA).

2.3 Solutions

1. 0.2 M phosphate buffer A: Mix 1 L of double distilled water (DDW) with 24 g of sodium phosphate monobasic anhydrous (NaH_2PO_4 FW 120.0).
2. 0.2 M phosphate buffer B: Mix 1 L of DDW with 28.4 g of sodium phosphate dibasic anhydrous (Na_2HPO_4 FW 141.96).
3. 0.2 M phosphate buffer (1 L) by mixing 230 mL of buffer A with 770 mL of buffer B. If not used immediately, filter solution using 0.22-μm filter.
4. Tissue collecting solution (TCS, 50 mM phosphate buffer containing 28.6% ethylene glycol and 23.8% glycerol): Mix 1 L of 0.1 M phosphate buffer with 600 mL of ethylene glycol and 500 mL of glycerol.
5. 1 M phosphate buffer C: Mix 142 g of sodium phosphate dibasic anhydrous (Na_2HPO_4) with DDW to complete volume to 1 L.
6. 1 M phosphate buffer D: Mix 120 g of sodium phosphate monobasic anhydrous (NaH_2PO_4) with DDW to complete volume to 1 L.
7. 4% paraformaldehyde solution. A series of steps are needed to prepare this solution. First, using a glass beaker, prewarm 1 L of DDW to 60 °C. Mix 900 mL of prewarmed water with 77 mL of phosphate buffer C, and then warm the solution to 60 °C. Add 40 g of paraformaldehyde to yield a 4% solution into the glass beaker on a stir plate in a ventilated hood. The paraformaldehyde will not dissolve unless the pH is basic (*see* **Note 3**). Add 22.6 mL of phosphate buffer D. Slowly adjust to pH 6.9–pH 7.4 by adding 1 N NaOH dropwise using a pipette until the solution clears and all powder had completely dissolved. Once dissolved, the solution should be cooled and

filtered. This solution can be aliquoted and frozen or stored at 2 °C–8 °C for up to one month, although it is best to prepare it fresh for every use.

8. 30% sucrose solution: Add 30 g of sucrose to DDW and complete volume to 100 mL.
9. PBS × 1: Mix 10 mL of PBS × 10 with 90 mL of DDW.
10. Blocking buffer (0.5% fish gelatin in PBST): Mix 49.45 mL of PBS pH 7.4, 50 μL of Tween-20 (0.1%), 500 μL of cold-water fish gelatin (fish gelatin stock is 40%–50%). Both Tween-20 and fish gelatin are viscous solutions; hence, cut the edge of the pipettor tip before collecting the solutions, and then wash the tip thoroughly with the PBS solution for proper application of these materials.
11. 40 mM Neu5Ac (MW = 309.273 g/mol) in PBS at pH 7 (12.37 mg/mL, ×2 solution): Dissolve 61.85 mg of Neu5Ac with 4.95 mL of DDW, adjust to pH 7 with NaOH (confirm pH with pH paper), and then Q.S. to 5 mL with DDW (*see* **Notes 4–6**).
12. 40 mM Neu5Gc (MW = 325.2693 g/mol) in PBS at pH 7 (13 mg/mL; ×2 solution): Dissolve 65 mg of Neu5Gc with 4.95 mL of DDW, adjust to pH 7 with NaOH (confirm pH with pH paper), and then Q.S. to 5 mL with DDW (*see* **Notes 4–6**).

2.4 Materials and Equipment

1. Dry ice.
2. 1.5-mL microtubes.
3. Peel-away disposable plastic tissue-embedding molds truncated size 12 × 12 mm.
4. Super frost PLUS slides white 76 × 26 mm 90 ground 90 corners (Cat. No.: BN9308C, BAR-NAOR Ltd., Ramat-Gan, Israel).
5. Microscope cover glass, thickness #1.0, 0.13–0.17 mm, 24 × 60 mm, 10 × 100 pcs. (Cat. No.: BN1052441C, BAR-NAOR Ltd., Ramat-Gan, Israel).
6. Microtome such as MICROM HM 450 (Cat. No.: 22-050-856, Thermo Scientific, MA, USA).
7. 96-well cell culture cluster, flat bottom with lid, polystyrene (Cat. No.: 3596, Corning Incorporated, NY, USA).
8. Disposable wipers such as Kimwipes.
9. Petri dish plastic 90 mm.
10. Hirschmann Pasteur pipettes 150 mm.
11. Costar® 48-well clear TC-treated multiple well plates (Cat. No.: CC-3548, Corning Incorporated, NY, USA).

12. *da Vinci* painting brush, series 373, size 2.
13. Blunt-nosed thumb forceps with serrated tips for increased grip.
14. Blunt-end forceps.

3 Methods

3.1 Preparation of Mouse Liver Tissues for Analysis (See Note 1)

1. Prepare 100 mL of PBS × 1 from the ×10 solution and cool it on ice.
2. Sacrifice C57BL/6 mouse in a CO_2 chamber, and excise the liver out.
3. Wash liver three times in 30 mL cold PBS in a 100 mL glass beaker. Cut the liver into small pieces (~10 × 10 × 2 mm).
4. Fix cut pieces of mouse liver tissues in 4% paraformaldehyde. Put each tissue section in 1 mL of 4% paraformaldehyde in a 1.5-mL microtube and keep overnight at 4 °C.
5. Using forceps (blunt-nosed thumb forceps with serrated tips for increased grip), transfer tissue into a new Eppendorf containing 1 mL 30% sucrose, and incubate for 2–3 h at 4 °C, until tissue settles down at the bottom of the tube (*see* **Note 7**).

3.2 Tissue Sectioning with Microtome

3.2.1 Prepare Microtome for Tissue Sectioning

1. Sectioning of tissues is done in a microtome instrument. The microtome tissue-holding platform must be leveled and adjusted before each sectioning (use bubble level to adjust). After leveling the surface, set the microtome to cut tissues into 30-μm-thick sections.
2. On the tissue mounting stand, set a square block of liquid OCT compound for each tissue to be sectioned (*see* **Note 8**).
3. Place crushed dry ice around the mounting stand to maintain cold temperature on the stand and allow OCT squares to freeze until it becomes opaque (*see* **Note 9**).
4. Replenish dry ice around the mounting stand throughout the tissue sectioning procedure (*see* **Note 10**).

3.2.2 Prepare Tissues for Cutting

1. Place 3–5 mL of PBS in a sterile Petri dish.
2. Place one drop of liquid OCT per tissue on another sterile Petri dish.
3. In order to prepare tissue for sectioning on OCT blocks, use blunt-end forceps (blunt-nosed thumb forceps with serrated tips for increased grip) and gently lift one fixed tissue piece (after overnight incubation in sucrose; *see* Subheading 3.1, **Step 5**) to gently dry the tissue with a Kimwipe.

4. Using a Pasteur pipet, place one drop of PBS on the center of one frozen OCT block on the microtome mounting stand. Then simultaneously take the dried tissue with forceps, dip it quickly and gently into the liquid OCT drop, and then immediately place it on top of the PBS drop on the frozen OCT block with the dipped OCT tissue facing down toward the PBS drop (*see* **Note 11**).
5. Flatten the tissue on the frozen OCT block using blunt-end forceps and then cover it locally with finely crushed dry ice (*see* **Note 12**).
6. Allow tissue to completely freeze before sectioning.

3.2.3 Sectioning of Tissue

1. Prepare a 96-well plate with TCS solution for collecting sections of tissues: place 200 μL of TCS per well in 96-well plates.
2. Place 3–5 mL of PBS in a sterile Petri dish.
3. Using a brush, remove the dry ice from the tissue to be sectioned, and then cut the OCT-embedded tissue sections into 30-μm-thick sections using the microtome.
4. Using a brush, pick up each cut section and place it in PBS in the Petri dish (*see* **Note 13**).
5. Using a brush, pick up each cut section from the PBS and place it individually into a well in the 96-well plate containing TCS solution.
6. Repeat **Steps 3–5** for each tissue block to be sectioned.
7. Properly label the 96 wells to identify the tissue sections.
8. Cover the sectioned-tissue plates with a cling film or plastic nylon cover.
9. Store at −20 °C until further use (up to 2 weeks; *see* **Note 14**).

3.3 Immuno-spi1; histochemistry Protocol

Tissue immunohistochemistry staining protocols involve the following steps: fixed tissue sections are washed, blocked, exposed to primary antibody, washed again, exposed to labeled secondary antibody, then washed, and staining allowed to develop with substrate. For tissue Neu5Gc-staining and specificity assessment, the primary antibody chicken anti-Neu5Gc IgY is used and pre-complexed with selected specific (Neu5Gc or Neu5Gc-glycoproteins) or non-specific (Neu5Ac or Neu5Ac-glycoproteins) inhibitors, as detailed in Table 1 (*see* **Note 15**).

3.3.1 Washing of Tissues

1. For tissue staining, wash the tissue sections with PBS as detailed below.
2. Prepare a 48-well plate with 300 μL of PBS pH 7.4 per well.
3. Using a brush, pick up each cut tissue section from the 96-well plate (tissues that were stored in TCS in −20 °C) and transfer it gently into the PBS in the 48-well plate.

Table 1
Description of tissue staining scheme

No.	Primary antibody	Inhibitor	Secondary antibody	Expected reactivity on a Neu5Gc-positive tissue
1.	Control IgY	–	HRP-donkey anti-chicken IgY	Negative
2.	Anti-Neu5Gc IgY	–	HRP-donkey anti-chicken IgY	Positive
3.	Anti-Neu5Gc IgY	Neu5Ac	HRP-donkey anti-chicken IgY	Positive
4.	Anti-Neu5Gc IgY	Neu5Gc	HRP-donkey anti-chicken IgY	Negative
5.	Anti-Neu5Gc IgY	Neu5Ac-glycoproteins	HRP-donkey anti-chicken IgY	Positive
6.	Anti-Neu5Gc IgY	Neu5Gc-glycoproteins	HRP-donkey anti-chicken IgY	Negative

4. Place the 48-well plate with tissue sections on top of horizontal shaker for 5 min.
5. Repeat the PBS wash two more times by aspirating the PBS with a pipette and re-adding 300 μL of PBS pH 7.4 per well (*see* **Notes 16** and **17**).

3.3.2 Blocking of Tissues

1. Prepare blocking buffer (*see* **Note 18**).
2. Aspirate PBS from tissue wells, and then add 300 μL of blocking buffer and incubate at room temperature for 1.5 h with shaking (*see* **Note 19**).

3.3.3 Preparation of Primary Antibody Solutions

Primary antibody is pre-complexed with the specific/non-specific inhibitors, and then incubated on ice for 1.5 h during the tissues blocking step. To prepare primary antibody–inhibitor solutions, the primary antibody and the inhibitors are first prepared at ×2 of their final concentration, and then mixed 1:1 to reach the appropriate concentration, as detailed in Table 2.

1. Prepare control primary antibody at 1:500 dilution in blocking buffer (×2 solution): mix 0.3 μL of control chicken IgY in 150 μL of blocking buffer.
2. Prepare primary antibody at 1:500 dilution in blocking buffer (×2 solution): mix 1.6 μL of chicken anti-Neu5Gc IgY in 800 μL of blocking buffer.
3. Prepare the non-specific inhibitor, Neu5Ac-glycoproteins (human serum or chicken serum), at 20% in blocking buffer (×2 solution): add 40 μL of human serum into 160 μL of blocking buffer (*see* **Note 20**).

Table 2
Description of tissue staining scheme

No.	Compound	Final concentration	Working concentration	Buffer
1.	Control IgY	1:1000	1:500	[a]Blocking
2.	Anti-Neu5Gc IgY	1:1000	1:500	Blocking
3.	Neu5Ac	20 mM	[b]40 mM	PBS
4.	Neu5Gc	20 mM	[b]40 mM	PBS
5.	Neu5Ac-glycoproteins (Human or chicken serum)	10%	20%	Blocking
6.	Neu5Gc-glycoproteins (chimpanzee serum)	10%	20%	Blocking

[a]Blocking buffer (0.5% fish gelatin in PBST)
[b]The sialic acid solutions must be titrated to pH 7 (*see* Note 4)

4. Prepare the specific inhibitor, Neu5Gc-glycoproteins (chimpanzee serum), at 20% in blocking buffer (×2 solution): add 40 μL of chimpanzee serum into 160 μL of blocking buffer (*see* **Note 21**).
5. Prepare as described or have ready after thawing the carbohydrate solutions (40 mM Neu5Ac as non-specific inhibitor and 40 mM Neu5Gc as specific inhibitor).
6. Then, label 1.5-mL microtubes as in Table 2 and prepare the antibody–inhibitor (×1) solutions as described below. All solutions are used at 300 μL per well for each tissue section.
7. Tube No. 1: add 150 μL of blocking buffer into 150 μL of control IgY diluted 1:500.
8. Add 150 μL of 1:500 diluted chicken anti-Neu5Gc IgY into five 1.5-mL microtubes (tubes no. 2–6).
9. Tube no. 2: add 150 μL of blocking buffer.
10. Tube no. 3: add 150 μL of 40 mM Neu5Ac in PBS pH 7.
11. Tube no. 4: add 150 μL of 40 mM Neu5Gc in PBS pH 7.
12. Tube no. 5: add 150 μL of 20% human serum in blocking buffer.
13. Tube no. 6: add 150 μL of 20% chimpanzee serum in blocking buffer.
14. Gently vortex all primary antibody tubes and then incubate on ice (without additional mixing) for 1.5 h during the tissues blocking (*see* **Note 22**).
15. Aspirate the blocking buffer from the 48-well plate with tissues, then add primary antibody solutions (tubes no. 1–6), and properly label the 48-well plate for the different tubes/treatments.

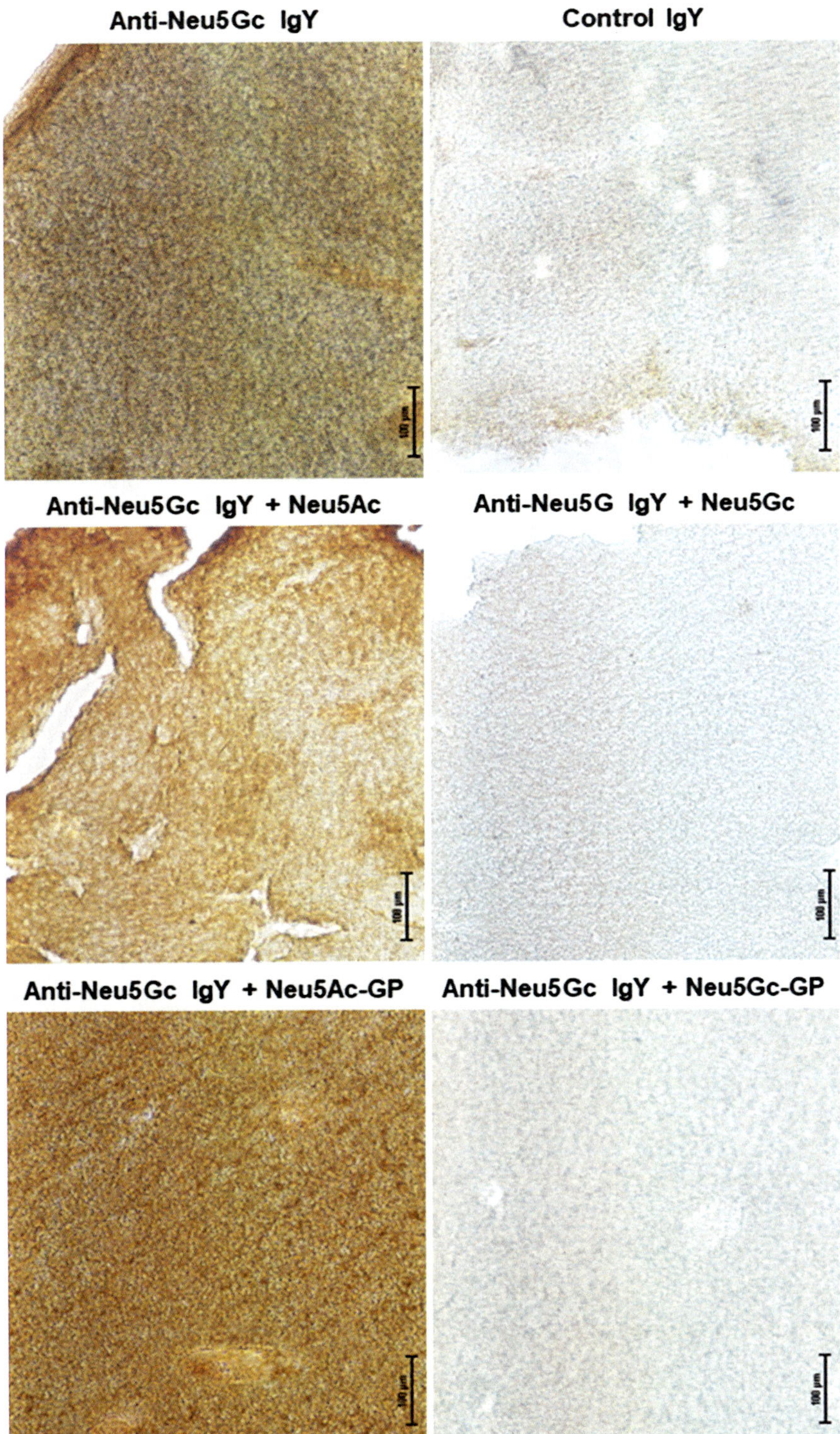

Fig. 1 Immunohistochemical determination of Neu5Gc in wild-type mouse liver tissue. Mouse liver tissue sections stored in −20 °C in 96-well plates in TCS were washed in PBS, and then incubated in blocking buffer (0.1% fish gelatin in PBST) at room temperature for 1.5 h. Subsequently, pre-complexed primary antibody

16. Incubate while shaking on a horizontal shaker for 4 h at room temperature (*see* **Note 23**).
17. Aspirate the primary antibody, then wash with 300 μL of PBST per well, twice, then finally wash with 300 μL of PBS per well.
18. Prepare secondary antibody at 1:500 dilution: 4 μL of HRP-conjugated donkey anti-chicken IgY-HRP in 2 mL of PBS.
19. Aspirate PBS from tissue wells and add diluted secondary antibody at 300 μL per well, and then incubate at room temperature for 1 h.
20. Aspirate then secondary antibody and wash with 300 μL of PBS per well. To this end, incubate while shaking for 5 min. Repeat the washing step two more times.
21. For generating the read out of the reactions, prepare substrate developing solution: use DAB substrate kit (*see* **Note 24**) according to manufacturer instructions: 1 drop of DAB chromogen substrate in 1 mL of kit buffer (prepare 2 mL), cover to keep away from light.
22. Aspirate PBS from all wells and then add 300 μL of DAB reagent to each well.
23. Allow DAB staining to develop for 1 min, and tissue becomes brown (allow 2 min if staining is faint). Once brown precipitate is seen on the tissues, aspirate the substrate solution from all the wells.
24. Wash stained tissue sections in 1 mL of PBS per well, three times, with intermittent shaking for 5 min per wash.
25. Place stained tissues on slides. To this end, use positively charged adhesion slides (adhesion microscope slides). To place tissue sections without folding, first place a drop of DDW (~50 μL) on the glass slide, and then, using a brush, place the tissue section gently on top of the water drop. Flatten tissue section on the slide by tilting the slide until water slides out and tissue sticks flat on the slide slightly dried.
26. Once the sections are slightly dried (up to 3 min, don't leave for over drying), place ~100 μL of Fluoromount-G (*see* **Note 25**). Then, gently place a microscope cover glass over the slide using forceps. Make sure that there are no bubbles near the tissue sections once the coverslip is placed.

Fig. 1 (continued) solutions were added (control IgY or chicken anti-Neu5Gc IgY in blocking buffer, in the presence of specific/non-specific inhibitors, as described in Table 2), and tissues incubated on for 4 h at room temperature. Binding was probed with HRP-donkey-anti-chicken IgY in PBS incubated at room temperature for 1 h. Tissue staining was developed with DAB reagent for 3 min, followed by PBS washing, and then observed in a Nikon Eclipse T*i*-S microscope under ×10 bright field objective. Images were analyzed using the imaging software NIS-Elements version 4.0 (representative of at least two independent experiments)

27. Observe stained tissue section in a microscope such as a Nikon Eclipse T*i*-S under 10× Bright field objective. The refractive index of 10× objective (MRL00042) is 0.30.
28. Capture images using a camera such as a Nikon DS-U3 DS-Filc-U3.
29. Analyze images using an imaging software such as NIS-Elements version 4.0.

4 Notes

1. Any other wild-type mouse strain can be used as a source for -Neu5Gc-positive tissues. In fact, any other non-human mammal can be used as a source for Neu5Gc-positive tissues, and any tissue can be used except for brain that lacks Neu5Gc [22–24].
2. The sera were obtained in our case from the local zoo only during routine maintenance procedures and kindly provided by Dr. Gillad Goldstein, curator of the Zoological Center Tel Aviv, Safari Park (Israel), and Dr. Nili Avni-Magen, Head Veterinarian and Zoological Director of The Tisch Family Zoological Gardens in Jerusalem (Israel).
3. Read the MSDS before working with paraformaldehyde. This chemical is toxic, flammable, and health hazardous; hence, prepare paraformaldehyde solution in a fume hood. (Use personal protective equipment. Ensure adequate ventilation. Avoid dust formation. Remove all sources of ignition. Take precautionary measures against static discharges). Also note that immunohistochemistry stains will be compromised by over-fixation.
4. Sialic acids have a net negative charge at physiological pH (pKa of 2.6) because their carboxylate group is deprotonated, a property that dominates their physiochemical properties. Therefore, the initial pH of 40 mM Neu5Ac/Neu5Gc in PBS is around pH2 and must be titrated with NaOH to reach ~ pH7. Without titration, the non-specific inhibitor Neu5Ac solution will also inhibit anti-Neu5Gc IgY binding, just due to the low pH of the solution.
5. 40 mM Neu5Ac/Neu5Gc pH adjustment: Add small drops of NaOH, mix the tube, allow pH to adjust, and then measure pH with a pH paper. Repeat until pH7 is reached.
6. The 40 mM Neu5Ac/Neu5Gc solutions can be aliquoted and stored in −20 °C.

7. Settling of tissue in the sucrose solution will occur within 1–2 h and up to overnight incubation. It is critical that tissue is settled down before starting the staining protocol.
8. In the microtome tissue mounting stand, while preparing for tissue sectioning, up to 6 OCT blocks can be made at a time.
9. OCT becomes opaque in nature once it is cooled. Dry ice must be put *after* arranging the OCT blocks in the microtome tissue mounting stand.
10. Throughout the tissue sectioning procedure, it is critical to make sure the dry ice is kept around the mounting stand.
11. To glue the tissue on top of the OCT frozen block, it is important to dip one side of the tissue in OCT drop within the Petri dish, and then immediately place it with OCT-laden side on the PBS on the frozen OCT block.
12. It is important that the dry ice is finely crushed.
13. Expect 15–20 tissue sections per tissue block.
14. Tissue sections in TCS do not freeze and remain in liquid solution and can be stored for up to two weeks before staining.
15. Specificity is demonstrated by competitive inhibition with free Neu5Gc or Neu5Gc-glycoproteins that is expected to inhibit anti-Neu5Gc IgY reactivity and with Neu5Ac or Neu5Ac-glycoproteins as controls that are not expected to inhibit anti-Neu5Gc IgY reactivity if the antigen is indeed Neu5Gc.
16. To aspirate the PBS, place the 48-well plate on top of a dark surface to visualize the tissue, tilt the plate, and aspirate the PBS with a pipet (make sure the tissue remains in the well), and gently add 300 μL of PBS.
17. If tissue is sturdy, it is possible to complete the whole tissue washing and staining procedures in a 96-well plate, with 100 μL of washing/blocking solution volumes (enough to completely cover the tissue).
18. Note that both Tween-20 and fish gelatin are viscous solutions.
19. Tissue can be incubated with blocking solution overnight (1.5 h at the minimum).
20. Humans and chicken have a dysfunctional *CMAH* encoding gene; hence, serum glycoproteins are decorated only with the Neu5Ac type of sialic acids and lack expression of Neu5Gc [11, 20, 25, 26].
21. Chimpanzee serum contains both Neu5Gc-conjugated glycoproteins (~14%) and Neu5Ac-glycoproteins (~86%) [13, 20].
22. After antibody–inhibitor solutions are placed on ice for incubation, do not mix the tubes again to avoid disruption of proper pre-complexing.

23. Incubation with primary antibody can be done at room temperature for 4 h or alternatively at 4 °C overnight while shaking.
24. Adjust DAB reagent to room temperature during the incubation with the secondary antibody for about 30 min.
25. Fluoromount-G is a viscous solution; hence, cut the edge of the pipettor 200 μL tip before collecting the solutions. Mounting media volume needs to be sufficient to cover the slides and coverslip, use ~100 μL. Place the mounting media around the tissues, not directly on the tissue. While placing the cover slip, it will cover the top of tissue as well.

Acknowledgment

This work was supported by the 7th Framework Program FP7-Health-2013-INNOVATION-1-603049 of the European Commission (to V.P-K).

References

1. Ekser B, Li P, Cooper DKC (2017) Xenotransplantation: past, present, and future. Curr Opin Organ Transplant 22:513–521
2. Vadori M, Cozzi E (2015) The immunological barriers to xenotransplantation. Tissue Antigens 86:239–253
3. Cooper DKC, Gaston R, Eckhoff D, Ladowski J, Yamamoto T, Wang L, Iwase H, Hara H, Tector M, Tector AJ (2018) Xenotransplantation-the current status and prospects. Br Med Bull 125:5–14
4. Byrne GW, McGregor CGA, Breimer ME (2015) Recent investigations into pig antigen and anti-pig antibody expression. Int J Surg 23:223–228
5. Galili U (2013) Anti-Gal: an abundant human natural antibody of multiple pathogeneses and clinical benefits. Immunology 140:1–11
6. Cooper DK (2016) Modifying the sugar icing on the transplantation cake. Glycobiology 26:571–581
7. Padler-Karavani V, Varki A (2011) Potential impact of the non-human sialic acid N-glycolylneuraminic acid on transplant rejection risk. Xenotransplantation 18:1–5
8. French BM, Sendil S, Pierson RN, Azimzadeh AM (2017) The role of sialic acids in the immune recognition of xenografts. Xenotransplantation 24
9. Salama A, Evanno G, Harb J, Soulillou JP (2015) Potential deleterious role of anti-Neu5Gc antibodies in xenotransplantation. Xenotransplantation 22:85–94
10. Reuven EM, Leviatan Ben-Arye S, Marshanski T, Breimer ME, Yu H, Fellah-Hebia I, Roussel JC, Costa C, Galiñanes M, Mañez R, Le Tourneau T, Soulillou JP, Cozzi E, Chen X, Padler-Karavani V (2016) Characterization of immunogenic Neu5Gc in bioprosthetic heart valves. Xenotransplantation 23:381–392
11. Varki A (2010) Colloquium paper: uniquely human evolution of sialic acid genetics and biology. Proc Natl Acad Sci U S A 107(Suppl 2):8939–8946
12. Nguyen DH, Tangvoranuntakul P, Varki A (2005) Effects of natural human antibodies against a nonhuman sialic acid that metabolically incorporates into activated and malignant immune cells. J Immunol 175:228–236
13. Padler-Karavani V, Yu H, Cao H, Chokhawala H, Karp F, Varki N, Chen X, Varki A (2008) Diversity in specificity, abundance, and composition of anti-Neu5Gc antibodies in normal humans: potential implications for disease. Glycobiology 18:818–830
14. Gao B, Long C, Lee W, Zhang Z, Gao X, Landsittel D, Ezzelarab M, Ayares D, Huang Y, Cooper DKC, Wang Y, Hara H (2017) Anti-Neu5Gc and anti-non-Neu5Gc

antibodies in healthy humans. PLoS One 12: e0180768

15. Okerblom J, Varki A (2017) Biochemical, Cellular, Physiological, and Pathological Consequences of Human Loss of N-Glycolylneuraminic Acid. Chembiochem 18:1155–1171
16. Leviatan Ben-Arye S, Yu H, Chen X, Padler-Karavani V (2017) Profiling anti-Neu5Gc IgG in human sera with a sialoglycan microarray assay. J Vis Exp 125
17. Paul A, Padler-Karavani V (2018) Evolution of sialic acids: implications in xenotransplant biology. Xenotransplantation 25(6):e12424
18. Burdorf L, Azimzadeh AM, Pierson RN (2018) Progress and challenges in lung xenotransplantation: an update. Curr Opin Organ Transplant 23:621–627
19. Zhang R, Wang Y, Chen L, Wang R, Li C, Li X, Fang B, Ren X, Ruan M, Liu J, Xiong Q, Zhang L, Jin Y, Zhang M, Liu X, Li L, Chen Q, Pan D, Li R, Cooper DKC, Yang H, Dai Y (2018) Reducing immunoreactivity of porcine bioprosthetic heart valves by genetically-deleting three major glycan antigens, GGTA1/β4GalNT2/CMAH. Acta Biomater 72:196–205
20. Diaz SL, Padler-Karavani V, Ghaderi D, Hurtado-Ziola N, Yu H, Chen X, Brinkman-Van der Linden EC, Varki A, Varki NM (2009) Sensitive and specific detection of the non-human sialic Acid N-glycolylneuraminic acid in human tissues and biotherapeutic products. PLoS One 4:e4241
21. Padler-Karavani V, Song X, Yu H, Hurtado-Ziola N, Huang S, Muthana S, Chokhawala HA, Cheng J, Verhagen A, Langereis MA, Kleene R, Schachner M, de Groot RJ, Lasanajak Y, Matsuda H, Schwab R, Chen X, Smith DF, Cummings RD, Varki A (2012) Cross-comparison of protein recognition of sialic acid diversity on two novel sialoglycan microarrays. J Biol Chem 287:22593–22608
22. Naito-Matsui Y, Davies LR, Takematsu H, Chou HH, Tangvoranuntakul P, Carlin AF, Verhagen A, Heyser CJ, Yoo SW, Choudhury B, Paton JC, Paton AW, Varki NM, Schnaar RL, Varki A (2017) Physiological exploration of the long term evolutionary selection against expression of N-glycolylneuraminic acid in the brain. J Biol Chem 292:2557–2570
23. Davies LR, Varki A (2015) Why is N-glycolylneuraminic acid rare in the vertebrate brain. Top Curr Chem 366:31–54
24. Springer SA, Gagneux P (2016) Glycomics: revealing the dynamic ecology and evolution of sugar molecules. J Proteome 135:90–100
25. Peri S, Kulkarni A, Feyertag F, Berninsone PM, Alvarez-Ponce D (2018) Phylogenetic distribution of CMP-Neu5Ac hydroxylase (CMAH), the enzyme synthetizing the proinflammatory human xenoantigen Neu5Gc. Genome Biol Evol 10:207–219
26. Varki A (2001) Loss of N-glycolylneuraminic acid in humans: mechanisms, consequences, and implications for hominid evolution. Am J Phys Anthropol Suppl 33:54–69

Chapter 5

Xenoantibodies and Complement Activity Determinations by Flow Cytometry in Pig-to-Primate Xenotransplantation

Nieves Doménech and Pilar Sánchez-Corral

Abstract

In pig-to-primate xenotransplantation, flow cytometry assays allow the examination of antibody reactivity to intact antigens in their natural conformation and location on cell membranes. Here we describe in detail the procedures of two flow cytometry assays to measure the antibody-mediated complement-dependent cytotoxicity (CDC) response or serum levels of IgG and IgM xenoantibodies. This information is key for understanding the rejection process of vascularized xenografts and finding strategies to overcome it.

Key words Primates, Pig, Serum, Xenoantibodies, Complement, Cytotoxicity, Flow cytometry

1 Introduction

A major objective during pig-to-primate xenotransplantation is to avoid acute xenograft rejection (AXR) to prolong graft survival [1]. AXR occurs within days after graft reperfusion and is mainly provoked by the presence of primate antibodies with specificities against xenoantigens on the surface of the pig endothelial cells. These xenoantibodies subsequently activate the recipient's complement, which then contributes to graft damage and loss by antibody-mediated complement-dependent cytotoxicity (CDC).

The main pig endothelial xenoantigen is the oligosaccharide galactose-α1,3 galactose (αGal or Gal), which is responsible for the hyperacute rejection of pig organs transplanted into nonhuman primates [2]. For many years, a major strategy in xenotransplantation was the pretransplantation removal of primate xenoantibodies by polymers containing multiple Gal epitopes [3]. However, antibodies to non-Gal antigens (NGal-Ab) also contribute to acute humoral xenograft rejection of pig organs in nonhuman primates [4]. One of these non-Gal antigens is a glycan produced by the pig β1,4 *N*-acetylgalactosaminyltransferase (B4GALNT2) gene [5]. Humans also have antibodies reacting with *N*-

Cristina Costa (ed.), *Xenotransplantation: Methods and Protocols*, Methods in Molecular Biology, vol. 2110,
https://doi.org/10.1007/978-1-0716-0255-3_5,

glycolylneuraminic acid (Neu5Gc)-modified glycans, because they are deficient in the enzyme cytidine monophosphate-*N*-acetylneuraminic acid hydroxylase (CMAH), required for Neu5Gc synthesis [5]. Also, class I swine leukocyte antigens (SLA) could be another target for genome editing in xenotransplantation, as some HLA antibodies in sensitized patients cross-react with class I SLA [6].

Current approaches in xenotransplantation are aimed at either xenoantigen reduction or transgene expression of human complement regulators that eliminate/limit the human-anti-porcine CDC response. To this end, the first approach developed when swine genetic modification became possible sought to avoid rejection by means of introducing complement regulatory proteins (CD55, CD59, CD46) to overcoming primate complement and coagulation activation [7]. Later, pigs lacking one, two, or three xenoantigens have been generated since 2000 by using different methodologies for gene knock out (KO) [8]. The first of these KO pigs lacked the α-1,3-glycosyltransferase-1 (GGTA1) gene, an enzyme necessary to generate the Gal epitope; these pigs are referred to as GalT KO (GTKO) or Gal KO [9]. Double (GGTA1/CMAH) and triple (GGTA1/CMAH/β4GalNT2) KO pigs lacking Gal and non-Gal antigens have also been generated taking advantage of the new techniques based on *clustered regularly interspaced short palindromic repeats* (CRISPR). The availability of these "minimally antigenic" pigs has highly decreased, although not completely abrogated, the adverse response of natural or induced xenoantibodies toward pig cells [10, 11]. Another promising strategy to limit the graft endothelial damage due to the presence of xenoantibodies is to incorporate transgenes of complement surface regulators (CD55, CD46, CD59) into GTKO pigs [12], although their advantages over the triple (GGTA1/CMAH/β4GalNT2) KO pigs are not clear enough [13]. In the last years, the CRISPR technology has made possible the production of pigs incorporating five or six genetic modifications [8]. The combination of these "multiple-gene pigs" with an adequate immunosuppressive regimen to minimize the induction of T-dependent xenoantibodies [14] has contributed to control the xenogeneic immune response and prolong pig graft survival in nonhuman primate recipients from a few days to many months.

To evaluate the efficacy of all these xenotransplantation strategies, several methods have been developed to assay the cytotoxicity mediated by antibody and complement or to determine the levels of the recipient functional xenoantibodies before the transplantation procedure and during posttransplant serum antibody monitorization. In the first years, the most utilized assays were the standard hemolytic assay CH50, which assesses activation of the complement classical pathway, and a specific ELISA to measure anti-Gal antibodies, which represent more than 85% of circulating natural anti-pig antibodies [15]. Nowadays, however, peripheral antibody response in pig-to-primate xenotransplantation is mostly determined by flow cytometry. This assay is of "predictive value" that is at least as sensitive as the CH50 or ELISA methods [16]. Moreover, flow cytometry approaches use unfixed cells and thus have the advantage of determining reactivity toward native non-denatured cellular antigens, which can also detect xenoantibodies that do not activate complement and xenoantibodies recognizing non-Gal antigens [17].

Based on the foregoing, in this chapter we describe in detail two different flow cytometry protocols currently used for measuring the CDC response and for quantifying circulating xenoantibodies in primate sera in xenotransplantation experiments: (1) a two-color flow cytometric complement-dependent cytotoxicity assay [18] that measures antibodies capable of lysing xenogeneic peripheral blood mononuclear cells (PBMC) in the presence of heterologous baby rabbit complement (BRC) and (2) a method to measure the binding of IgG and IgM xenoantibodies reacting with primary/immortalized xenogeneic endothelial cells (EC). The cells used in these assays as a source of xenoantigens could be either from wild-type (WT) or from genetically modified pigs.

2 Materials

2.1 Two-Color Flow Cytometric Complement-Dependent Cytotoxicity Assay

1. Serum test samples (*see* **Notes 1** and **2**).
2. Peripheral blood mononuclear cells (PBMC) obtained from anticoagulated blood samples collected from WT or genetically modified pigs after separation on discontinuous Ficoll gradients (*see* **Note 3**).
3. Selected baby rabbit complement (BRC) (*see* **Note 4**) as a source of external complement. For the BRC preparation, follow the supplier's instructions. It is presented by most suppliers as a lyophilized preparation and usually requires the sample to be dissolved in a small volume of distilled water.
4. Hanks' balanced salt solution (HBSS).
5. Phosphate-buffered saline (PBS) tablets without Ca^{++} and Mg^{++} (follow manufacturer's instructions).
6. 1 mg/mL fluorescein diacetate (FDA) stock solution in acetone, stored at 4 °C. On the day of the experiment, it is diluted to 1 μg/mL in PBS.
7. 50 μg/mL propidium iodide (PI) stock solution in HBSS. Stored at 4 °C.
8. Microcentrifuge tubes of 1.5 or 2.0 mL.
9. Variable micropipettes for 200–1000 μL and 20–200 μL and appropriate tips.

10. Multichannel pipette.
11. 96-well rigid V-bottom microtiter plates.
12. Water bath at 56 °C.
13. Standard laboratory centrifuge with microtiter plate holders.
14. High-speed bench centrifuge (e.g., microfuge).
15. Fluorescence-activated cell sorter (FACS) tubes.
16. Flow cytometer and software for analysis (i.e., Becton Dickinson FACScan flow cytometer and CellQuest software).

2.2 Quantification of IgG and IgM Anti-Pig Antibodies by Flow Cytometry

1. Serum test samples (*see* **Notes 1** and **2**).
2. Primary or immortalized porcine aortic endothelial cells (pAEC) from WT or genetically modified pigs (*see* **Note 5**).
3. Penicillin/streptomycin (P/S): 5000 units/mL penicillin and 5000 μg/mL streptomycin.
4. Complete DMEM culture medium: Dulbecco's modified Eagle's medium (DMEM) supplemented with 10% fetal bovine serum (FBS) and 1% P/S.
5. Trypsin-ethylenediaminetetraacetic acid (EDTA) solution: 0.25% trypsin/0.04% EDTA.
6. Sterile physiologic saline solution.
7. Fluorescein isothiocyanate (FITC)-labeled rabbit antihuman IgG (γ-chain specific) or rabbit antihuman IgM (μ-chain specific). These stock solutions should be kept at 4 ° C in the dark.
8. PBS tablets without Ca^{++} and Mg^{++} (follow manufacturer's instructions).
9. Fluorescence-activated cell sorter (FACS) buffer: PBS + 1% bovine serum albumin (BSA) + 0, 01% sodium azide.
10. 50 μg/mL PI stock solution in PBS. Stored at 4 °C.
11. Sterile centrifuge tubes and coated culture flasks (*see* **Note 6**).
12. Incubated at 37 °C and humidified, 5% CO_2 atmosphere.
13. Laminar air flow cabinet.
14. Inverted optical microscope.
15. Microcentrifuge tubes of 1.5 or 2.0 mL.
16. Sterile graduated pipettes.
17. Variable micropipettes for 200–1000 μL and 20–200 μL and appropriate tips.
18. Multichannel pipette.
19. 96-well rigid V-bottom microtiter plates.
20. Water bath at 56 °C.

21. Standard laboratory centrifuge with tubes and/or microtiter plate holders.
22. High-speed bench centrifuge (e.g., microfuge).
23. FACS tubes.
24. Flow cytometer and software for analysis (i.e., Becton Dickinson FACScan flow cytometer and CellQuest software).

3 Methods

3.1 Two-Color Flow Cytometric Complement-Dependent Cytotoxicity Assay

1. Prepare serial twofold dilutions of test sera for each assay (*see* **Note 7**) with HBSS in a 96-well V-bottom plate at 100 μL/well (serum concentration varied by dilution 50%, 25%, 12.5%, 6.25%, 3125%, and 1.56%).
2. Add 100 μL of pig PBMC (at 3×10^6 cells/mL) to all wells. A negative control (only cells, not serum) should be included.
3. Incubate the assay plate for 30 min at 4 °C.
4. After incubation, the plate is centrifuged at 500 × *g* for 5 min, decanted, and washed with 100 μL of cold HBSS. To resuspend the PBMC pellet, use a multichannel pipette and pipette up and down several times for each column. This step is repeated twice.
5. Add 50 μL of optimal dilutions of BRC in HBSS to all wells, including the negative control wells (optimal dilution of BRC from Serotec UK is usually 1/5).
6. Incubate the assay plate for 30 min at 37 °C.
7. After incubation, wash the cells three times with 100 μL of cold HBSS.
8. Incubate the cells in the dark for 15 min with 20 μL of 1 μg/mL FDA (in PBS) and 10 μL of 50 μg/mL PI (in PBS), reaching a final volume of 30 μL (*see* **Note 8**).
9. After incubation, transfer the cells from 96-well assay plate to FACS tubes containing 100 μL of HBSS.
10. Acquire and analyze results with a flow cytometer such as Becton Dickinson FACScan flow cytometer and CellQuest software. Adjust voltage values for forward scatter (FSC) and side scatter (SSC) to visualize the entire cell population. Acquire 10,000 gated events. Quadrant analysis is performed, with cells stained green but not red (PI- FDA+) counted as viable, cells stained red but not green (PI+ FDA-) counted as dead, and double-positive cells (PI+ FDA+) counted as having membrane damage. Double-negative cells (PI- FDA-) are excluded from the analysis. Background spontaneous cytotoxicity is determined by analyzing cells not exposed to serum (negative control).

11. For each serum dilution, the percentage of cell cytotoxicity (PC) is calculated from the PI label determined according to the formula: PC =100 × (% of cells dead under experimental conditions – % of spontaneously dead cells)/(100 – % spontaneously dead cells).
12. Serum antibody-mediated cytotoxicity is analyzed by generating a curve in which the percentage of dead cells is plotted against serum dilution. For each serum sample, the serum dilution that destroyed 50% of the target cells (SLU50%) is calculated and expressed as serum lytic units (*see* **Note 9**).

3.2 Quantification of IgG and IgM Anti-pig Antibodies by Flow Cytometry

1. Culture pAEC as monolayer in the complete DMEM culture medium in an appropriate gelatin-coated culture flask. The cells are maintained in an incubator at 37 °C in 5% CO_2.
2. On the day of the assay, detach cells (at 80%–95% confluency) from the culture flask. For this purpose, remove culture medium, wash the cell monolayer with physiologic saline solution, and dissociate the cells by treatment at 37 °C for 2–3 min with a trypsin-EDTA solution (1×) (add enough volume to cover the cell monolayer). Block enzyme activity by adding complete DMEM.
3. Transfer the cell suspension to a 50 mL centrifuge tube and wash carefully the culture flask in order to remove all the remaining cells. Centrifuge pAEC at 300 × *g* for 10 min.
4. Remove the supernatant, and resuspend cells in complete DMEM and count. Harvest sufficient cells for 1×10^5 cells/test.
5. Bring final volume of cell suspension up to 20 mL with complete DMEM and spin at 300 × *g* for 10 min.
6. Discard the supernatant and resuspend the pellet in the appropriate volume of DMEM (no FBS from now on) to have 2×10^6 cell/mL.
7. Pipette 50 μL (1×10^5 cells) into each well of a 96-well plate, and add 50 μL of the appropriate solution for staining as follows:
   - Control of cells only: 50 μL of DMEM.
   - Controls of secondary antibody (two wells, one for FITC-rabbit antihuman IgG and one for FITC-rabbit antihuman IgM): 50 μL of DMEM.
   - Test sera samples (two wells for each, one for detecting IgG and one for IgM): dispense 50 μL of test sera diluted 1:4 in FACS buffer (25% serum final concentration).
8. Incubate the assay plate for 30 min at 4 °C.

9. After incubation, wash the cells with 150 μL of FACS buffer and spin at 500 × *g* for 5 min. This step is repeated twice.
10. Resuspend the appropriate pellets (in all the wells except in the control of cells only) in 20 μL of 1/20 dilution of FITC-labeled rabbit antihuman IgG or rabbit antihuman IgM in FACS buffer.
11. Incubate for 30 min at 4 °C in the dark.
12. After incubation, wash the cells with 150 μL of FACS buffer and spin at 500 × *g* for 5 min. This step is repeated twice.
13. Resuspend the final pellets in 200 μL of FACS buffer, and transfer the cells from the 96-well assay plate to FACS tubes immediately before acquisition.
14. Optional: add 10 μL of 50 μg/mL PI (in PBS) solution to each tube in order to include a dead/live gating.
15. Acquire and analyze results with a flow cytometer such as a Becton Dickinson FACScan flow cytometer and CellQuest software. Adjust voltage values for FSC and SSC to visualize the entire cell population. Collect 5000 events in the live cell gate, and plot the specific median fluorescence intensity (MFI), which is generated by "test MFI—control (secondary antibody only) MFI," for each tube.

4 Notes

1. Serum primate samples are obtained from coagulated blood (allowed to clot for 30 min at room temperature) after centrifugation (800 × *g* for 10 min). Samples should be immediately distributed in aliquots and frozen at −80 °C, and they should be thawed immediately before the experiments.
2. For all assays, serum primate samples are heated at 56 °C for 30 min in microfuge tubes to inactivate complement, and then cooled and spun in the microfuge at 17,000 × *g* to remove fine particulates (precipitate) and lipids (surface) before use in the assay. Heat inactivation is performed on the same day of the assay.
3. PBMC are acceptable for the assay only when viability >80%, and a cell preparation that has been treated with Percoll must not be used; Percoll increases background binding, which then reduces the sensitivity of the assay and may result in a false-negative result.
4. It is recommended to test a small number of BRC samples to select one or more with maximum complement activity and minimum intrinsic cytotoxicity (i.e., cytotoxicity in the absence of serum) toward PBMC targets.

5. Primary porcine aortic endothelial cells (pAECs) could be isolated from donor pigs by collagenase digestion and/or mechanical scrapping and seeded in gelatin-coated culture flasks. Primary cells are used in early passages and collected just prior to confluency. However, there are some inherent problems with using primary cells to measure primate antibodies on a regular basis, such as having sufficient cells, spontaneous dedifferentiation of the cultures, mycoplasma contaminations, among others. Alternatively, immortalized pAEC lines cultured in the same conditions as the primary cells can be used.
6. It is recommended that gelatin-coated culture flasks are used to enhance attachment of EC. Make a 1% gelatin solution in PBS and sterilize by autoclaving (20–10 min on liquid cycle). Then just coat your flasks under the hood, let gelatin to polymerize on the incubator (37 °C) for 1 h, and then, discard the excess under the hood and let the flasks open to dry, at least 1 h, before using. Alternatively, you can wrap the flask in parafilm and store the coated flasks at 4 °C for up to 1 month.
7. The use of serial dilutions of the serum is required to accurately assess antibody binding, and to measure (or exclude the presence of) prozone effects.
8. Although discrimination between live and dead cells by flow cytometry analysis of immunofluorescently stained cells does not require two-color detection, it is recommended the simultaneous use of two stains for better background exclusion because it permits to detect counts derived from artifacts such as the formation of cells doublets.
9. Because linear curves are not obtained from experimental data, it is recommended to use the Von Krogh equation (VK) as the mathematical method to calculate SLU50% in assays of lytic complement activity.

References

1. Ekser B, Rigotti P, Gridelli B et al (2009) Xenotransplantation of solid organs in the pig-to-primate model. Transpl Immunol 21:87–92. https://doi.org/10.1016/j.trim.2008.10.005
2. Sandrin MS, McKenzie IFC (1994) Gal (1,3) Gal, the major xenoantigen(s) recognised in pigs by human natural antibodies. Immunol Rev 141:169–190
3. Mañez R, Lopez-Pelaez E, Centeno A et al (2004) Failure to deplete anti-Galα1-3Gal antibodies after pig-to-baboon organ xenotransplantation by immunoaffinity columns containing multiple Galα1-3Gal oligosaccharides. Xenotransplantation 11:408–415
4. Harnden I, Kiernan K, Kearns-Jonker M (2010) The anti-nonGal xenoantibody response to alpha1,3-galactosyltransferase gene knockout pig xenografts. Curr Opin Organ Transplant 15:207–211. https://doi.org/10.1097/MOT.0b013e328336b854
5. Byrne GW, McGregor CGA, Breimer ME (2015) Recent investigations into pig antigen and anti-pig antibody expression. Int J Surg 23:223–228. https://doi.org/10.1016/j.ijsu.2015.07.724

6. Barreau N, Godfrin Y, Bouhours JF et al (2000) Interaction of anti-HLA antibodies with pig xenoantigens. Transplantation 69:148–156
7. Klymiuk N, Aigner B, Brem G et al (2010) Genetic modification of pigs as organ donors for xenotransplantation. Mol Reprod Dev 77:209–221. https://doi.org/10.1002/mrd.21127
8. Ekser B, Li P, Cooper DKC (2017) Xenotransplantation: past, present, and future. Curr Opin Organ Transplant 22:513–521. https://doi.org/10.1097/MOT.0000000000000463
9. Cooper DK, Dorling A, Pierson RN 3rd et al (2007) Alpha1,3-galactosyltransferase gene-knockout pigs for xenotransplantation: where do we go from here? Transplantation 84:1–7
10. Estrada JL, Martens G, Li P et al (2015) Evaluation of human and non-human primate antibody binding to pig cells lacking GGTA1/CMAH/β4GalNT2 genes. Xenotransplantation 22:194–202. https://doi.org/10.1111/xen.12161
11. Martens GR, Reyes LM, Butler JR et al (2017) Humoral reactivity of renal transplant-waitlisted patients to cells from GGTA1/CMAH/B4GalNT2, and SLA class I knockout pigs. Transplantation 101:e86–e92. https://doi.org/10.1097/TP.0000000000001646
12. Mohiuddin MM, Singh AK, Corcoran PC et al (2014) Role of anti-CD40 antibody-mediated costimulation blockade on non-Gal antibody production and heterotopic cardiac xenograft survival in a GTKO.hCD46Tg pig-to-baboon model. Xenotransplantation 21:35–45. https://doi.org/10.1111/xen.12066
13. Butler JR, Paris LL, Blankenship RL et al (2016) Silencing porcine CMAH and GGTA1 genes significantly reduces xenogeneic consumption of human platelets by porcine livers. Transplantation 100:571–576. https://doi.org/10.1007/s11248-016-9958-0
14. Adams AB, Kim SC, Martens GR et al (2018) Xenoantigen deletion and chemical immunosuppression can prolong renal xenograft survival. Ann Surg 268:564–573. https://doi.org/10.1097/SLA.0000000000002977
15. Doménech N (2012) Xenoantibodies and complement activity determinations in pig-to-primate xenotransplantation. Methods Mol Biol 885:57–69. https://doi.org/10.1007/978-1-61779-845-0_5
16. Diaz TM, Moscoso I, Centeno A et al (2004) Flow cytometry complement-mediated cytotoxicity assay detects baboon xenoantibodies directed to porcine epitopes undetected by hemolytic assay. Transpl Immunol 13:313–317
17. Azimzadeh AM, Byrne GW, Ezzelarab M et al (2014) Development of a consensus protocol to quantify primate anti-non-Gal xenoreactive antibodies using pig aortic endothelial cells. Xenotransplantation 21:555–566. https://doi.org/10.1111/xen.12125
18. Díaz TM, Pertega S, Ortega D et al (2004) An Fda/Pi flow cytometry assay of complement-mediated cytotoxicity of antibodies generated during xenotransplantation. Cytometry 62:54–60

Chapter 6

3D Cell-Culture Models for the Assessment of Anticoagulant and Anti-Inflammatory Properties of Endothelial Cells

Riccardo Sfriso and Robert Rieben

Abstract

Endothelial cells (EC) play a crucial role in the pathophysiology of cardiovascular diseases, ischemia/reperfusion injury, and graft rejection in (xeno-)transplantation. In such nonphysiological conditions, EC are known to lose their quiescent phenotype and switch into an actively pro-inflammatory, procoagulant, and anti-fibrinolytic state. This case happens essentially because the endothelial glycocalyx—a layer of proteoglycans and glycoproteins covering the luminal surface of the endothelium—is shed. Heparan sulfate, one of the main components of the endothelial glycocalyx, contributes to its negative charge. In addition, many plasma proteins such as antithrombin III, superoxide dismutase, C1 inhibitor, and growth factors and cytokines bind to heparan sulfate and by this scenario contribute to the establishment of an anticoagulant and anti-inflammatory endothelial surface. Shedding of the glycocalyx results in a loss of plasma proteins from the endothelial surface, and this phenomenon causes the switch in phenotype. Particularly in xenotransplantation, both hyperacute and acute vascular rejection are characterized by coagulation dysregulation, a situation in which EC are the main players.

Since many years, EC have been used in vitro in 2D flatbed cell culture models, with or without the application of shear stress. Such models have also been used to assess the effect of human transgenes on complement- and coagulation-mediated damage of porcine EC in the context of xenotransplantation. The methods described in this chapter include the analysis of endothelial cell–blood interactions without the necessity of using anticoagulants as the increased EC surface-to-volume ratio allows for natural anticoagulation of blood. Furthermore, this chapter contains the description of a novel microfluidic in vitro model carrying important features of small blood vessels, such as a 3D round-section geometry, shear stress, and pulsatile flow—all this in a closed circuit, recirculating system aiming at reproducing closely the in vivo situation in small vessels.

Key words 3D cell culture, Coagulation, Complement, Endothelial cells, Genetic modifications, Microcarrier beads, Microfluidics, Vascular biology, Xenotransplantation

1 Introduction

The endothelium consists of a monolayer of endothelial cells (EC) which constitute the inner lining of blood vessels. Given their location, EC are directly in contact with blood and circulating cells. Blood fluidity, the vascular tone, and platelet aggregation are all actively regulated by EC which carry also a physical

Cristina Costa (ed.), *Xenotransplantation: Methods and Protocols*, Methods in Molecular Biology, vol. 2110, https://doi.org/10.1007/978-1-0716-0255-3_6, © Springer Science+Business Media, LLC, part of Springer Nature 2020

semi-selective barrier function and are responsible for the maintenance of blood vessel homeostasis by surface expression and secretion of different key molecules [1, 2]. The endothelial glycocalyx—a carbohydrate-rich layer covering the endothelial surface—serves several functions acting as a mechanotransducer and influencing blood–EC interactions [3].

EC activation is characterized by a change of the endothelial phenotype which loses its anti-inflammatory, anticoagulant, and pro-fibrinolytic features as a consequence of endothelial glycocalyx shedding. Coagulation dysregulation is a hallmark of acute vascular rejection in xenotransplantation, leading to thrombotic microangiopathy and systemic consumptive coagulopathy which culminate in organ failure [4]. This case is mainly due to genetic differences between the donor and recipient.

Recent developments in genetic engineering techniques (CRISPR-Cas9) paved the way to quick and efficient production of new, multi-transgenic source pigs (over)expressing human regulators of both the complement and coagulation systems [5]. The effects of different combinations of transgenes and knockouts in preventing rejection need to be tested before preclinical pig to nonhuman primate xenotransplantation experiments are performed. Because the graft endothelium is the first site in contact with the recipient blood, many in vitro models use porcine endothelial cells to analyze the effect of human transgenes on complement and coagulation. However, the majority of in vitro experiments involving EC are still carried out in flat bed tissue culture systems, which are far from the physiological environment of EC and have only limited validity for the assessment of the anticoagulant, anti-inflammatory, and pro-fibrinolytic functions of EC. We therefore developed a 3D cell culture model of EC on microcarrier beads, in which the increased cell surface-to-volume ratio allows for the study of the natural anticoagulant properties of EC [6, 7]. In addition, culturing of porcine EC on the inner surface of cylindrical microchannels (Fig. 1) and exposing them to pulsatile flow of cell culture medium (or human serum/plasma during the experiments) allows to assess the function of anticoagulant and anti-inflammatory properties of the EC under near-natural conditions.

This chapter focuses on the description of 3D cell culture techniques to assess anticoagulant and anti-inflammatory properties of EC. It does not directly cover techniques for measuring coagulation markers in blood samples because most of them have been extensively described in the previous book by Bulato et al. [8]. However, the whole blood collected after incubation of EC-coated microcarrier beads and the perfusate (plasma) collected from the microfluidic system can be easily analyzed to detect coagulation activation markers such as thrombin–antithrombin complex (TAT), prothrombin fragments (F1 + F2), and tPA-PAI-1 complexes using commercial ELISA kits.

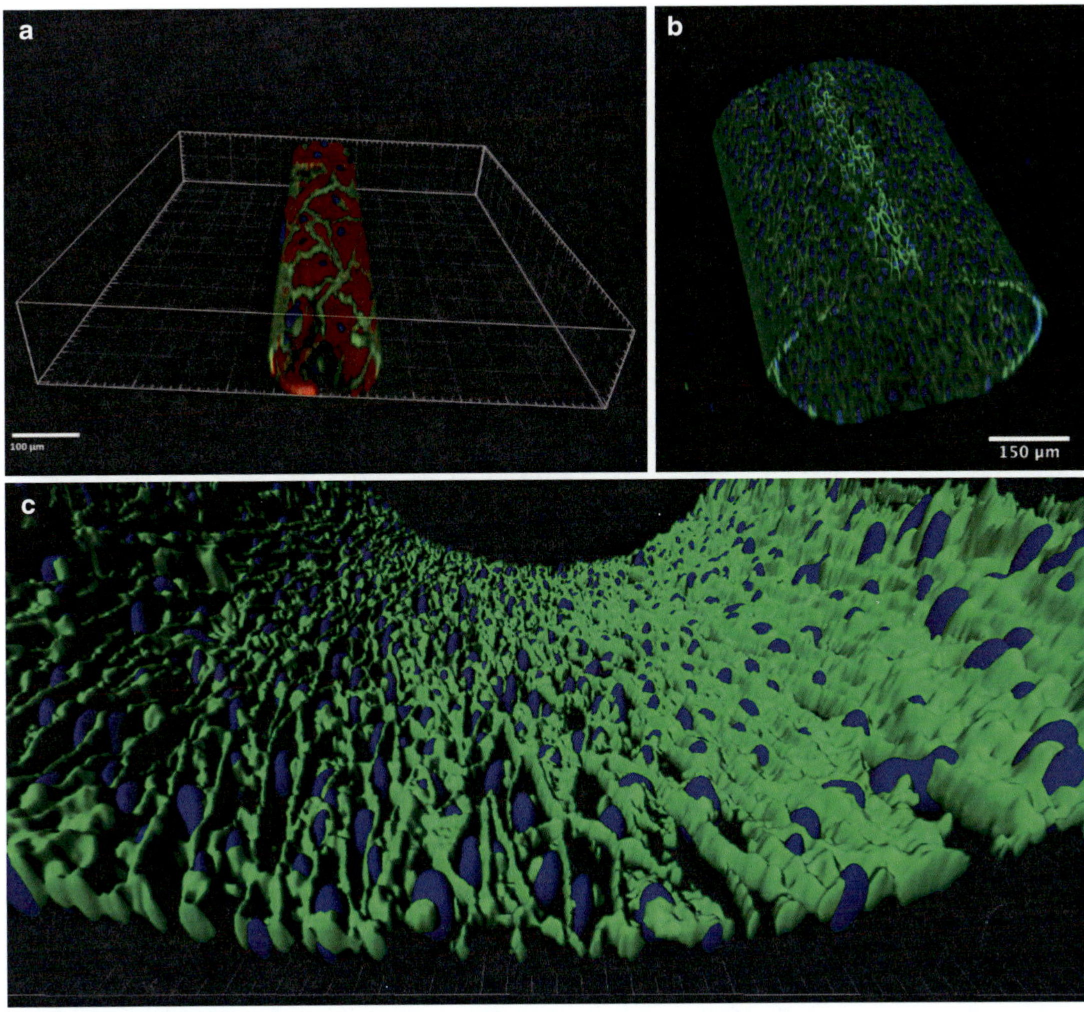

Fig. 1 Confocal images of EC-coated microchannels. (**a**) 3D rendering of the 120 μm round section channel. EC monolayer was stained for VE-cadherin (green) and F-actin (red). Nuclei were stained with DAPI (blue). (**b**) 3D z-stack of the 550 μm round section channel. EC monolayer was stained for VE-cadherin (green). Nuclei were stained with DAPI (blue). (**c**) 3D rendering of the 550 μm round section channel, detail. EC were stained or VE-cadherin (green). Nuclei were stained with DAPI (blue)

2 Materials

2.1 Microcarrier Bead Assay

2.1.1 Coating of Microcarrier Beads

The following materials must be of cell culture grade. A cell culture laminar flow hood is required:

1. Microcarrier beads (Biosilon, polystyrene beads with a diameter of 160–300 μm, a density of 1.05 g/cm^3 and carry a negative surface charge).
2. Ultra-pure water.

3. 0.2% acetic acid solution (prepared in a 50 mL falcon tube by adding 57 μL of acetic acid (60.05 g/mol) in 50 mL of sterile ultrapure water).
4. 100 μg/mL bovine collagen I in 0.2% acetic acid.
5. Phosphate-buffered saline (PBS): 137 mM NaCl, 2.6 mM KCl, 8 mM Na_2HPO_4, 1.4 mM KH_2PO_4, pH 7.4.
6. Medium 199.
7. Endothelial cell growth medium supplement mix and 25 μL of heparin (5000 IU/mL).

2.1.2 Collecting Cells

The following materials must be of cell culture grade. A cell culture laminar flow hood is required.

1. Subconfluent (85%–90%) T175 flask with EC.
2. PBS 1× at 37 °C.
3. Trypsin-0.05% EDTA.
4. Cell culture incubator at 37 °C with 5% CO_2 in a humid atmosphere.
5. Medium 199 supplemented with 10% FBS, 1% penicillin/streptomycin, 0.4% endothelial cell growth medium supplement mix.
6. 50 mL falcon tube.

2.1.3 Seeding Cells into the Spinner Flask

The following materials must be of cell culture grade. A cell culture laminar flow hood is required:

1. Subconfluent (85%–90%) T175 flask with EC.
2. 500 mL spinner flasks.
3. 10 mL and 25 mL serological pipettes.
4. Medium 199.
5. Colorless RPMI medium supplemented with 10% FBS, 1% penicillin/streptomycin, 1% L-glutamine, 0.4% endothelial cell growth medium supplement mix, and 25 μL of heparin.

2.1.4 Confluence Verification

The following materials do not need to be of cell culture grade unless otherwise specified:

1. 200 μL pipette.
2. PBS 1× at room temperature.
3. Fixative: Parapicric acid.
4. Nuclear labeling solution: 1 μg/mL of 4,6′-diamidino-2-phenylindole (DAPI) in PBS 1×.
5. Glass slides.
6. Mounting media (Glycergel).
7. Confocal microscope.

2.1.5 Experimental Procedure

1. 10 mL serological pipette and pipetboy.
2. Colorless RPMI not supplemented.

2.1.6 Incubation with Non-anticoagulated Blood

1. Butterfly sterile syringe needle.
2. 9 mL neutral polypropylene tubes (*see* **Note 1**).
3. 10 mL sterile serological pipette.
4. 12 mL polypropylene tubes.
5. Parafilm.
6. Horizontal tilting table.
7. 37 °C incubator.
8. Timer.
9. 2 mL tubes for serum/plasma collection.

2.1.7 Immuno-fluorescence Staining of EC-Coated Microcarrier Beads

1. PBS at room temperature.
2. PBS containing 3% BSA.
3. Fixative: Parapicric acid.
4. Antibody dilution buffer: 1% BSA in PBS (*see* **Note 2**).
5. 1 μg/mL of 4,6′-diamidino-2-phenylindole (DAPI) in PBS.
6. Mounting medium: Glycergel.
7. Microscope slides (26 × 76 mm) and sterile coverslips.
8. Confocal microscope.

2.2 Microfluidic Model

2.2.1 Microfluidic Chip Preparation

1. PDMS silicon elastomer.
2. Curing agent.
3. Ø 90 mm Petri dish.
4. Supporting needles (Ø 120 μm).
5. Mold needles (Ø 550 μm).
6. Vacuum chamber.
7. 60 °C incubator.
8. Ø 2 mm biopsy punch.
9. Scalpel.

2.2.2 PDMS-Glass Bonding

1. Scotch tape.
2. Plasma oxygen cleaner.
3. 100% Isopropanol.
4. Soap water.
5. Ultrapure water.
6. Nitrogen gun.
7. Glass slides (24 × 60 mm).

2.2.3 Coating of Microchannels

The following materials must be of cell culture grade. A cell culture laminar flow hood is required:

1. Ultrapure water.
2. 5% (3-Aminopropyl)triethoxysilane (APTES) (prepared by adding 250 μL of APTES to 4.75 mL of ultrapure water, corresponding to a 1:20 dilution, in a 5 mL tube and mixing well).
3. 0.1% glutaraldehyde in ultrapure water (prepared by diluting 40 μL of the 25% glutaraldehyde stock solution in 9′960 μL of ultrapure water, corresponding to a 1:250 dilution).
4. 15 mL falcon tubes.
5. 50 μg/mL human fibronectin in PBS (dilute 75 μL of the 1 mg/mL fibronectin stock solution in 1425 μL of sterile PBS 1×). The volume of this solution is calculated for the coating of four microchannels. Please adjust the volume accordingly if more microchannels need to be coated.
6. 100 μg/mL bovine collagen I in 0.2% acetic acid.
7. DMEM Glutamax supplemented with 10% FBS, 1% penicillin/streptomycin, 0.4% endothelial cell growth medium supplement mix.

2.2.4 Cell Seeding

The following materials must be of cell culture grade. A cell culture laminar flow hood is required:

1. DMEM Glutamax supplemented with 10% FBS, 1% penicillin/streptomycin, 0.4% endothelial cell growth medium supplement mix, and 4% dextran.
2. Subconfluent (85%–90%) T75 flask with EC.
3. Trypsin-0.05% EDTA.
4. Phosphate-buffered saline (PBS): 137 mM NaCl, 2.6 mM KCl, 8 mM Na_2HPO_4, 1.4 mM KH_2PO_4, pH 7.4.

2.2.5 Peristaltic Pump Connection

The following materials must be of cell culture grade. A cell culture laminar flow hood is required:

1. Peristaltic pump.
2. Silicon extension tubings (inner diameter 1 mm).
3. Silicon pump head tubings with stoppers (Gilson).
4. Adapters (Bio-Rad).
5. Autoclaved ultrapure water.
6. Autoclaved PBS 1×.
7. DMEM Glutamax supplemented with 10% FBS, 1% penicillin/streptomycin, 0.4% endothelial cell growth medium supplement mix, and 4% dextran.

8. 15 mL falcon tubes (reservoir tubes) with two holes in the cap (Ø 2 mm made with a driller).

3 Methods

Carry out all procedures under cell culture laminar flow hood unless otherwise specified.

3.1 Microcarrier Bead Assay [6]

3.1.1 Coating of Microcarrier Beads

1. Mix 7 mL of microcarrier beads with 42 mL of the 100 μg/mL collagen solution in a 50 mL tube and incubate for 1 h at room temperature.
2. Wash beads two times with 25 mL of PBS pH 7.4 (add 25 mL of PBS, mix well with the pipet, and wait until the beads are settled down then discard the supernatant and repeat) and one time with 25 mL of medium 199.
3. Cover the beads in the 50 mL tube with 10 mL of the supplemented medium 199 and allow equilibration for 10 min before further use.

3.1.2 Collecting Cells

1. Remove the cell culture medium from the T175 flask containing PAEC and add 5 mL of PBS.
2. Remove PBS from the T175 flask.
3. Add 5 mL of trypsin-0.05% EDTA and incubate for 3–4 min at 37 °C.
4. Collect the cells by rinsing the flask with 15 mL of cell culture medium and transfer the suspension to a 50 mL tube.
5. Centrifuge cells at 1200 × *g* for 8 min at room temperature, remove excess medium, and resuspend the pellet in 5 mL of cell culture medium.

3.1.3 Seeding Cells into the Spinner Flask

1. Add 20 mL of cell culture medium to the cell suspension and resuspend.
2. Add 20 mL of cell culture medium (w/o cells) into the 500 mL magnetic spinner flask.
3. Add the cells to the washed microcarrier beads and mix carefully with a 25 mL serological pipette.
4. Transfer the beads/cell mixture into the magnetic spinner flask.
5. Rinse the 50 mL tube with 10 mL of cell culture medium to collect the remaining cells.
6. Add an additional 85 mL of cell culture medium into the spinner flask and place it into the incubator overnight at

37 °C on a shaker (100 × *g*, mixing interval: 3 min every 45 min).

7. Add 50 mL of cell culture medium (total volume 200 mL) and continue stirring for an additional 24 h at 37 °C on a shaker (100 × *g*, mixing interval: 3 min every 45 min).
8. Add the colorless supplemented RPMI medium up to 320 mL of total volume.
9. Replace the medium every 48 h: Remove 100 mL of old medium and add 100 mL of fresh supplemented colorless RPMI.
10. Culture the cells for 5–7 days. The time depends on the confluence state of the cell-coated beads.

3.1.4 Confluence Verification

1. Collect 200 μL of cell-coated beads using a pipette and transfer them into a polypropylene tube (*see* **Note 3**).
2. Wash the beads three times with 600 μL of PBS (add PBS, tilt the tube, and mix gently to avoid detachment of the cells from the beads, wait for the beads to settle down, discard the PBS, and repeat).
3. Fix the beads for 10 min by adding 200 μL of parapicric acid.
4. Wash three times with 600 μL of PBS as in **Step 2**.
5. Add DAPI diluted in PBS and incubate for 10 min.
6. Transfer the beads on a glass slide and apply a coverslip using glycerol-based mounting medium (*see* **Note 4**).
7. Visualize the beads under a confocal microscope (Fig. 2).

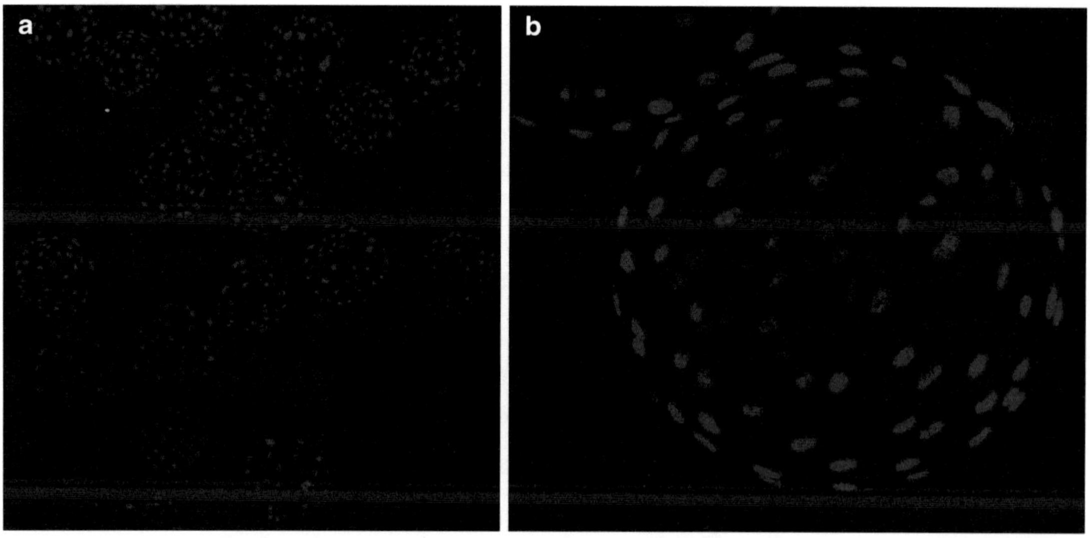

Fig. 2 Confluence verification of endothelial cells grown on microcarrier beads. Confocal microscopy picture of endothelial cell nuclei (stained with DAPI, blue) after 7 days of culture

3.1.5 Experimental Procedure

1. Remove the cell-coated beads from the magnetic spinner flask with a 10 mL serological pipette and transfer them into 12 mL round-bottom polypropylene tubes.
2. Let the beads settle down (around 1–2 mins) and remove excess medium.
3. Add more beads to the tubes until every tube contains exactly 2 mL of beads (*see* **Note 5**).
4. Add 5 mL of clear RPMI to each tube and mix carefully using a 10 mL serological pipette.
5. Let the beads settle down and remove excess medium.
6. Repeat the washing procedure one more time with RPMI and remove all excess medium.

3.1.6 Incubation with Non-anticoagulated Blood

1. Carefully and slowly (using neither jet nor vacutainers) draw blood from a healthy volunteer and collect it in 9 mL neutral polypropylene tubes (without anticoagulant).
2. Slowly transfer 8 mL of blood with a 10 mL serological pipette into each of the polypropylene tubes containing 2 mL of cell-coated beads (the total volume will be 10 mL). Always avoid rough handling of blood or beads to avoid premature EC activation. The procedure takes 1–2 min.
3. Carefully tilt the blood/bead mixture to ensure equal mixing and seal the cap with paraffin film.
4. Place the tubes on a horizontal tilting table (with gentle tilting settings only) inside a 37 °C incubator and record clotting times.
5. At set time intervals (e.g., after 10, 20, 30, 50, 70, 90 min), remove 1.5–2 mL of blood–bead mixture for serum or plasma analysis (*see* **Note 6**).
6. For collection of serum, leave the blood to coagulate (*see* **Note 7**). To collect the plasma, add EDTA or citrate to 2 mL tubes before adding blood samples.

3.1.7 Immuno<!–[spi] break--?>fluorescence Staining of EC-Coated Microcarrier Beads

1. Collect 200 μL of cell-coated beads using a pipette and transfer them into a polypropylene tube (*see* **Note 3**).
2. Wash the beads three times with 600 μL of PBS (add PBS, tilt the tube, and mix gently to avoid detachment of the cells from the beads, wait for the beads to settle down, discard the PBS, and repeat).
3. Fix the beads for 10 min by adding 200 μL of parapicric acid.
4. Wash three times with 600 μL of PBS as in **Step 2**.
5. Block using 600 μL of PBS-3% BSA for 30 min at room temperature.

6. Add primary antibodies diluted in PBS-1%BSA and incubate for 1 h at room temperature.
7. Wash three times with 600 μL of PBS as in **Step 2**.
8. Add fluorescent secondary antibodies diluted in PBS-1% BSA and incubate for 1 h at room temperature.
9. Wash three times with 600 μL of PBS as in **Step 2**.
10. Add DAPI diluted in PBS and incubate for 10 min.
11. Wash three times with 600 μL of PBS as in **Step 2**.
12. Transfer the beads on a glass slide and apply a coverslip using glycerol-based mounting medium (*see* **Note 4**).
13. Visualize the beads under a confocal microscope.

3.2 Microfluidic Model

3.2.1 Microfluidic Chip Preparation

1. Mix silicon elastomer and curing agent in proportion 10:1 (*see* **Note 8**).
2. Mix well the two components with a plastic spoon for 3–5 min.
3. Apply vacuum to remove air bubbles.
4. Clean the needles (Ø 550 μm and Ø 120 μm) with isopropanol and leave them to dry on a tissue.
5. Transfer liquid PDMS into a Petri dish (Ø 90 mm; *see* **Note 9**) and vacuum it to remove further air bubbles.
6. Place support needles (Ø 120 μm) on the bottom of the Petri dish.
7. Place mold needles (Ø 550 μm) orthogonally on top of support needles (Fig. 3).
8. Carefully transfer the Petri dish into the incubator (60 °C).
9. Cure overnight at 60 °C.
10. Remove the solidified PDMS and cut four equal chips.

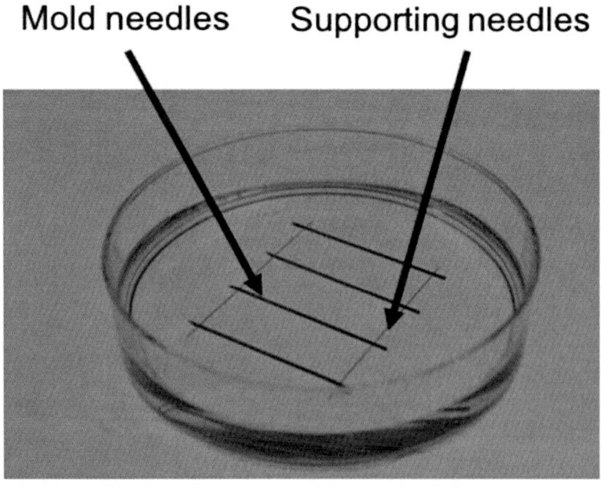

Fig. 3 Position of supporting needles and mold needles

11. Remove the needles with forceps (Ø 550 μm and Ø 120 μm). Store the 550-μm needles (reusable) and discard the small needles.
12. Punch holes, distance 1 cm (use a ruler), as inlets and outlets using a Ø 2.0 mm biopsy puncher.
13. Seal the side holes left by the needles with liquid PDMS (*see* **Note 10**).
14. Cure at 60 °C for at least 2 h.
15. Chips can be stored at this point, and then tape the structures to protect them from dust.

3.2.2 PDMS-Glass Bonding (Fig. 4)

1. Cut a chip in four parts with a scalpel. Every single part must contain a microchannel.
2. Tape a PDMS chip with scotch tape on the bottom side while keeping the channel side on top and leave a small space between the channels.
3. Clean a glass slide with 70% ethanol and soap water and rinse with ultrapure water.
4. Dry the glass slide with a nitrogen gun (*see* **Note 11**).
5. Place PDMS-chip and glass slide into the oxygen plasma cleaner.
6. Turn on the oxygen tank, the pressure indicator, the plasma cleaner, and the vacuum pump.
7. Wait for the pressure to decrease until approximately 300 mTorr, and then turn the O_2 valve into position.
8. Wait until the pressure stabilizes at approximately 650 mTorr.
9. Turn on plasma to a high level for glass-PDMS bonding. Leave 3 min under oxygen plasma.
10. Open the valve to let the pressure normalize by room air (*see* **Note 12**).

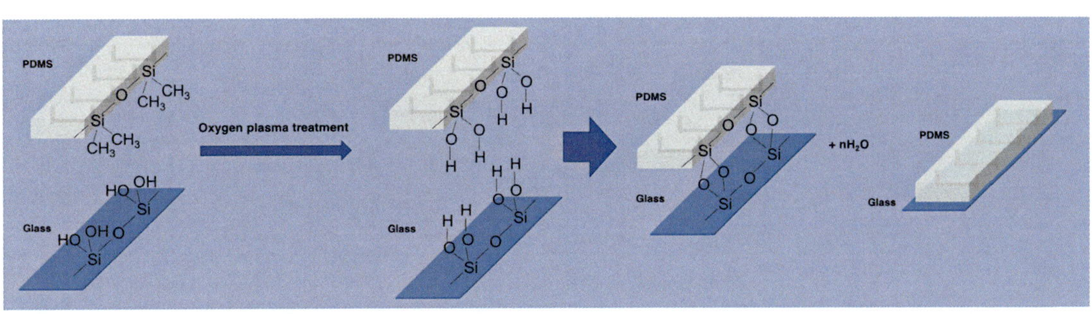

Fig. 4 Schematic representation of the bonding procedure. The plasma oxygen treatment removes hydrocarbon groups (CxHy), leaving behind silanol groups on the PDMS and OH groups on the glass substrate, respectively. This process allows strong Si–O–Si covalent bonds to form between the two materials

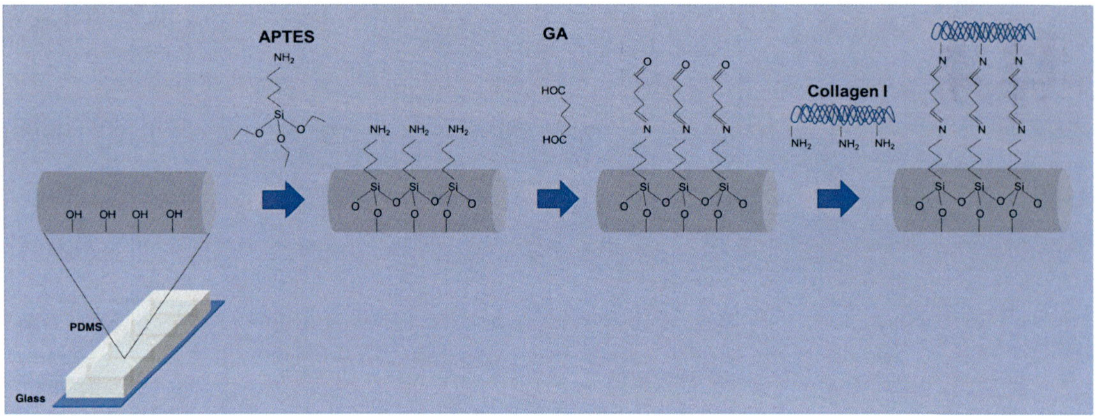

Fig. 5 Schematic representation of PDMS surface modification to cross-link extracellular matrix proteins. *GA* glutaraldehyde

11. Place the PDMS chip in the center of the glass slide and gently press them together to allow a covalent bonding formation.

3.2.3 Coating of Microchannels (Fig. 5)

1. After oxygen-plasma bonding, modify the PDMS surface immediately by filling the microchannels with 5% APTES (the shorter the time between plasma oxygen treatment and APTES, the more APTES-groups will bind to the surface), and leave 20 min (*see* **Note 13**).
2. Wash the channels three times with distilled water.
3. Replace the water with 0.1% glutaraldehyde three times. Leave it for 30 min.
4. Wash three times with distilled water.
5. Replace the water with 50 μg/mL fibronectin (three times), incubate 60 min at 37 °C, or place it in the laminar flow overnight at room temperature under UV light.
6. Fill the channels with the 100 μg/mL collagen-I solution (three times) and leave it in the laminar flow at room temperature for 1.5 h.
7. Add cell culture medium and place the Petri dish with the chip in the 37 °C incubator for 30 min or longer before cell seeding.

3.2.4 Cell Seeding

1. Harvest the cells of interest and adjust cell density at 10^6/mL by adding cell culture medium supplemented with dextran (4% final dextran concentration).
2. Inject 10^6 cells/mL into the channels from both ends using 200 μL pipette (make sure using a single cell suspension, avoid cell clumps!).

3. Place the microfluidic chip upside down in the incubator and leave it for 15 min. The cells should have enough time to attach to the surface (block the glass slide with a tape to the Petri dish).
4. Aspirate the unattached cells and add new cells. Incubate the microfluidic chip upright for 30 min.
5. Verify cell attachment using a microscope.
6. Wash the channels with 4% dextran medium to get rid of unattached cells.
7. Wait until the cells are confluent (normally the next day) and change the medium regularly (every 2 h).

3.2.5 Peristaltic Pump Connection (Four Channels)

1. Connect the peristaltic pump only when the endothelial cells are confluent.
2. Flush tubings with 30 mL of autoclaved ultrapure water and then with 30 mL of autoclaved PBS.
3. Check the tubings for leakage and continue flushing with 20 mL 4% dextran medium.
4. Be sure that the medium runs through all the channels with the same speed. If not, adjust the flow with the screws on the back of the pump.
5. Fill 15 mL falcon tubes (reservoir tubes) with 10 mL of 4% dextran medium and insert the inlet and outlet of the tubings through the cap. Adjust the inlet at the level of 8 mL and the outlet at 2 mL so that mixing of the medium is ensured.
6. Insert the ends with the adapters in the inlet and outlet of the microchannels.
7. Start the pump at 7 rpm (flow 600 μL/min).
8. Increase to desired rpm or leave the cells at 7 rpm.
9. Keep the desired rpm for 48 h or longer if you change medium reservoir every 2 days.

4 Notes

1. For example, Sarstedt blood collection system—do not use open systems to reduce blood–air contact and don't use vacutainers because of blood activation due to formation of a blood jet.
2. Always use fresh PBS-1% BSA solution for antibody dilution. Because it does not contain antibiotics, contamination may occur if stored for long time.
3. Carefully pipet and handle the cell-coated microbeads to avoid cell detachment and/or activation.

4. Use a tissue to absorb excess PBS after pipetting the microbeads on the glass slide. Afterward, apply the mounting medium and place the coverslip without applying excessive pressure.
5. Pipet the cell-coated microbeads carefully into the polypropylene tubes without stirring them to prevent cell detachment.
6. For six time points, we suggest having at least three replicates within each group of cells, as the blood sampling will be done in different tubes.
7. Appropriate collection of serum samples is necessary to prevent complement degradation. The blood must be allowed to coagulate at room temperature for 30 min, immediately after placing the tubes in a 4 °C centrifuge and centrifuging the samples at 2000 × *g* for 10 min. After centrifugation, place the tubes on ice and collect the serum in 1.5 mL tubes and store them immediately at −80 °C for further analysis.
8. Example for one Petri dish (four microchips): 35 g of silicon elastomer, 3.5 g of curing agent.
9. Use a nitrogen gun to clean the surface of the Petri dish before pouring the PDMS.
10. Prepare 5 g of PDMS: 5 g of silicon elastomer, 0.5 g of curing agent.
11. Clean the glass slide carefully and extensively. If not properly cleaned, then a good bonding cannot be achieved and the microchannels will leak.
12. Slowly open the valve letting the pressure normalize. Avoid opening the valve too fast, as the microchips and the glass slides might be blown against the walls of the plasma oxygen cleaner chamber. After oxygen plasma treatment, the PDMS and the glass slide are strongly reactive, and if they come in contact with other surfaces inside the chamber, they could stick to them.
13. Fill the microchannel by pipetting the solution on the inlet until it reaches the outlet. Remove the excess solution from the outlet with a pipet paying attention not to dry the channel. Add more solution on the inlet and repeat the partial removal from the outlet two times more. This step has to be followed throughout the whole coating protocol.

References

1. Félétou M (2011) The Endothelium. Morgan & Claypool, San Rafae
2. Kása A, Csortos C, Verin AD (2015) Cytoskeletal mechanisms regulating vascular endothelial barrier function in response to acute lung injury. Tissue Barriers 3:e974448. https://doi.org/10.4161/21688370.2014.974448
3. Reitsma S, Slaaf DW, Vink H et al (2007) The endothelial glycocalyx: composition, functions, and visualization. Pflugers Arch 454:345–359. https://doi.org/10.1007/s00424-007-0212-8
4. Lin CC, Cooper DKC, Dorling A (2009) Coagulation dysregulation as a barrier to xenotransplantation in the primate. Transpl Immunol

21:75–80. https://doi.org/10.1016/j.trim.2008.10.008

5. Hryhorowicz M, Zeyland J, Słomski R, Lipiński D (2017) Genetically modified pigs as organ donors for xenotransplantation. Mol Biotechnol 59:435–444. https://doi.org/10.1007/s12033-017-0024-9
6. Sfriso R, Bongoni A, Banz Y et al (2017) Assessment of the anticoagulant and anti-inflammatory properties of endothelial cells using 3D cell culture and non-anticoagulated whole blood. J Vis Exp:e56227–e56227. https://doi.org/10.3791/56227
7. Sfriso R, Zhang S, Bichsel CA et al (2018) 3D artificial round section micro-vessels to investigate endothelial cells under physiological flow conditions. Sci Rep 8:5898. https://doi.org/10.1038/s41598-018-24273-7
8. Bulato C, Radu C, Simioni P (2012) Studies on coagulation incompatibilities for xenotransplantation. Methods Mol Biol 885:71–89. https://doi.org/10.1007/978-1-61779-845-0_6

Chapter 7

Cell-Based Assays for Modeling Xenogeneic Immune Responses

Kelly Casós, Roberta Sommaggio, Magdiel Pérez-Cruz, and Cristina Costa

Abstract

Research in xenotransplantation implies a high experimental complexity comprising molecular, cellular, and in vivo studies to investigate the mechanisms of xenograft immune rejection and functional failure, as well as the strategies to counteract them. After major advances associated with the identification of the carbohydrate xenoantigens and their elimination through genomic edition of the source pigs, the study of the cellular immune response against the xenograft is gaining particular attention. Xenogeneic cell-based assays that put together pig cells and human leukocytes such as monocytes, NK cells, and T cells are relevant to address this hurdle. Thus, we describe here coculture, co-stimulatory, and cytotoxicity assays for investigating the cellular and molecular mechanisms of xenograft rejection. These techniques allow elucidating the key pathways that take place during the xenogeneic immune response in a simplified setting. Treatment with either pro-inflammatory or anti-inflammatory cytokines can be used for studying the regulation of adhesion, co-stimulatory molecules, and receptors involved in triggering the immune response under various conditions. Furthermore, these assays can be used for the follow-up of the immune response of in vivo studies as well as for the development of tolerogenic approaches that promote xenograft survival.

Key words Pig, Human, Monocytes, T cells, NK cells, Cytokines, Immune response

1 Introduction

Xenotransplantation involves the transplantation of animal cells, tissues, or organs to recover function of the corresponding part of the human recipient. At present, the domestic pig (Sus scrofa domestica) is considered the best source of biological material for xenotransplantation. Unfortunately, the molecular incompatibility between the donor and the recipient, resulting from the phylogenetic distance between pigs and humans, entails a range of immune complications following transplantation, leading to xenograft rejection [1]. The introduction of new technologies based on in vitro and in vivo studies has allowed the identification of molecular

Kelly Casós and Roberta Sommaggio contributed equally to this work.

Cristina Costa (ed.), *Xenotransplantation: Methods and Protocols*, Methods in Molecular Biology, vol. 2110,
https://doi.org/10.1007/978-1-0716-0255-3_7,

targets, as well as the generation of genetically engineered pigs designed to control cross-species incompatibilities and xenograft rejection by means of multiple genetic modifications, i.e., knockouts and transgenes [2–4]. In particular, the cellular assays are being used to model in a simplified setting the rejection process led by immune cells. These tools help elucidating the cellular and molecular mechanisms of xenograft rejection and set the bases for the development of the strategies that counteract them.

Cell-, tissue-, and organ-based xenografts are subjected to distinct immune rejection processes that share humoral and cellular mechanisms. In solid organs, a classification of the rejection processes has been established based on the time of occurrence: (1) hyperacute rejection, produced within minutes after transplantation by natural xenoreactive antibodies and complement activation; (2) acute xenograft rejection (AXR), taking place within weeks post-transplantation, being mediated by activated endothelium, antibody binding, complement and coagulation factors, as well as a variety of immune cells; and (3) chronic rejection (with similarities to allografts) [3, 5]. Interestingly, non-encapsulated islets infused by injection into the portal vein suffer instant blood-mediated inflammatory reaction also involving innate immune mechanisms that lead to substantial loss of islets [6]. As for other cellular and tissue-based xenografts, subsequent rejection events involve cellular and humoral mechanisms.

AXR comprises a cellular component that is mainly mediated by a T-cell response to pig antigens and is triggered within days after transplantation. Different studies have demonstrated the presence of a mononuclear cell infiltrate containing primarily $CD4^+$ T cells, CD8+ T cells, macrophages, and some natural killer (NK) cells in solid organs transplanted into nonhuman primates (NHP) without immunosuppression [7, 8]. Nevertheless, current immunosuppression protocols markedly decrease the cellular immune infiltrate and prolong survival of solid organ xenografts in NHP models [7, 9–11]. Notably, recent work has established the critical role of $CD4^+$ T cells in rejection of xenografted kidneys in NHP selected for low anti-pig antibody titers prior to transplantation, obtaining very long survivals (>400 days) when specifically depleting this T-cell population [12]. In this accordance, carbohydrate remodeling of pig cells to reduce human natural antibody reactivity through genetic modifications does not affect the cellular immune response when confronted with human T cells [13]. In fact, cell-mediated rejection seems highly relevant in cell-based xenografts as the use of immunotherapies that suppress T-cell activation results in long-term survival of porcine islets in diabetic monkeys [14–16]. Thus, understanding the underlying mechanisms of how xenografts trigger cellular rejection is currently of high interest to develop pharmacological and genetic engineering approaches that promote engraftment.

The study of transplant rejection and tolerance was traditionally focused on cells of the adaptive immune system, especially T cells. T cells in xenotransplantation are activated through direct and indirect pathways. During direct activation, recipient (NHP or human) T-cell receptors (TCR) bind swine leukocyte antigen (SLA) class I and class II on porcine antigen-presenting cells (APC). Porcine APC include passenger dendritic cells or endothelial cells that constitutively express CD80/86 (as opposed to human APC) [17, 18]. This interaction results in T-cell-mediated cytotoxicity directed against the xenograft. The indirect pathway of T-cell activation involves NHP/human recipient $CD4^+$ T cells recognizing porcine donor peptide presented on NHP/human recipient major histocompatibility complex (MHC) class II leading to activated T cells producing cytokines that prime B cells and the innate immune system, including macrophages and NK cells, a process which ultimately leads to xenograft dysfunction [18–20]. Most of what is known about the molecular interactions that trigger the T-cell response to xenografts results from cellular studies that coculture porcine aortic endothelial cells (PAEC) with human peripheral blood mononuclear cells (PBMC) or purified lymphocyte populations. PAEC function as authentic professional APC promoting direct activation of CD8+ and $CD4^+$ T cells through respective binding of SLA I and II together with co-stimulatory molecules (porcine CD86 and CD40) [18, 21, 22].

AXR also comprises an innate immune cellular response that has received much attention because it could be responsible of relevant rejection mechanisms resistant to immunosuppression. Monocytes/macrophages are thus important in AXR, being present from an early stage in the xenograft of nonhuman primates [7, 23] and contributing to xenograft rejection of solid organs in small animal models [24]. A major contribution of macrophages to rejection has also been demonstrated in cellular xenografts such as pancreatic islets. Different studies demonstrated that macrophages exhibit an activated cytokine profile and mediate their effects through NO and/or TNFα, as well as through cellular interactions with the adaptive immune system [24]. A target considered key for the control of these responses is that of the signal-regulatory protein-α (SIRPα), a receptor expressed on macrophages which interacts with ubiquitously expressed CD47 and prevents the phagocytosis of cells [25]. As the porcine CD47 does not trigger the desired inhibitory signals across species, transgenic expression of the human homolog in the source pig is probably a good solution [26]. Nevertheless, other strategies are being considered as well with this purpose such as expression of human CD200, PDL-1, or HLA-G [27–29].

The initiation and regulation of adaptive immune responses after solid organ transplantation by natural killer (NK) cells, promoting either rejection or tolerance, has only been recognized more recently [30–32]. NK cells function under a tightly regulated balance of activating and inhibitory signals mediated by a variety of NK cell receptors and their respective ligands on potential target cells. Upon recognition of "altered or abnormal cells" by one or more pathways, the NK cells will trigger lysis of the target cells and secrete cytokines such as IFNγ. In agreement with these functions, NK cells play a prominent role in rejection of xenografts. The incapacity of SLA class I molecules to signal through human NK cell inhibitory receptors is one of the reasons why pig cells are susceptible to human NK-cell-mediated lysis [33]. Furthermore, human NK cells can adhere and lyse porcine endothelial cells both directly and by antibody-dependent cell-mediated cytotoxicity (ADCC) in the presence of human serum [34]. The cellular studies have also allowed to identify some key NK cell activating receptors and pig ligands implicated in triggering cytotoxicity [35]. Thus, a variety of potential strategies to reduce the observed endothelial damage caused by NK cells during rejection of vascularized xenografts have been developed including the generation of HLA-E transgenic pigs [36], which have been studied in different xenotransplantation models [37–39]. Furthermore, it is expected that all the genetic modifications recently developed to reduce human antibody reactivity toward pig cells will confer additional protection by controlling ADCC.

Major progress has thus been made in the last decades to understand the immunobiology of pig-to-primate xenotransplantation. However, the cellular assays remain a key tool to elucidate the molecular mechanisms of cellular rejection and test the strategies to counteract them. In this chapter, we describe the methodology of some basic cellular assays that can be easily adapted depending on the purpose of the study. We have used various of these techniques for a study characterizing the cellular immune response against porcine chondrocytes [40], and it could be applied to other cells of interest for xenotransplantation. The adhesion assays are used to assess the capacity of various human immune cells to adhere to porcine cells and determine the adhesion molecules involved. The coculture assays allow studying the response of both human and porcine cells under conditions that favor cell contact. Particularly, the co-stimulation assays are performed to identify the cell surface pig molecules that provide functional co-stimulatory signals to human T cells. Finally, the cytotoxicity assay is a fundamental tool for the study of NK cell function that can be applied to identify the triggering ligands and receptors.

2 Materials

2.1 Cell Culture and Cytokine Treatments

1. Porcine cells of interest for xenotransplantation such as PAEC or PK-15 (*see* **Note 1**).
2. Multiple pig and human cytokines, as well as combinations of them, can be used (*see* **Note 2**).
3. Sterile cell culture supplies: 15-mL conical tubes, culture flasks, pipettes, Pasteur pipettes, and tips.
4. Culture medium appropriate for the chosen cell: Most established cell lines do well in Dulbecco's modified Eagle's medium (DMEM) containing high glucose, 10% fetal bovine serum (FBS) (*see* **Note 3**), 200 U/mL penicillin, and 200 μg/mL streptomycin. Primary cultured cells usually have special requirements such as media supplements and surface coating (*see* **Note 4**).
5. Solution of trypsin (0.05%) and ethylenediaminetetraacetic acid (EDTA) (1 mM) for detachment of adherent cells. The reagent TrypLE™ Express (Invitrogen, Carlsbad, CA, USA) works well.
6. Tissue culture equipment: Pipettors, laminar-flow cabin, aspiration pump or source of vacuum, 5% CO_2 incubator, centrifuge inverted microscope, and Neubauer chamber.

2.2 Static Adhesion Assays

1. Porcine cells of interest for xenotransplantation such as PAEC (*see* **Note 1**). These can be tested in resting conditions and after cytokine stimulation.
2. Human leukocytes. It is preferable to assess adhesion of a purified population such as NK cells or $CD4^+$ T cells. The activation state of the cells should be taken into account as many can be assayed either freshly isolated or after culture under controlled conditions (e.g., IL-2-activated NK cells, monocyte-derived macrophages). We recommend assessing both settings whenever possible. A simpler approach is to assess adhesion of an established cell line such as the NK92 cells or the U937 monoblastic cells. The assay we describe here should work for all of them.
3. Sterile tissue culture supplies: 15-mL conical tubes, culture flasks, pipettes, Pasteur pipettes, tips, and 96-flat-bottom-well plates.
4. Cell culture medium appropriate for the chosen pig cells: *See* Subheading 2.1. for description of culture media of pig cells.
5. Phytohemagglutinin (PHA-P). A solution of 100 μg/mL can be prepared and stored at −20 °C in aliquots. Do not filter to prevent the reagent loss.

6. Cell culture medium appropriate for the chosen human leukocytes: It has to be adjusted to each cell type. U937 do well in DMEM/10% FBS, but cultured NK cells will need Iscove's modified Dulbecco's medium (IMDM) supplemented with 10% (v/v) human serum, 10%(v/v) nonessential amino acids, 10% (v/v) sodium pyruvate, 200 U/mL of penicillin/streptomycin, and 200 U/mL recombinant human IL-2 (hIL-2). For 3-week culturing, the NK cells are seeded in the same medium in 96-round-bottom-well plates with irradiated JY feeder cells, 0.5 μg/mL PHA-P, and 5 ng/mL recombinant human IL-15. Fresh feeder cells are added also on day 5 of culture to restimulate NK cells.
7. TrypLE™ Express or equivalent reagent.
8. Hanks' balanced salt solution (HBSS) 1% BSA for all incubations and washes (*see* **Note 5**).
9. Labeling solution: Calcein AM (10 μg/mL, Invitrogen). It has to be prepared fresh for each assay. Prepare a tube with 5 mL of HBSS/1% BSA wrapped in aluminum foil. Dissolve the content of a calcein AM 50 μg vial in 10 μL of dimethyl sulfoxide (DMSO), mix well with about 0.5 mL of the HBSS/1% BSA, and transfer back to the wrapped tube. This amount usually suffices for an assay, but it can be scaled up if needed. Store at room temperature until use.
10. Equipment: tissue culture equipment as described in Subheading 2.1 and cytofluorometer. The cytofluorometer is to be used at a 485-nm excitation filter and a 520-nm emission filter.

2.3 Coculture Assay

1. Porcine cells of interest for xenotransplantation such as PAEC (*see* **Note 1**). These can be tested in resting conditions and after cytokine stimulation.
2. Human leukocytes: preferably a purified population such as monocytes to determine their response in front of pig cells. Nevertheless, more complex settings can be assayed as well such as peripheral blood mononuclear cells or combinations of T cells and monocytes/macrophages. A simpler approach is to use an established cell line such as the U937 monoblastic cells. We describe a basic assay for a single population that can be modified for more complex settings.
3. Sterile tissue culture supplies: 15-mL conical tubes, culture flasks, pipettes, Pasteur pipettes, 1.5-mL tubes, tips, and 96-flat-bottom-well plates. The assay can also be conducted in other multi-well plates such as 24- or 48-well plates depending on the amount of supernatant desired.
4. Cell culture medium: DMEM/10% FBS with antibiotic will work for PAEC and monocyte cocultures.
5. DMEM alone for washes.

6. ELISA kits for detection of pig and human IL-8 and/or IL-6 (*see* **Note 6**).
7. Equipment: tissue culture equipment as described in Subheading 2.1, multichannel pipettor, basins, and microplate reader.

2.4 Co-stimulation Assay

1. Porcine cells of interest for xenotransplantation such as PAEC.
2. Human T cells: preferably a purified population such as $CD4^+$ T cells or $CD8^+$ T cells. Nevertheless, a simpler approach is to use an established cell line such as Jurkat-T cells that can provide relevant information about the presence on pig cells of functional co-stimulatory molecules for human receptors.
3. Sterile tissue culture supplies: 15-mL conical tubes, culture flasks, pipettes, Pasteur pipettes, 1.5-mL tubes, tips, and 96-flat-bottom-well plates.
4. Culture medium for PAEC: DMEM/10% FBS with antibiotic.
5. Culture medium for Jurkat/T cells: RPMI/10% FBS.
6. PHA-P. Remember that a solution of 100 μg/mL can be prepared and stored at −20 °C in aliquots. Do not filter to prevent the reagent loss.
7. Purified antihuman CD28 (hCD28) mAb CD28.2 (BD Biosciences, San José, CA, USA). This is an agonist antibody to be used as positive control.
8. ELISA kit for detection of hIL-2.
9. Equipment: tissue culture equipment as described in Subheading 2.1, multichannel pipettor, basins, and microplate reader.

2.5 NK Cytotoxicity Assay

1. Porcine cells of interest for xenotransplantation such as PAEC.
2. NK cells. These can be tested either freshly isolated or after culture with hIL-2. The NK92 cell line can also be used as parallels well the activity of hIL-2-activated NK cells on pig cells, but it does not reflect the donor variability.
3. Sterile tissue culture supplies: 15-mL conical tubes, culture flasks, pipettes, Pasteur pipettes, tips, and round-bottom 96-well plates.
4. Culture medium for PAEC: DMEM/10% FBS with antibiotic. Use the appropriate medium for culturing other pig cells.
5. Culture medium for the cytotoxicity assay: IMDM/10% FBS with 200 U/mL penicillin/streptomycin. The IMDM/10% human serum can only be used when the effect of xenogeneic antibodies is assessed.
6. TripLE Express or cell dissociation buffer.
7. Optional: Blocking antibodies for NK receptors such as NKG2D (Clone 149810, R&D Systems, Minneapolis, MN,

USA), NKp44 (p44/8, Beckman Coulter GmbH, Krefeld, Germany), and NKp46 (Clone BAB28, Beckman Coulter GmbH).

8. Equipment: tissue culture equipment as described in Subheading 2.1 and a room equipped for radioactive tissue culture with gamma counter.

3 Methods

3.1 Cell Culture and Cytokine Treatments

1. Use primary cultured pig cells at low passages with the recommended media and monitor for morphologic and phenotypic changes when necessary. This is not necessary for well-established cell lines.
2. Split primary cultured cells such as PAEC when they just reached confluence. Do not dilute the PAEC more than 1–2 for subculture as low-density culture causes fast senescence. However, other primary cultured cells may have other particular requirements that need to be taken into account for each specific cell type.
3. Treat PAEC, pig chondrocytes, or established cell lines with cytokines after they reach confluence. We routinely preincubate with either 10 ng/mL hTNFα for 24 h or 5–10 ng/mL hIL-1α for 8 h before other assays to activate the cells as a model of a pro-inflammatory setting.

3.2 Static Adhesion Assay

1. Seed the pig cells to reach confluence in 96-flat-bottom-well plates overnight. Distribute evenly the cells within the well by carefully applying cross motions.
2. Leave these cells untreated or preincubate with cytokines as described in preparation for the adhesion assay. Prepare triplicates for each assay condition.
3. First, wash the chosen effector cells by centrifugation (*see* **Note 7**) and resuspend in 10 μg/mL calcein AM. Incubate for 20 min at 37 °C for loading. The amount of cells to be used can be estimated in advance following the calculations described below. Prepare always in excess. Use HBSS/1% BSA for all the assay incubations and washes.
4. In the case of assessing the contribution of specific molecules to adhesion, one good option is the utilization of blocking antibodies just prior to the addition of the human leukocytes. For blocking pig molecules, wash the plate and keep at room temperature for 20 min with the blocking antibody and the isotype control in designated wells at 2× concentration (10 μg/mL is usually sufficient). This procedure can be conducted simultaneously to the labeling of the effector cells.

5. Wash the calcein-loaded cells by centrifugation, resuspend in HBSS/1% BSA, and count (*see* **Note 8**). For blocking human molecules, preincubate a sufficient amount of labeled cells (such as NK92) (*see* **Note** 7) with the blocking or the isotype control antibody for 10 min at room temperature.
6. Add the effector cells in equal volume (e.g., 100 μL for a 200-μL final volume) to the confluent pig cells at a specific effector/target (E/T) ratio in triplicates (*see* **Note 9**). We recommend to work in a range of 1:1–5:1 E/T ratio (*see* **Note 10**), but it may be of interest to assess higher ratios for establishing the amount needed to saturate the system.
7. Optional: Centrifuge the plate at 16 × *g* for 2 s without brake.
8. Optional: Determine the start fluorescence with cytofluorometer at 485/520 nm.
9. After an incubation of 30–45 min at 37 °C, remove non-adherent cells with thorough washes. When testing the effect of antibodies, these can be kept during the coculture time (a final concentration of 5 μg/mL is enough).
10. Lyse the remaining cells with 1% SDS.
11. Measure fluorescence intensity with a cytofluorometer at 485/520 nm. The percentage of adhesion can be calculated by dividing the end value with its corresponding start values (if taken) and multiplying by 100.

3.3 Coculture Assay

1. Seed the pig cells in multi-well plates to reach confluence overnight.
2. Leave them untreated or preincubate with cytokines as described.
3. Wash the cells with DMEM alone just prior to the assay.
4. Harvest and count the effector cells (*see* **Note 8**).
5. Initiate the coculture by adding the effector cells to the pig cells at the desired E/T ratio. We recommend to work in a range of 1:1–5:1 E/T ratio (*see* **Note 10**). Assay each condition at least in duplicate.
6. Maintain the coculture for the desired length of time. The coculture time should be established for each study, but we find that cocultures of 18–24 h usually lead to a clear response. Shorter times such as 8 h can be used for kinetics, but may result in a partial response for changes in protein expression or secretion.
7. Collect and centrifuge the supernatants at 260 × *g* for 5 min.
8. Harvest the cell-free supernatants. These can be stored in 1.5-mL tubes or 96-well plates (depending on the volume) at −20 °C until use. Regarding the cellular pellet, it can be discarded or processed for further analyses (*see* **Note 11**).

9. Determine the levels of human and porcine IL-8 in supernatants by ELISA using a commercialized kit following the instructions of the manufacturer. Use a microplate reader at 450 nm.

3.4 Co-stimulation Assay

1. Seed the pig cells in 96-flat-bottom-well plates to reach confluence overnight in preparation for the coculture.
2. On the day of the assay, harvest and count the Jurkat or human T cells after some pipetting up and down to disperse the cells (which usually grow in clusters) (*see* **Note 8**).
3. At this time point, the media of the pig cells may be removed and changed to RPMI/10% FBS (e.g., 100 μL). In experiments determining the contribution of specific molecules such as pCD86, preincubate (prior to the coculture) selected wells in triplicate with the specific blocking antibody or the isotype-matched control (10 μg/mL in culture media) for 20 min at room temperature. The antibodies can be conveniently maintained during the coculture (5 μg/mL).
4. Add the effector cells in RPMI/10% FBS in equal volume (e.g., 100 μL for a 200-μL final volume) to the confluent pig cells. Prepare the first cocultures with at least two different E/T ratios such as 2:1 and 5:1 in duplicates or triplicates. A multichannel pipettor and sterile basins can be used for this purpose.
5. Add the PHA to selected wells to a final concentration of 10 μg/mL at the time of initiating the cocultures and incubate at 37 °C for the desired times (*see* **Note 12**) and conditions. All experiments should include a positive control such as T cells incubated with PHA and the anti-hCD28 agonist antibody (5 μg/mL).
6. Collect cell-free supernatants at the end of the assay by transferring a fixed amount of media to 96-V-bottom-well plates and centrifuging at 260 × *g*. The harvested supernatants can be assayed instantly and/or frozen for storage at −20 °C.
7. Assay for hIL-2 by ELISA following the instructions of the manufacturer. Final determinations are done in a microplate reader.

3.5 NK Cytotoxicity Assay

1. Detach the adherent pig cells from the culture flask using TripLE Express, Accumax™ (Thermo Fisher Scientific), or cell dissociation buffer (*see* **Note 13**) and wash by centrifugation. Resuspend the cells in IMDM/10% FBS with antibiotic and count (*see* **Note 8**).
2. Incubate 5×10^5 target (pig) cells with 100 μCi (3.7 MBq) ^{51}Cr as sodium chromate in a 15-mL tube for 1 h at 37 °C. The tube is placed on a rotating system (in a way that the sample cannot be spilled) inside the 5%CO_2 incubator with the tap loose.

3. Harvest and wash the NK cells to avoid any traces of human serum and resuspend in IMDM/10% FBS with antibiotic. When multiple E/T ratios are used, the effector cells are plated in serial dilution before the labeled target cells are added. The E/T ratios 5:1, 10:1, 20:1, and 40:1 are usually tested for this purpose. Have ready the appropriate amount of cells in 100 μL/well in a round-bottom 96-well plate.
4. Wash the labeled pig cells with IMDM/10% FBS with antibiotic twice and resuspend in the appropriate amount of media to have 5×10^4 cells/mL. Load 100 μL per well for the assay.
5. For experiments assaying the contribution of specific molecules with blocking antibodies, preincubate the NK cells for 30 min at room temperature with 10 μg/mL blocking antibodies specific for NK cell receptors such as NKG2D, NKp30, NKp44, and NKp46. The antibodies are later left for the duration of the coculture at a concentration of 5 μg/mL.
6. Add the labeled target cells to the NK cells into the round-bottom 96-well plate to have a final volume of 200 μL/well and coculture at 37 °C and 5% CO_2. Triplicate each assay condition.
7. Stop the assay after 4 h of coculture and collect the supernatants.
8. Determine the maximum release by incubating the labeled target cells with 1% Triton X-100 and the spontaneous release by incubating the labeled target with medium alone.
9. Analyze the ^{51}Cr release on a gamma counter. To this end, thick filters are placed on each well to absorb all the supernatants and transferred to the appropriate tubes for measures in the gamma counter.
10. Calculate the percentage of specific lysis as ((experimental release-spontaneous release)/(maximum release-spontaneous release)) $\times$ 100.

4 Notes

1. There are many pig cells that are of interest for xenotransplantation and can be studied in cellular assays such as chondrocytes, hepatocytes, and fetal mesencephalon cells between others. In general, primary cultured cells such as PAEC are better models for cellular assays. Nevertheless, it is also common to use immortalized and established cell lines because they are easier to grow in culture and genetically engineer.
2. In xenogeneic cellular models, it is especially interesting to assess the effect of human cytokines. However, at least one of

them has been reported not to be functional on pig cells, and that is human IFNγ. On the contrary, human TNFα displays a very strong effect on pig cells.

3. Remember to heat-inactivate the FBS before use to prevent undesired complement activation. We usually place a bottle with 1 L FBS recently thawed in a 56 °C water bath for 1 h for heat inactivation. Then, let it cool down and prepare 50-mL aliquots for storage at −20 °C. For growing primary cultures, it is advisable to test different batches and providers of FBS and reserve various bottles of a good batch.
4. To enhance the growth and survival of primary cultured pig cells, we supplement the basal media DMEM/10% FBS with a bovine hypothalamus extract such as Endothelial Cell Growth Supplement (Merck Millipore). We use 50 mg/L for PAEC and 25 mg/L for chondrocytes with good results. We remove this supplement at least 24 h before initiating any assay. For culturing PAEC, it can also be of help to pre-coat the culture plates or flasks with porcine gelatin following standard procedures. It is wise to optimize the culture conditions prior to starting the experimental assays.
5. Note that HBSS can lose its properties when stored for a long time. Renew your stock often or have a concentrated supplement of calcium/magnesium as it is needed for good results.
6. Note that some ELISA antibodies cross-react with homologues of other species. Thus, it is advisable to include always controls of cells without coculture in the same quantities assessed in the coculture. Nevertheless, we have good experience with kits from R&D Systems to detect the pig cytokines and from RayBiotech for the human counterparts. Other cytokines could be tested as well taking into account this limitation and the capabilities of the effector cell chosen.
7. Human NK cells, monocytes, and macrophages have Fc receptors. In the case of assaying NK cells cultured with human serum, these should be washed at least twice before the assays to avoid any traces of the human serum that could affect adhesion. In the case of using blocking antibodies in the assay, optimize the conditions to minimize the effect of Fc receptors unless these are the subject of study. For instance, assess the effect of molecules with NK92 cells that lack CD16 or use mouse IgG1. In any case, we advise to include an isotype-matched control antibody when including antibodies in the assay.
8. The cells can be simply counted with the microscope using a Neubauer chamber and an appropriate dilution. It is especially advisable to also assess viability (with trypan blue, for instance) to make sure the cells are in good condition for the assay.

9. Be especially careful during washes and the removal of the media. We recommend inverting the plate over a container for waste collection to avoid touching the cells or leaving media. The remaining drops are removed by patting the plate on a clean paper towel. We also recommend to be very thorough with the pipetting when adding the effector cells. To determine the number of target (pig) cells, it can be appropriate to count cells after harvest from a larger culture surface such as a T25 flask in the previous culture step. The number of confluent cells per well in a 96-well plate can then be estimated by correcting for culture surface (e.g., $y = 0.32\times\ (n/25)$ in which n is the number of cells isolated from a confluent T25).
10. We recommend working at low E/T ratios such as 2:1 for assessing the effect of specific molecules.
11. It may be possible to determine changes in the cells after coculture at the level of mRNA or protein that could be assessed respectively by RT-PCR and western blotting.
12. Cocultures of 18–24 h should suffice to detect the T-cell response. Evaporation of the media may be a problem for longer cocultures such as 3–5 days.
13. The choice of buffer for harvesting the adherent pig cells can be important for determining their susceptibility to NK cell-mediated lysis, as some ligands can be destroyed by enzymatic treatments. Thus, this should be evaluated in the first experiments if it has not been previously reported in the literature.

Acknowledgments

This work was supported by the European Union Seventh Framework Programme (FP7/2007–2013) under Grant Agreement no. 603049 (TRANSLINK), the SLT002/16/00445 grant funded by the Department of Health of the Generalitat de Catalunya, Ministerio de Economía y Competitividad-ISCIII (PI15/00181), and Ministerio de Educación y Ciencia (SAF2008-00499) (all granted to CC), and co-funded by FEDER funds, a way to build Europe. IDIBELL benefits from the support of CERCA. R.S. was supported by a one-year fellowship from Fundació Catalana de Trasplantament and M.P.-C. by an IDIBELL fellowship.

References

1. Cooper DK, Gollackner B, Sachs DH (2002) Will the pig solve the transplantation backlog? Annu Rev Med 53:133–147
2. Cooper DK, Ekser B, Ramsoondar J, Phelps C, Ayares D (2016) The role of genetically engineered pigs in xenotransplantation research. J Pathol 238:288–299
3. Niemann H, Petersen B (2016) The production of multi-transgenic pigs: update and perspectives for xenotransplantation. Transgenic Res 25:361–374
4. Yum SY, Yoon KY, Lee CI, Lee BC, Jang G (2016) Transgenesis for pig models. J Vet Sci 17:261–268
5. Cooper DKC, Ekser B, Tector AJ (2015) Immunobiological barriers to xenotransplantation. Int J Surg 23:211–216
6. Goto M, Tjernberg J, Dufrane D et al (2008) Dissecting the instant blood-mediated inflammatory reaction in islet xenotransplantation. Xenotransplantation 15:225–234
7. Ashton-Chess J, Roussel JC, Manez R et al (2003) Cellular participation in delayed xenograft rejection of hCD55 transgenic pig hearts by baboons. Xenotransplantation 10:446–453
8. Davila E, Byrne GW, LaBreche PT et al (2006) T-cell responses during pig-to-primate xenotransplantation. Xenotransplantation 13:31–40
9. Tseng YL, Kuwaki K, Dor FJ et al (2005) alpha1,3-Galactosyltransferase gene-knockout pig heart transplantation in baboons with survival approaching 6 months. Transplantation 80:1493–1500
10. Higginbotham L, Mathews D, Breeden CA et al (2015) Pre-transplant antibody screening and anti-CD154 costimulation blockade promote long-term xenograft survival in a pig-to-primate kidney transplant model. Xenotransplantation 22:221–230
11. Mohiuddin MM, Singh AK, Corcoran PC et al (2016) Chimeric 2C10R4 anti-CD40 antibody therapy is critical for long-term survival of GTKO.hCD46.hTBM pig-to-primate cardiac xenograft. Nat Commun 7:11138
12. Kim SC, Mathews DV, Breeden CP et al (2019) Long-term survival of pig-to-rhesus macaque renal xenografts is dependent on CD4 T cell depletion. Am J Transplant 19 (8):2174–2185
13. Butler JR, Wang ZY, Martens GR et al (2016) Modified glycan models of pig-to-human xenotransplantation do not enhance the human-anti-pig T cell response. Transpl Immunol 35:47–51
14. Cardona K, Korbutt GS, Milas Z et al (2006) Long-term survival of neonatal porcine islets in nonhuman primates by targeting costimulation pathways. Nat Med 12:304–306
15. Hering BJ, Wijkstrom M, Graham ML et al (2006) Prolonged diabetes reversal after intraportal xenotransplantation of wild-type porcine islets in immunosuppressed nonhuman primates. Nat Med 12:301–303
16. Liu Z, Hu W, He T et al (2017) Pig-to-primate islet xenotransplantation: past, present, and future. Cell Transplant 26:925–947
17. Griesemer A, Yamada K, Sykes M (2014) Xenotransplantation: immunological hurdles and progress toward tolerance. Immunol Rev 258:241–258
18. Scalea J, Hanecamp I, Robson SC, Yamada K (2012) T-cell-mediated immunological barriers to xenotransplantation. Xenotransplantation 19:23–30
19. Satyananda V, Hara H, Ezzelarab MB, Phelps C, Ayares D, Cooper DK (2013) New concepts of immune modulation in xenotransplantation. Transplantation 96:937–945
20. Yamada K, Sachs DH, DerSimonian H (1995) Human anti-porcine xenogeneic T cell response. Evidence for allelic specificity of mixed leukocyte reaction and for both direct and indirect pathways of recognition. J Immunol 155:5249–5256
21. Murray AG, Khodadoust MM, Pober JS, Bothwell AL (1994) Porcine aortic endothelial cells activate human T cells: direct presentation of MHC antigens and costimulation by ligands for human CD2 and CD28. Immunity 1:57–63
22. Rollins SA, Kennedy SP, Chodera AJ, Elliott EA, Zavoico GB, Matis LA (1994) Evidence that activation of human T cells by porcine endothelium involves direct recognition of porcine SLA and costimulation by porcine ligands for LFA-1 and CD2. Transplantation 57:1709–1716
23. Ezzelarab M, Garcia B, Azimzadeh A et al (2009) The innate immune response and activation of coagulation in alpha1,3-galactosyltransferase gene-knockout xenograft recipients. Transplantation 87:805–812
24. Li S, Waer M, Billiau AD (2009) Xenotransplantation: role of natural immunity. Transpl Immunol 21:70–74
25. Ide K, Wang H, Tahara H et al (2007) Role for CD47-SIRPalpha signaling in xenograft rejection by macrophages. Proc Natl Acad Sci U S A 104:5062–5066

26. Watanabe H, Sahara H, Nomura S et al (2018) GalT-KO pig lungs are highly susceptible to acute vascular rejection in baboons, which may be mitigated by transgenic expression of hCD47 on porcine blood vessels. Xenotransplantation 25:e12391
27. Yan JJ, Koo TY, Lee HS et al (2018) Role of human CD200 overexpression in pig-to-human xenogeneic immune response compared with human CD47 overexpression. Transplantation 102:406–416
28. Plege-Fleck A, Lieke T, Romermann D et al (2014) Pig to rat cell transplantation: reduced cellular and antibody responses to xenografts overexpressing PD-L1. Xenotransplantation 21:533–542
29. Esquivel EL, Maeda A, Eguchi H et al (2015) Suppression of human macrophage-mediated cytotoxicity by transgenic swine endothelial cell expression of HLA-G. Transpl Immunol 32:109–115
30. Hirohashi T, Chase CM, DellaPelle P et al (2014) Depletion of T regulatory cells promotes natural killer cell-mediated cardiac allograft vasculopathy. Transplantation 98:828–834
31. Harmon C, Sanchez-Fueyo A, O'Farrelly C, Houlihan DD (2016) Natural killer cells and liver transplantation: orchestrators of rejection or tolerance? Am J Transplant 16:751–757
32. Benichou G, Yamada Y, Aoyama A, Madsen JC (2011) Natural killer cells in rejection and tolerance of solid organ allografts. Curr Opin Organ Transplant 16:47–53
33. Sullivan JA, Oettinger HF, Sachs DH, Edge AS (1997) Analysis of polymorphism in porcine MHC class I genes: alterations in signals recognized by human cytotoxic lymphocytes. J Immunol 159:2318–2326
34. Seebach JD, Comrack C, Germana S, LeGuern C, Sachs DH, DerSimonian H (1997) HLA-Cw3 expression on porcine endothelial cells protects against xenogeneic cytotoxicity mediated by a subset of human NK cells. J Immunol 159:3655–3661
35. Forte P, Lilienfeld BG, Baumann BC, Seebach JD (2005) Human NK cytotoxicity against porcine cells is triggered by NKp44 and NKG2D. J Immunol 175:5463–5470
36. Weiss EH, Lilienfeld BG, Muller S et al (2009) HLA-E/human beta2-microglobulin transgenic pigs: protection against xenogeneic human anti-pig natural killer cell cytotoxicity. Transplantation 87:35–43
37. Puga Yung G, Bongoni AK, Pradier A et al (2018) Release of pig leukocytes and reduced human NK cell recruitment during ex vivo perfusion of HLA-E/human CD46 double-transgenic pig limbs with human blood. Xenotransplantation 25
38. Laird CT, Burdorf L, French BM et al (2017) Transgenic expression of human leukocyte antigen-E attenuates GalKO.hCD46 porcine lung xenograft injury. Xenotransplantation:24
39. Bongoni AK, Kiermeir D, Schnider J et al (2015) Transgenic expression of human CD46 on porcine endothelium: effect on coagulation and fibrinolytic cascades during ex vivo human-to-pig limb xenoperfusions. Transplantation 99:2061–2069
40. Sommaggio R, Manez R, Costa C (2009) TNF, pig CD86, and VCAM-1 identified as potential targets for intervention in xenotransplantation of pig chondrocytes. Cell Transplant 18:1381–1393

Chapter 8

Determination of Redox Status in Serum

Kelly Casós, Cristina Costa, and Manuel Galiñanes

Abstract

Free radicals of oxidative and nitrosative stress can trigger both pro-inflammatory and anti-inflammatory responses. In the transplant setting, reactive oxygen species (ROS) and reactive nitrogen species (RNS) are produced at the rejection site by different cell types including endothelial cells and macrophages. In particular, production of nitric oxide (NO) by inducible nitric oxide synthase (iNOS) seems to play an important role in promoting inflammation after exposure to inflammatory stimuli. In xenotransplantation, NO produced by iNOS upregulate multiple vasoactive substances, cytokines, chemokines, and growth factors, whereas production of NO by endothelial nitric oxide synthase (eNOS) could confer a protective effect to the graft. Accordingly, further research is needed to better understand the associated mechanisms in order to enhance protection and prevent tissue damage. Here, we describe simple methods to determine the redox state in serum that could be applied to animal models such as for xenotransplantation studies, as well as to clinical samples. Notably, caution should be taken when interpreting results of ROS and RNS measurements due to this dual role of free radicals in protecting and injuring the graft.

Key words Oxidative stress, Nitric oxide, Malondialdehyde, Nitrotyrosine, Xenotransplantation, Graft rejection, Immune response, Inflammation

1 Introduction

Oxidative stress occurs as a result of an imbalance between reactive oxygen species (ROS) production and antioxidant activity in favor of ROS. When the antioxidant defenses are unable to protect the organism from the amount of free radicals generated, there is an increase in lipid peroxidation and oxidative stress [1]. Reactive nitrogen species (RNS) are derived from the reaction of nitric oxide (NO) with superoxide anion produced via the enzymatic activities of inducible nitric oxide synthase (iNOS) and nicotinamide adenine dinucleotide phosphate oxidase. RNS induces nitrosative stress, which adds to the pro-inflammatory burden of ROS. Although ROS and RNS play an important homeostatic role as

Cristina Costa and Manuel Galiñanes contributed equally to this work.

Cristina Costa (ed.), *Xenotransplantation: Methods and Protocols*, Methods in Molecular Biology, vol. 2110,
https://doi.org/10.1007/978-1-0716-0255-3_8, © Springer Science+Business Media, LLC, part of Springer Nature 2020

regulatory mediators in signaling processes, they are harmful for living organisms at moderate or high concentrations, inactivating important cellular molecules [2, 3]. Elevated ROS levels can influence cellular proliferation and survival and induce inflammatory signaling cascades to mediate cellular damage [4]. The dysregulation of ROS synthesis in an autoimmune setting can contribute to inappropriate activation of the immune system that may induce to recognize healthy tissue as foreign. This problem is particularly dangerous if an elevated level of ROS production overwhelms antioxidant defenses, which can result in oxidative stress, ROS-mediated damage, and eventual cell death [5]. Thus, oxidative stress has been widely implicated in many inflammatory diseases, varying from ischemia-reperfusion injuries to neurological disorders and aging. The increase of ROS can trigger immune responses via two main mechanisms: (1) by incorrectly oxidizing proteins, lipids, or DNA, which are no longer recognized as self by immune cells triggering an autoimmune response, or (2) by activating redox-sensitive proteins inside the cells which can then participate in pro-inflammatory signaling.

Acute allograft rejection is a complex and incompletely understood process. The complexity arises from the interaction of several cell types as well as a variety of mediators of inflammation. In this context, antigen-presenting cells interact with T lymphocytes in the initial stages of organ rejection [6]. $CD4^+$ T cells stimulated by alloimmune activation release inflammatory cytokines such as interleukin-2 (IL-2), interferon-γ (IFN-γ), IL-4, and other mediators. The synthesis of IL-2 is a pivotal event that serves to stimulate T cells and NK cells to proliferate and produce IFN-γ. IFN-γ, in turn, activates macrophages to produce IL-1, IL-6, TNF-α, and NO. NO is produced following upregulation of iNOS which occurs initially predominately in macrophage infiltrates and later on within graft parenchymal tissue [7]. Xenograft rejection shows similarities to allotransplantation, but the innate immune response plays a more prominent role in xenotransplantation [8]. In the xenogeneic setting, neutrophil sequestration, which occurs within minutes, coupled with the release of damaging enzymes and ROS can induce reperfusion injury contributing to hyperacute and acute xenograft rejection [9, 10]. In addition, the generated ROS bind endogenous NO to produce peroxynitrate, accelerating NO removal and altering the homeostatic endogenous effects of NO [11].

NO generated by iNOS is considered to play a significant detrimental role in a variety of allotransplanted cells and tissues including lung [12, 13], kidney [14], pancreatic islets [15, 16], cornea [17], and aorta [18]. However, a beneficial effect of iNOS acting as an anti-inflammatory agent is revealed by the genetic deletion of iNOS in mice. Thus, iNOS limits chronic transplant-induced atherosclerosis and vasculopathy of the aorta [19], heart [20, 21], and kidney [22] in the corresponding transplant models conducted in the knockout mice. The mechanism of this protective action has been explained by the ability of NO to limit adhesion of platelets and leukocytes to vascular endothelium and to induce apoptosis of macrophages and proliferating vascular smooth muscle cells [23]. Regardless of the enzymatic source of NO, there are also discrepancies of the NO involvement in xenotransplantation. Administration of SIN-1, a donor drug of NO, to the endothelium of ex vivo perfused pulmonary xenografts improved graft function by inhibiting vWF secretion, vasoconstriction, and thrombin generation and indirectly through inhibition of complement activation [24]. On the contrary, Ngo et al. [25] demonstrated that nitrosative stress contributed to hyperacute rejection of rat livers perfused with human blood.

Although there is a close association between oxidative stress, inflammation, and tissue injury, the actual role of ROS and RNS in xenotransplantation remains unclear due to their dual role. Therefore, additional research is needed to investigate the redox state in preclinical models and patients to elucidate the underlying mechanisms induced by the graft during acceptance and rejection. A better understanding of these processes would allow exploiting their potential beneficial effect in this setting. Accordingly, we describe in this chapter simple colorimetric methods to measure free radicals or their effects including nitrotyrosine, an indirect marker of peroxynitrate; malondialdehyde, a lipoperoxidation index; NO bioavailability; and total antioxidant capacity.

2 Materials

Prepare all solutions with deionized water and analytical-grade reagents.

2.1 Sample Preparation

1. Cryopreservation or microfuge sterile tubes.
2. Micropipettes and micropipette tips.
3. Microcentrifuge.
4. −80 °C Freezer.

2.2 NO Bioavailability

There are various commercial kits available for this determination. Those based on the Griess reaction, as the one we use and describe here, generate highly reproducible and robust results. These kits provide a convenient colorimetric method for determination of total nitrate/nitrite concentration. Nitrates and nitrites are the natural products of NO, which is processed very rapidly (within a few seconds) in the organism. The procedure for their quantification comprises two steps: the first one converts nitrate to nitrite using nitrate reductase; the second step involves the addition of the Griess reagents to convert nitrite into a deep-purple azo compound. The materials needed are:

1. Serum samples.
2. Nitrate standard.
3. Assay reagents: assay buffer, cofactor and enzyme reaction, and Griess reagents.
4. Micropipettes, multichannel micropipette, micropipette tips, and reservoir.
5. 96 Flat-bottom well plates and 1.5 mL tubes.
6. Spin filter cups for 30 kDa protein concentration.
7. Microcentrifuge.
8. Microplate reader with filter of 540–550 nm.

2.3 Nitrotyrosine Determination

The OxiSelect™ Nitrotyrosine ELISA Kit (Cell Biolabs, Inc., San Diego, CA, USA) is the only commercial kit we found fully developed for measuring nitrotyrosine. It is based on a competitive ELISA. The samples and nitrated BSA standards are first added to an EIA plate coated with nitrated BSA. After a brief incubation, an anti-nitrotyrosine antibody is added, followed by thorough washes and an HRP-conjugated secondary antibody. The protein nitrotyrosine content in the samples is determined by comparing with the standard curve generated with the predetermined nitrated BSA standards. The signal intensity will correlate inversely with the amount of nitrated protein in each sample or standard as the nitrated protein will bind to the anti-nitrotyrosine antibody in solution and prevent (by competition) its binding to the nitrated BSA on the plate. The materials needed are:

1. Serum samples.
2. Nitrated bovine serum albumin (BSA) standard and nitrotyrosine-coated EIA plate, provided with the kit.
3. Assay reagents: assay diluents, wash buffer, primary anti-nitrotyrosine rabbit IgG, HRP-conjugated secondary antibody, substrate solution, and stop solution, all provided with the kit.
4. Micropipettes, multichannel micropipette, micropipette tips, and reservoir.
5. 96 Flat-bottom well plate (optional for serum samples) and 1.5 mL tubes.
6. Microplate reader with 450 nm filter.

2.4 Total Antioxidant Capacity

When assessing oxidative stress in a transplant setting, the measurement of the total antioxidant capacity is more informative than measuring the levels of specific natural antioxidants. Thus, we use the OxiSelect™ Total Antioxidant Capacity (TAC) Assay Kit (Cell Biolabs), which is a commercial kit that measures the total antioxidant capacity within a sample through the reduction of copper ions from Cu^{2+} to Cu^{+}. Samples and standards are diluted with a reaction reagent, and upon the addition of copper, the reaction proceeds for just a few minutes. Antioxidant capacity is then determined by comparing it to the reducing activity of uric acid standards. The materials needed are:

1. Serum samples.
2. Uric acid standard, provided with the kit.
3. Assay reagents: reaction buffer, cupper ion reagent, and stop solution (all provided with the kit).
4. 1N NaOH.
5. PBS 1×.
6. Micropipettes, multichannel micropipette, micropipette tips, and reservoir.
7. 96 Flat-bottom well plates and 1.5 mL tubes.
8. Microplate reader with 490 nm filter.

2.5 Malondialdehyde Determination

For this determination, we have set up an assay in our laboratory following the method described by the team of Yagi [26]. The materials needed are:

1. Serum samples.
2. 97% 1,1,3,3-Tetraethoxypropane (TEPP).
3. Stock solution: 10 mM TEPP (protect from light).
4. Working solution: 10 μM TEPP.
5. 0.083N Sulfuric acid (1/12 N H_2SO_4).
6. 10% Phosphotungstic acid.
7. 0.14 mM Ethylenediaminetetraacetic acid (EDTA).
8. 0.2% Butylated hydroxytoluene (BHT) in ethanol.
9. 1% Thiobarbituric acid in 50% (v/v) acetic acid (heat at 95 °C to solubilize).
10. Butanol.
11. Micropipettes, multichannel micropipette, micropipette tips, and reservoir.
12. 96 Flat-bottom well plates and 1.5 mL tubes.
13. Microcentrifuge.
14. Microplate reader with filter of 540–620 nm.

3 Methods

3.1 Sample Preparation

1. After clotting, centrifuge blood samples at 2300 × *g* in the microfuge for 10 min at room temperature.
2. Transfer the serum (upper level) to a clean tube distributing in aliquots as needed (*see* **Note 1**).
3. Freeze aliquots of serum at −80 °C until analysis.

3.2 NO Bioavailability

3.2.1 Preparation of Reagents

1. Assay buffer preparation: dilute the contents of the assay buffer vial with deionized water up to 100 mL of final volume. This assay buffer should be used for sample dilution. Store at 4 °C.
2. Nitrate reductase enzyme preparation: reconstitute the contents of the vial with 1.2 mL of assay buffer. Keep on ice during use, and store at −20 °C when not in use. Freezing and thawing of this solution should be limited to one time.
3. Nitrate reductase cofactor preparation: Reconstitute the contents of the vial with 1.2 mL of assay buffer. Keep on ice during use, and store at −20 °C when not in use. Freezing and thawing of this solution should be limited to one time.
4. Nitrate standard preparation: reconstitute the contents of the vial with 1.0 mL of assay buffer. Store at 4 °C when not in use (do not freeze). The reconstituted standard will be stable for about four months when stored at 4 °C.
5. Griess reagents R1 and R2: do not add any water or assay buffer to these reagents, as they are ready for use. These reagents should be stored at 4 °C (do not refreeze).
6. The nitrated standard curve must be performed in order to quantify the sample nitrate + nitrite concentrations. To this end, place 0.9 mL of assay buffer, and add 0.1 mL of reconstituted nitrate standard. Mix with vortex. The concentration of this stock standard is 200 μM. Use this standard (200 μM) for the preparation of the nitrate standard curve (*see* **Note 2**, Table 1).
7. Optional: the serum samples could be submitted prior to the measurements to ultracentrifugation at 16000 × *g* in the microfuge for 10 min (*see* **Note 3**).

3.2.2 Assay

1. To start the assay, add 200 μL of assay buffer to the blank wells of the plate. Do not add any other reagents to these wells.
2. Use samples in a particular dilution such as 1/4 (*see* **Note 4**). For this particular dilution, add 20 μL of each sample and 60 μL of assay buffer to each of the sample wells in a pattern you choose (80 μL final volume).
3. Add 10 μL of the enzyme cofactor mixture to each of the wells (standards and samples).

Table 1
Nitrate standard concentrations[a]

Well	Nitrate standard (μL)	Assay buffer (μL)	Nitrate concentration (mM)
A1, A2	0	80	0
B1, B2	5	75	5
C1, C2	10	70	10
D1, D2	15	65	15
E1, E2	20	60	20
F1, F2	25	55	25
G1, G2	30	50	30
H1, H2	35	45	35

[a]The concentration is calculated for the final 200 μL assay volume after addition of the Griess reagents

4. Add 10 μL of the nitrate reductase mixture to each of the wells (standards and samples).
5. Cover the plate with the plate cover or aluminum paper, and incubate at room temperature for 3 h (*see* **Note 5**).
6. After the incubation time, add 50 μL of Griess reagent R1.
7. Immediately add 50 μL of Griess reagent R2 to each of the wells (standards and samples).
8. The color will develop in about 10 min at room temperature. It is not necessary to cover the plate (*see* **Note 6**).
9. Read the absorbance at 540 or 550 nm using a plate reader.
10. For calculations, subtract the blanks first. To this end, average the absorbance value of the blank wells, and subtract this from the absorbance values of all the other wells. Then, plot the standard curves. Make a plot of absorbance at 540–550 nm as a function of nitrate or nitrite concentration. The nitrate standard curve is used for determination of total nitrate+nitrite concentration (*see* **Note 7**).

3.3 Nitrotyrosine Determination

3.3.1 Preparation of Reagents

1. Wash buffer 1× preparation: dilute the 10× wash buffer 1:10 with deionized water. Stir to homogeneity.
2. Anti-nitrotyrosine primary and secondary antibody preparation: immediately before use, dilute the antibodies 1:1000 with assay diluent. Do not store the diluted solutions.
3. For the preparation of the nitrotyrosine BSA standard curve, prepare a dilution series of nitrated BSA standards in the nitrotyrosine concentration range of 0–8000 nM by diluting the nitrated BSA stock solution in assay diluents (*see* **Note 8**, Table 2).

Table 2
Preparation of nitrated BSA standards

Standard tubes	Nitrated BSA standard (μL)	Assay diluent (μL)	Nitrated BSA (μg/mL)	Nitrotyrosine (nM)
1	60	240	200	8000
2	100 of tube #1	300	50	2000
3	100 of tube #2	300	12.5	500
4	100 of tube #3	300	3.125	125
5	100 of tube #4	300	0.78	31.25
6	100 of tube #5	300	0.195	7.81
7	100 of tube #6	300	0.049	1.95
8	0	300	0	0

3.3.2 Assay

1. Prepare and mix all reagents thoroughly before use. Each protein sample including nitrated BSA and blank should be assayed in duplicate.
2. To start the assay, add 50 μL of serum sample (*see* **Note 9**) or nitrated BSA standard to the wells of the EIA plate. Incubate at room temperature for 10 min on an orbital shaker.
3. Next, add 50 μL of the diluted anti-nitrotyrosine antibody to each well; incubate at room temperature for 1 h on an orbital shaker.
4. Wash microwell strips three times with 250 μL wash buffer 1× per well with thorough aspiration between each wash. After the last wash, empty wells and tap microwell strips on absorbent pad or paper towel to remove the wash buffer (*see* **Note 10**).
5. Add 100 μL of the diluted secondary antibody-enzyme conjugate to all wells.
6. Incubate at room temperature for 1 h on an orbital shaker.
7. Wash microwell strips three times according to **step 4**. Proceed immediately to the next step.
8. Warm substrate solution to room temperature. Add 100 μL of substrate solution to each well, including the blank wells. Incubate at room temperature on an orbital shaker for 10 min (*see* **Note 11**).
9. Stop the enzyme reaction by adding 100 μL of stop solution into each well, including the blank wells. Results should be read immediately (color will fade over time).

10. Read absorbance of each microwell on a spectrophotometer using 450 nm as the primary wave length.
11. For calculations, plot the standard curve read at 450 nm absorbance in function of nitrotyrosine concentration (*see* **Note 12**).

3.4 Total Antioxidant Capacity

3.4.1 Preparation of Reagents

1. Reaction buffer 1× preparation: dilute the reaction buffer 1:100 with PBS (hydrophilic). Mix to homogeneity. Store the reaction buffer 1× at 4 °C for up to three months.
2. Copper ion reagent 1×: dilute the copper ion reagent 1:100 with deionized water (hydrophilic). Mix to homogeneity. Store the copper ion reagent 1× at 4 °C for up to three months.
3. Stop solution 1×: dilute the stop solution 1:10 with deionized water (hydrophilic). Mix to homogeneity. Store the stop solution 1× at 4 °C for up to three months.
4. Prepare fresh the uric acid standards. Start by weighing out the uric acid powder for a 10 mg/mL solution in 1N NaOH, which corresponds to a concentration of 60 mM. Use the 60 mM uric acid solution to prepare a 2 mM solution of uric acid (e.g., add 100 μL of the 60 mM uric acid solution to 2.900 mL of deionized water). This concentration is suitable for the preparation of the uric acid standard curve (*see* **Note 13**).

3.4.2 Assay

1. To start the assay, add 20 μL of the uric acid standards or diluted serum samples 1/20 (*see* **Note 14**) to the 96-well microtiter plate.
2. Add 180 μL of the reaction buffer 1× to each well using either a multichannel pipette or a plate liquid handling system. Mix thoroughly.
3. Obtain an initial absorbance by reading the plate at 490 nm.
4. To initiate the reaction, add 50 μL of the copper ion reagent 1× into each well. Incubate 5 min on an orbital shaker.
5. Add 50 μL of stop solution 1× to each well to terminate the reaction.
6. Read the plate again at 490 nm.
7. Calculate the net absorbance by subtracting the initial absorbance readings for samples and standards (**step 3**) from the final readings taken for each one (**step 6**).
8. Plot the net absorbance against the uric acid concentration for the uric acid standard curve.
9. Calculate the antioxidant capacity of samples by using the uric acid standard curve.

3.5 Malondialdehyde Determination

1. Use different concentrations of TEPP in order to obtain a standard curve (*see* **Note 15**).
2. For the precipitation of lipids and proteins, mix 50 μL of serum, 400 μL of 1/12N H_2SO_4, and 50 μL of 10% phosphotungstic acid in a microfuge tube.
3. Keep the mixture during 5 min at room temperature in gentle agitation.
4. Centrifuge at 2300 × *g* during 10 min at room temperature in the microcentrifuge. Discard the supernatant.
5. Add to the pellet 400 μL of 1/12 N H_2SO_4 and 30 μL of 10% phosphotungstic acid.
6. Keep in gentle agitation during 5 min at room temperature.
7. Repeat spinning as in **step 4**. Discard the supernatant, and resuspend the pellet in 400 μL of deionized water.
8. To initiate the assay, add 400 μL of resuspendent pellet (**step 7**) or standards (*see* **Note 16**) to a microfuge tube.
9. Add 40 μL of 0.2% BHT to each well (samples and standards).
10. Add 5 μL of 0.14 mM EDTA to each well (samples and standards).
11. Add 100 μL of 1% TBA to each well (samples and standards).
12. Boil at 95 °C for 1 h (*see* **Note 17**).
13. Cool down at room temperature for 30 min approximately.
14. Add 550 μL of butanol to each well (samples and standards).
15. Mix gently by pipetting.
16. Centrifuge at 2300 × *g* in the microfuge for 15 min.
17. Recover the upper phase (butanolic), and read the absorbance at 540 and 620 nm.
18. Calculate the final absorbance by subtracting Abs620–Abs540. Plot the results using the TEPP standard curve.

4 Notes

1. We recommend preparing serum aliquots of sufficient volume for each determinations being considered in order to avoid thawing cycles.
2. The standard curve for nitrate is prepared by addition of nitrate standard and assay buffer to the plate wells as shown in Table 1.
3. To avoid residual hemoglobin and bilirubin, samples could be all subjected to ultracentrifugation using filters with a cutoff of 10–30 kDa. This is an optional step that could be avoided if hemolysis is not observed as it can reduce the overall

measurements. Before use, clean the filter with deionized water or PBS. The flow through is kept for further processing of the sample and measurements.

4. For human serum samples, we have used in the past a ¼ dilution in this assay (20 μL of serum + 60 μL assay buffer, total volume 80 μL). Some tests may be needed to find the appropriate dilution for other species.
5. It is not necessary to shake the plate during incubation.
6. The 10 min incubation is optimal for color development for this determination. Nevertheless, the plate can be left to develop for longer time periods and still generate valid data, provided that the Griess reagents have been added to the standard curve and samples at the same time. Thus, it is very important to assay the standard curve at the same time as the samples to ensure an accurate measurement.
7. The equation to calculate the NO concentration uses slope of the standard curve:

$$[\text{Nitrate} + \text{Nitrite}]\ (\mu\text{M}) = \left[\frac{\text{Abs}_{540} - y\ \text{intercept}}{\text{Slope}}\right] \times \left[\frac{200\ \mu\text{L} \times \text{dilution}}{\text{Volume of sample used}\ (\mu\text{L})}\right]$$

8. The standard curve for nitrated BSA is prepared by addition of nitrate standard and assay diluents to the plate wells as shown in Table 2.
9. Avoid using any hemolyzed serum or with traces of bilirubin and cholesterol as these can interfere in the reaction or the read out.
10. It is very important to dry well the wells to avoid diluting the subsequent solutions.
11. The development of the chromogen should be carefully monitored as it can change rapidly, requiring rapid intervention to stop the reaction and prevent saturation. In fact, the incubation time can vary between 2 and 30 min.
12. This competitive assay generates an inverted standard curve which follows a logarithmic profile. For calculating the sample nitrotyrosine, we use the analysis software solution from MyAssays (https://www.myassays.com) and apply the function "Four Parameter Logistic Regression." To this end, the standard data points (concentration vs. measurement) are plotted on semilog axes, and the abovementioned regression is applied through the points. This method is widely used and cited for ELISA data analysis.

Table 3
Preparation of uric acid standards

Tubes	2 mM Uric acid (μL)	Deionized water (μL)	Uric acid concentration (mM)
1	500	500	1
2	500 of tube #1	500	0.5
3	500 of tube #2	500	0.25
4	500 of tube #3	500	0.125
5	500 of tube #4	500	0.0625
6	500 of tube #5	500	0.03125
7	500 of tube #6	500	0.0156
8	500 of tube #7	500	0.0078
9	500 of tube #8	500	0.0039
10	0	500	0.0

Table 4
Preparation of TEPP standards

$[TEPP]_{final}$ (μL)	10 μM	8 μM	6 μM	4 μM	2 μM	1 μM
$[MDA]_{final}$ (μL)	10	8	6	4	2	1
TEPP 10 μM	1000	800	600	400	200	100
Deionized water (μL)	–	200	400	600	800	900

13. The standard curve for uric acid is prepared by addition of uric acid standard and deionized water to the plate wells as shown in Table 3.
14. For determinations in human serum, we dilute the samples 1/20 in PBS.
15. The standard curve for TEPP is prepared by addition of TEPP standards and deionized water to the plate wells as shown in Table 4.
16. According to Table 4, take 50 μL of the standard, and add 350 μL of deionized water in order to perform the assay.
17. To prevent the microfuge tube from opening during heating, make a hole in the tube cap.

Acknowledgments

The research leading to this work received funding from the European Union's Seventh Framework Programme (FP7/2007–2013) under Grant Agreement no. 603049, TRANS-LINK. This work was also supported by Ministerio de Economía y Competitividad-ISCiii (PI15/00181, granted to CC, and PI12/00119, granted to MG) and the SLT002/16/00445 grant funded by the Department of Health of the Generalitat de Catalunya and co-funded by FEDER funds, a way to build Europe. IDIBELL and VHIR benefit from the support of CERCA.

References

1. Campise M, Bamonti F, Novembrino C, Ippolito S, Tarantino A, Cornelli U et al (2003) Oxidative stress in kidney transplant patients. Transplantation 76:1474–1478
2. Di Meo S, Reed TT, Venditti P, Victor VM (2016) Role of ROS and RNS sources in physiological and pathological conditions. Oxid Med Cell Longev 2016:1245049
3. Sunil VR, Shen J, Patel-Vayas K, Gow AJ, Laskin JD, Laskin DL (2012) Role of reactive nitrogen species generated via inducible nitric oxide synthase in vesicant-induced lung injury, inflammation and altered lung functioning. Toxicol Appl Pharmacol 261:22–30
4. Dickinson BC, Chang CJ (2011) Chemistry and biology of reactive oxygen species in signaling or stress responses. Nat Chem Biol 7:504–511
5. Ramkumar KM, Sekar TV, Bhakkiyalakshmi E, Foygel K, Rajaguru P, Berger F et al (2013) The impact of oxidative stress on islet transplantation and monitoring the graft survival by non-invasive imaging. Curr Med Chem 20:1127–1146
6. Nagano H, Nadeau KC, Takada M, Kusaka M, Tilney NL (1997) Sequential cellular and molecular kinetics in acutely rejecting renal allografts in rats. Transplantation 63:1101–1108
7. Yang X, Chowdhury N, Cai B, Brett J, Marboe C, Sciacca RR et al (1994) Induction of myocardial nitric oxide synthase by cardiac allograft rejection. J Clin Invest 94:714–721
8. Harris DG, Quinn KJ, Dahi S, Burdorf L, Azimzadeh AM, Pierson RN (2014) 3rd. Lung xenotransplantation: recent progress and current status. Xenotransplantation 21:496–506
9. Kim YT, Lee HJ, Lee SW, Kim JY, Wi HC, Park SJ et al (2008) Pre-treatment of porcine pulmonary xenograft with desmopressin: a novel strategy to attenuate platelet activation and systemic intravascular coagulation in an ex-vivo model of swine-to-human pulmonary xenotransplantation. Xenotransplantation 15:27–35
10. Rao JN, Clark SC, Ali S, Kirby J, Flecknell PA, Dark JH (2003) Improvements in lung compliance after pulmonary transplantation: correlation with interleukin 8 expression. Eur J Cardiothorac Surg 23:497–502
11. Murad F (2006) Shattuck Lecture Nitric oxide and cyclic GMP in cell signaling and drug development. N Engl J Med 355:2003–2011
12. Shiraishi T, DeMeester SR, Worrall NK, Ritter JH, Misko TP, Ferguson TB Jr et al (1995) Inhibition of inducible nitric oxide synthase ameliorates rat lung allograft rejection. J Thorac Cardiovasc Surg 110:1449–1459. discussion 60
13. Worrall NK, Boasquevisque CH, Misko TP, Sullivan PM, Ferguson TB Jr, Patterson GA (1997) Inducible nitric oxide synthase is expressed during experimental acute lung allograft rejection. J Heart Lung Transplant 16:334–339
14. Vos IH, Joles JA, Schurink M, Weckbecker G, Stojanovic T, Rabelink TJ et al (2000) Inhibition of inducible nitric oxide synthase improves graft function and reduces tubulointerstitial injury in renal allograft rejection. Eur J Pharmacol 391:31–38
15. Brandhorst D, Brandhorst H, Zwolinski A, Nahidi F, Bretzel RG (2001) Prevention of early islet graft failure by selective inducible nitric oxide synthase inhibitors after pig to nude rat intraportal islet transplantation. Transplantation 71:179–184
16. Stevens RB, Ansite JD, Mills CD, Lokeh A, Rossini TJ, Saxena M et al (1996) Nitric

oxide mediates early dysfunction of rat and mouse islets after transplantation. Transplantation 61:1740–1749

17. Strestikova P, Plskova J, Filipec M, Farghali H (2003) FK 506 and aminoguanidine suppress iNOS induction in orthotopic corneal allografts and prolong graft survival in mice. Nitric Oxide 9:111–117
18. Ouyang J, Xu D, Zhang X, Qi S, Ma A, Jiang W et al (2005) Effect of a novel inducible nitric oxide synthase inhibitor in prevention of rat chronic aortic rejections. Transplantation 79:1386–1392
19. Shears LL, Kawaharada N, Tzeng E, Billiar TR, Watkins SC, Kovesdi I et al (1997) Inducible nitric oxide synthase suppresses the development of allograft arteriosclerosis. J Clin Invest 100:2035–2042
20. Koglin J, Glysing-Jensen T, Mudgett JS, Russell ME (1998) NOS2 mediates opposing effects in models of acute and chronic cardiac rejection: insights from NOS2-knockout mice. Am J Pathol 153:1371–1376
21. Qian Z, Gelzer-Bell R, Yang Sx SX, Cao W, Ohnishi T, Wasowska BA et al (2001) Inducible nitric oxide synthase inhibition of weibel-palade body release in cardiac transplant rejection. Circulation 104:2369–2375
22. Du C, Jiang J, Guan Q, Diao H, Yin Z, Wang S et al (2007) NOS2 (iNOS) deficiency in kidney donor accelerates allograft loss in a murine model. Am J Transplant 7:17–26
23. Weis M, Cooke JP (2003) Cardiac allograft vasculopathy and dysregulation of the NO synthase pathway. Arterioscler Thromb Vasc Biol 23:567–575
24. Park HS, Kim JE, You HJ, Gu J, Yoo B, Lee S et al (2015) Beneficial effect of a nitric oxide donor in an ex vivo model of pig-to-human pulmonary xenotransplantation. Xenotransplantation 22:391–398
25. Ngo BT, Beiras-Fernandez A, Hammer C, Thein E (2013) Hyperacute rejection in the xenogenic transplanted rat liver is triggered by the complement system only in the presence of leukocytes and free radical species. Xenotransplantation 20:177–187
26. Ohkawa H, Ohishi N, Yagi K (1979) Assay for lipid peroxides in animal tissues by thiobarbituric acid reaction. Anal Biochem 95:351–358

Chapter 9

Generation of Genetically Modified Mice Using CRISPR/Cas9

D. Muñoz-Santos, L. Montoliu, and A. Fernández

Abstract

Using CRISPR-based genome-editing techniques, we are able to generate a variety of new mouse models of several types of diseases. These animal models will be instrumental not only for enabling the comprehension of a particular disease and its underlying molecular mechanism but also as unique recipients for testing novel and innovative therapeutic approaches that are being currently explored. This chapter describes detailed step-by-step protocols, reagents, and equipment required for successful generation of genome-edited mice using CRISPR tools.

Key words CRISPR, Cas9, Knock-in mouse, Knockout mouse, Genotyping, Genome editing

1 Introduction

Clustered regularly interspaced short palindromic repeat (CRISPR) sequences cooperate with CRISPR-associated (Cas) proteins to form the basis of CRISPR–Cas adaptive immune systems in prokaryotes [1]. In 2012, the potential of CRISPR–Cas systems was uncovered, and these were presented as genome-editing tools with an outstanding capacity to trigger targeted genetic modifications that can be applied to virtually any organism, including genome-edited mouse models [2–4]. This nuclease-mediated system is the most robust genome-editing system known to date. It is based on three components: Cas9 protein, the CRISPR RNA (crRNA), and a trans-activating crRNA (tracrRNA) [5, 6]. Basically, it requires a 20-nucleotide guide sequence as part of the crRNA and tracrRNA partially complementary to the crRNA (or even, as it is mentioned in other approaches, a single-guide RNA (sgRNA) consisting of a fusion of crRNA and tracrRNA in vitro) and a Cas9 endonuclease protein to cleave the genomic DNA at specific sites producing a double-stranded DNA break (DSB), which is immediately repaired by the endogenous DNA repair mechanisms [7]. The nonhomologous end-joining (NHEJ) route is associated with the insertion and deletion (INDEL) of nucleotides, usually

Cristina Costa (ed.), *Xenotransplantation: Methods and Protocols*, Methods in Molecular Biology, vol. 2110, https://doi.org/10.1007/978-1-0716-0255-3_9,

resulting in gene disruption. The homology-directed repair (HDR) route, in the presence of single-strand DNA (ssDNA) or double-strand DNA (dsDNA) oligonucleotides with homology to sequences surrounding the target DSB, drives repair by introducing exogenous DNA sequences, which normally leads to knock-in/reporter insertion or gene editing.

Although most CRISPR–Cas applications derive from the CRISPR–Cas9 tools from *Streptococcus pyogenes*, other CRISPR–Cas-like systems are being isolated and characterized from other prokaryotes [8]. In the near future, more efficient tools could be discovered for improved genome modification. CRISPR/Cas9 technology has allowed for simple and rapid development of knock-outs and knock-ins, revolutionizing genome editing in mice. This protocol covers the basic techniques needed to rapidly generate knockout or knock-in mice with CRISPR tools via electroporation of Cas9 protein, the guide RNA, and possible donor DNA into mouse fertilized oocytes.

2 Materials

There are many web tools available to design the RNA guides required for using a CRISPR–Cas tool, [9] but we do recommend Breaking-Cas [10], developed at CNB-CSIC, for its ease and flexibility. Additional programs are used for engineering silent mutations (WatCut) and for designing oligonucleotides (Primer Blast NCBI or Primer3). All solutions should be prepared using nuclease-free water, of embryo-transfer quality, and store all reagents at −80 °C until use.

2.1 Guide RNA (gRNA) and Oligonucleotide Design

1. Breaking-Cas: online CRISPR tool (https://http://bioinfogp.cnb.csic.es/tools/breakingcas/) [10].
2. WatCut tool: http://watcut.uwaterloo.ca/template.php (web-based programs for designing silent mutations).
3. Primer Blast NCBI: https://www.ncbi.nlm.nih.gov/tools/primer-blast/.

2.2 CRISPR–Cas9 Components

1. Cas9 protein (PNABio Inc. or Integrated DNA Technologies, Inc.) (*see* **Note 1**).
2. 5X Cas9 buffer: 0.1 M HEPES, 0.75 M KCl, 2.5 mM DTT, 0.5 mM EDTA, and 0.05 M $MgCl_2$.
3. crRNA (Edit-R Mouse synthetic crRNA, Dharmacon).
4. tracrRNA (Edit-R CRISPR–Cas9 synthetic tracrRNA, Dharmacon).
5. Nuclease-free water embryo transfer.
6. Taq DNA polymerase.

7. 10 mM Solution of each nucleotide.
8. 2% TBE agarose gel.
9. Gel electrophoresis equipment.
10. PCR cleanup kit such as QIAquick PCR purification kit from QIAGEN.
11. Two oligo primers for PCR.
12. 0.2 mL PCR tubes.
13. Thermocycler.
14. Orbital shaker.

2.3 Electroporation of Mouse Fertilized Oocytes

1. 200 Mouse fertilized oocytes C57BL/6J.
2. 0.2 mL nuclease-free microcentrifuge tubes.
3. Stock gRNA(s) at ≥500 ng/μL in water or 10 mM Tris buffer, pH 7.5–8.
4. Stock oligonucleotide at 100 μM in embryo-transfer water, if needed.
5. Heat block set to 37 °C.
6. Pipettors and tips for volumes of 0.5–50 μL.
7. Microcentrifuge.
8. Embryo-transfer water.
9. 1 M Potassium chloride (KCl) solution.
10. 1M HEPES buffer solution, pH 7.2–7.4.
11. Cas9 protein (e.g., IDT 1074182).
12. 5× EP Cas9 buffer: 1 M potassium chloride (KCl) solution and 1 M HEPES buffer solution, pH 7.4; embryo-transfer water.
13. Opti-MEM.
14. M2 Medium (SIGMA M-7167).
15. KSOM Medium (Cosmo Bio CSR-R-B074).
16. Mineral Oil (SIGMA M-8410).
17. Stereomicroscopes with additional cold light systems (Leica MZ8, MZ9).
18. Mini-Incubator K-MINC-1000 (Cook).
19. CO_2 Incubator (C60 Labotech).
20. Tissue-culture dish 35 × 10 mm.
21. Tissue-culture dish 60 × 15 mm.
22. Gene Pulser Xcell electroporation System (Bio-Rad).
23. Glass slide with electrodes (BEX Co., Ltd. LF501PT1–10), 1 mm gap.

2.4 Genotyping Genome-Edited Mice

1. Proteinase K-containing lysis solution: 0.5 mg/mL proteinase K, 100 mM NaCl, 50 nM Tris–HCl pH 8, 100 mM EDTA pH 8, and 1% SDS.
2. Taq polymerase.
3. TE buffer: 10 mM Tris–HCl pH 7.5 and 1 mM EDTA pH 8.
4. 10 mM Solution of each nucleotide.
5. Two oligo primers for PCR.
6. TAE 1×: 40 mM Tris, 20 mM acetic acid, and 1 mM EDTA, prepared with ultrapure water and 0.2 μm filtered.
7. Thermocycler.
8. Gel electrophoresis equipment.
9. pGEM-T Easy Vector (Promega).
10. Gibson Assembly® Master Mix (New England Biolabs).
11. Competent bacteria (i.e., Top10) and reagents for transformation.

3 Methods

3.1 CRISPR Experimental and gRNA Design

3.1.1 Design of the gRNA

1. Use the Breaking-Cas program (http://bioinfogp.cnb.csic.es/tools/breakingcas/) because of its flexibility and capacity to use any genome deposited in ENSEMBL [10].
2. Select in the program "Ensembl collection of vertebrates genomes."
3. On the next screen, choose organism from Ensembl list, upload a file or paste the sequence(s) (up to 20,000 nucleotides in total) to be scanned in FASTA format, select the suitable endonuclease (Cas9, Cpf1. . .), and submit.
4. Download report or view online (Fig. 1).
5. Choose the gRNA whose position is the closest to your intended mutation and according to the arbitrary indicated scores (*see* **Note 2**).

3.1.2 Design of the ssDNA Oligonucleotide Template for Gene Edition/Repair

Design a single-strand (ss) DNA oligonucleotide (90–120 nucleotides in length) with your desired mutation centrally located, approximately (Fig. 2). You might wish to add some silent mutations to help you in genotyping the expected mutation, such as removing or creating a new restriction enzyme site for RFLP analysis (Fig. 2). You might also wish to add some silent mutations to inactivate the PAM sequence to prevent a reversion of a mutated allele by reedition. Web-based programs for designing silent mutations: *WatCut* (http://watcut.uwaterloo.ca/template.php).

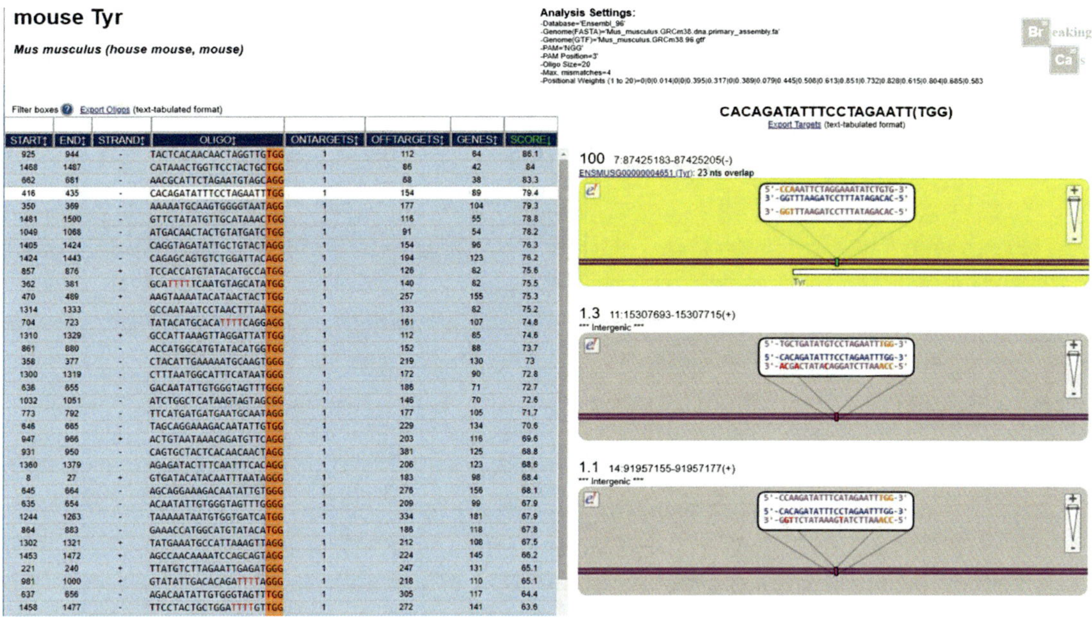

Fig. 1 Using Breaking-Cas with the mouse Tyr gene (exon 3). The page is divided into two panels. The left panel lists all the predicted guides found, with its associated sequence, on-target, off-target, and score. Upon hovering over any sgRNA from this panel, the right panel will show the corresponding on-target (yellow box) and all its potential off-targets, associated scores, mismatches with respect to the real target, and genomic locations (gray boxes)

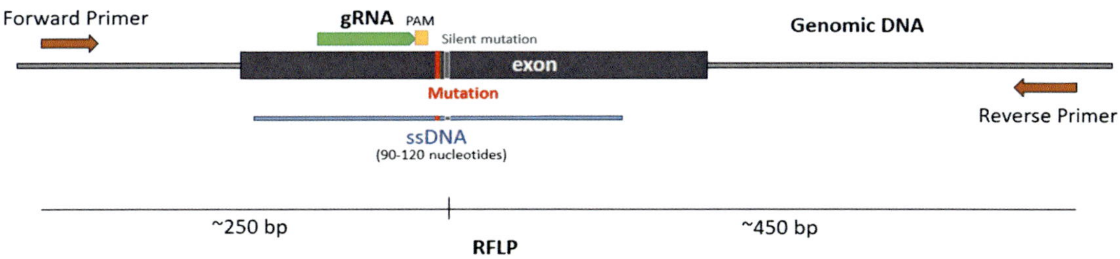

Fig. 2 Designing a knock-in strategy using CRISPR–Cas tools on a gene of interest. The designed crRNA (sgRNA) is shown in green; PAM sequence is indicated in orange; the ssDNA (in blue) has been designed to carry two-point mutations: the pathogenic mutation (in red) and the silent mutation (in gray) to generate a new cutting site for assisting the genotyping process. To amplify the region, two flanking oligonucleotides are designed (forward and reverse primer, in brown) to generate a PCR product of 700 bp size. The new cutting site produces two fragments of 250pb and 450pb, useful for RFLP analyses

3.1.3 Design of DNA Primers for Analyzing Your Gene Edition Event

Design convenient DNA primers, flanking your gene editing event, to analyze the gene edition process and be able to amplify the relevant DNA fragment and eventually sequence it. Design the primers in a way such that the final PCR product is around 500–800 bp. Considering your amplified fragment, place your mutation asymmetrically, leaving approximately ~200–300 bp and ~300–500 bp at either side. This will be useful for the T7 endonuclease assays and restriction fragment length polymorphism (RFLP) analyses (Fig. 2) [11]. The web-based programs recommended for designing DNA primers are *Primer Blast NCBI* (https://www.ncbi.nlm.nih.gov/tools/primer-blast/) and *Primer 3* (http://primer3.sourceforge.net/).

3.2 Preparing and Validating CRISPR–Cas9 Components

3.2.1 Resuspending crRNA and tracrRNA

1. Add embryo-transfer-tested water or standard microinjection buffer to the crRNA and tracrRNA lyophilized powders to a final concentration of 100 μM.
2. Place the solutions in an orbital shaker for 30 min at room temperature.
3. Prepare crRNA and tracrRNA aliquots and store at −80 °C.

3.2.2 Preparing the gRNA

1. Mix crRNA 5 μL (100 μM) and tracrRNA 5 μL (100 μM) (1:1 molar rate).
2. Anneal crRNA with the tracrRNA using a thermocycler using the following program: first 5 min at +95 °C and then ramp down to +25 °C at +5 °C/min.
3. Prepare crRNA–tracrRNA (gRNA) aliquots and store at −80 °C.

3.2.3 Preparing Recombinant Cas9 Protein

1. Select a Cas9 protein for your experiment (*see* **Note 1**). You may use recombinant Cas9 protein with 6-His and nuclear localization signal (NLS) from SV40.
2. Dissolve the lyophilized protein powder with nuclease-free water with 20% glycerol to obtain a final concentration of 1 mg/mL.

3.2.4 In Vitro Validation of gRNAs

There are several procedures available, including testing the CRISPR–Cas tools in cells or even in a subset of mouse embryos. However, the simplest method, totally run in vitro, uses the resulting ribonucleoprotein (RNP) particle to target and cut a test DNA fragment containing the expected site for DSB. This is equivalent to a restriction enzyme digestion. Currently, we routinely test (and recommend testing) all gRNAs through this option:

1. In preparation of the PCR template, run a standard PCR reaction for the DNA fragment of choice in a volume of 50 μL.
2. Check a small aliquot (5 μL) on a 2% agarose gel (in TBE buffer) to confirm the presence of a single DNA band.
3. Purify the rest of the PCR DNA band using a suitable method or kit (i.e., QIAquick PCR purification kit).
4. Quantify the purified DNA PCR product.
5. For in vitro digestion of the PCR template, set up the following reaction in an Eppendorf tube: 100–200 ng purified PCR

Table 1
Recommended controls to be included in the in vitro (IV) assay

	IV assay 1	IV assay 2	gRNA control	Cas9 control	Negative control
Cas9 protein	100 ng	100 ng	–	100 ng	–
gRNA	50 ng	100 ng	100 ng	–	–
5× Cas9 buffer	2 μL	2 μL	2 μL	2 μL	2 μL
Purified DNA PCR product	150 ng	200 ng	150 ng	150 ng	150 ng
mQ H_2O	Up to 10 μL	Up to 10 μL	Up to 10 μL	Up to 10 μL	Up to 10 μL

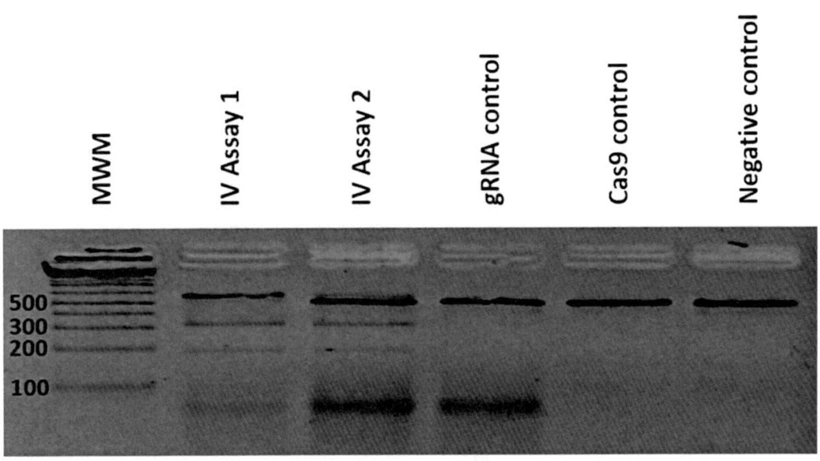

Fig. 3 In vitro testing CRISPR–Cas tools reagents. The ~500 bp PCR product is digested into the expected ~200 and ~300 bp bands in the in vitro assay lanes 1 and 2, where all reagents are present. The rest of lanes are included for control purposes

product, 100 ng of Cas9 protein, 50–100 ng of gRNA, and 2 μL of buffer 5× Cas 9. Add nuclease-free water to 10 μL final volume (Table 1).

6. Incubate the reaction for 3 h at +37 °C.
7. Subsequently incubate it for 10 min at +65 °C.
8. Run the digestions in a 2% agarose gel (TBE) (Fig. 3).

3.3 Electroporation of RNPs into Mouse Fertilized Oocytes

1. For preparation of the RNPs, calculate the volumes of gRNA and Cas9 protein needed for 11 μL of electroporation mix (Table 2).
2. Mix 300 ng/μL of gRNA, 600 ng/μL of Cas9, and 2.2 μL of EP Cas9 buffer (5×).
3. Incubate at 37 °C for 10 min. Pulse centrifuge to collect droplets after the incubation.

Table 2
Concentrations of RNA and Cas9 for electroporation into mouse fertilized oocytes

	ng/μL Recommended (final concentration to 11 μL)	Range ng/μL
crRNA + tracrRNA (1:1)	300	100–500
Cas9 protein	600	120–1000
ssDNA	400	100–400

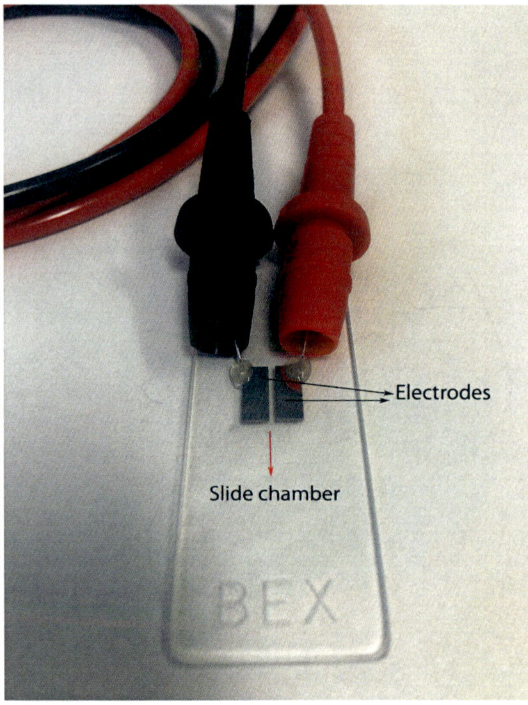

Fig. 4 Using a slide chamber for embryo electroporation. The Opti-MEM/RNP mix and a group of embryos (20–30) are transferred between the electrodes (black arrows) into the slide chamber (red arrow)

4. Optional: add ssDNA donor to 400 ng/μL.
5. Adjust the final volume to 11 μL using nuclease-free water.
6. For the electroporation, add 2.5 μL of RNP mix into the 2.5 μL of an Opti-MEM drop per slide.
7. Transfer the 5 μL of RNP/Opti-MEM into the slide chamber (Fig. 4).
8. Using the mouth pipette, transfer 20–30 mouse fertilized oocytes to the slide.
9. Electroporate the oocytes setting up the following conditions: 30 V, 3 ms, 10 pulses, and 100 ms interval.
10. Transfer the electroporated mouse fertilized oocytes into suitable foster females.

3.4 Genotyping Genome-Edited Mice to Detect Your Allele of Choice by Cloning and Sequencing

1. Isolation of genomic DNA from mouse tail. Mouse genomic DNA is extracted from tail biopsies (<1 cm). The tissue is digested with proteinase K-containing lysis solution; incubate overnight 56 °C. Thereafter, the DNA is purified by ethanol precipitation and resuspended in TE buffer.
2. Clone genome-edited fragments into pGEM-T Easy or any suitable plasmid vector. To this end, design convenient DNA primers (described in Subheading 3.1.3), flanking your gene editing event, to analyze the gene edition process and be able to amplify the relevant DNA fragment and eventually sequence it. As mentioned, design the primers in a way such that the final PCR product is around 500–800 bp and your mutation is placed asymmetrically (~200–300 bp and ~300–500 bp at either side). This will be useful for the T7 endonuclease assays and RFLPs analyses [11]. Use the Gibson Assembly® Cloning Kit for cloning the PCR products into pGEM-T Easy Vector. The primers used for the amplification of genomic sequences include a 5′ adapter sequence homologous to the ends of the vector cloning sites.
3. Sequence 10–20 independent clones. Your allele can be detected by sequencing of the PCR product cloned into the vector from the target region (*see* **Note 3**).

4 Notes

1. We recommend using recombinant Cas9 protein with 6-His and nuclear localization signal (NLS) from SV40 (PNABio, Inc.). Alternative Cas9 and other Cas-like proteins are available from a variety of commercial providers. Thus, you may wish to use recombinant *S. pyogenes* Cas9 nuclease, purified from an *E. coli* strain expressing the protein. It contains a nuclear localization sequence (NLS) and a C-terminal 6-His tag. Usually, it is provided in solution at 10 μg/μL (Integrated DNA Technologies, Inc.).
2. To generate knockout mutants of protein coding genes, non-coding genes, and repetitive sequences, the use of dual sgRNA-guided Cas9 nuclease is necessary. Co-injection of dual sgRNAs results in the removal of the interval between two sgRNAs and the loss-of-function phenotype of targeted genes.
3. You need to take into account that this procedure can be biased since not all alleles may exist in equivalent proportions plus not all alleles will be equally accepted by bacteria for cloning purposes. Next-generation sequencing techniques may be necessary to detect low-frequency alleles (10–20% or less).

Acknowledgment

Laboratory research grants used for this manuscript include MINECO (BIO2015-70978-R) and CIBERER-ISCIII (ACCI program).

References

1. Mojica FJM, Montoliu L (2016) On the origin of CRISPR-Cas technology: from prokaryotes to mammals. Trends Microbiol 24 (10):811–820
2. Seruggia D, Montoliu L (2014) The new CRISPR-Cas system: RNA-guided genome engineering to efficiently produce any desired genetic alteration in animals. Transgenic Res 23(5):707–716
3. Fernández A, Josa S, Montoliu L (2017) A history of genome editing in mammals. Mamm Genome 28(7–8):237–246
4. Burgio G (2018) Redefining mouse transgenesis with CRISPR/Cas9 genome editing technology. Genome Biol 19(1):27
5. Jinek M, Chylinski K, Fonfara I, Hauer M, Doudna JA, Charpentier E (2012) A programmable dual-RNA-guided DNA endonuclease in adaptive bacterial immunity. Science 337 (6096):816–821
6. Cong L, Ran FA, Cox D et al (2013) Multiplex genome engineering using CRISPR/Cas systems. Science 339(6121):819–823
7. Doudna JA, Charpentier E (2014) Genome editing. The new frontier of genome engineering with CRISPR-Cas9. Science 346 (6213):1258096
8. Shmakov S, Abudayyeh OO, Makarova KS et al (2015) Discovery and functional characterization of diverse class 2 CRISPR-Cas systems. Mol Cell 60(3):385–397
9. Torres-Perez R, Garcia-Martin JA, Montoliu L, Oliveros JC, Pazos F (2019) WeReview: CRISPR tools—Live repository of computational tools for assisting CRISPR/Cas experiments. Bioengineering 6(3):63
10. Oliveros JC, Franch M, Tabas-Madrid D et al (2016) Breaking-Cas-interactive design of guide RNAs for CRISPR-Cas experiments for ENSEMBL genomes. Nucleic Acids Res 44 (W1):W267–W271
11. Harms DW, Quadros RM, Seruggia D et al (2014) Mouse genome editing using the CRISPR/Cas system. Curr Protoc Hum Genet 83:15.7.1–15.727

Chapter 10

Genome-Wide PERV Inactivation in Pigs Using CRISPR/Cas9

Marc Güell

Abstract

The shortage of organs for transplantation is probably the biggest unmet medical need. A potential problem with the clinical use of porcine xenografts is the risk that porcine endogenous retroviruses (PERVs) could infect human cells. In the past, we determined the PERV copy number in the porcine kidney epithelial cell line PK15 and in primary fibroblasts. Using CRISPR-Cas9, we disrupted the catalytic center of *pol*, which is essential for virus replication. Next, we isolated cells in which 100% of the PERV elements had been inactivated. This method enables the possibility of eradicating PERVs in vitro for application to pig-to-human xenotransplantation. Here we describe the methodological bases of this work.

Key words CRISPR, Cas9, Xenotransplantation, PERVs, Pig genome editing

1 Introduction

Porcine xenografts derived from genetically modified swine have been shown to exhibit increased survival rates upon transplantation to nonhuman primates, compared to wild-type xenografts [1]. Despite progress in increasing immune and functional compatibility between pigs and humans, concerns surrounding the transmission of zoonotic pathogens have hindered the production of transgenic organs for clinical xenotransplantation [2]. Although many potential human pathogens can be eliminated from animals through therapeutics and bio-secure [3] breeding, porcine endogenous retroviruses (PERVs) cannot be eliminated in this way, since they reside in multiple copies in the genome of all pig strains and are transmitted vertically.

Endogenous retroviruses are present in the genome of most vertebrate species [4]. Typically, they do not cause disease in the native species, but may do so in other species [5]. Pig genomes contain from a few to several dozen copies of PERV elements [6]. Though no clinical data have demonstrated PERV transmission to humans in vivo [7, 8], several studies have described PERV infection of human cells in vitro [5, 9]. Three subgroups of

Cristina Costa (ed.), *Xenotransplantation: Methods and Protocols*, Methods in Molecular Biology, vol. 2110,
https://doi.org/10.1007/978-1-0716-0255-3_10,

infectious PERVs have been identified: PERV-A, PERV-B, and PERV-C [10].

Multiple strategies have been proposed to mitigate the risk of PERV transmission to human cells. These include using PERV-specific small interfering RNAs (RNAi) to silence PERV expression and developing vaccines to combat infection by PERVs [11–13]. RNAi reduces PERV expression but does not eliminate the risk of infection, especially in immunocompromised recipients [14]. Given the genotypic variability of PERVs, a vaccine strategy is attractive but similarly unlikely to completely eliminate risk of infection. Additionally, there have been various unsuccessful attempts to eliminate PERVs using genome engineering tools such as zinc finger nucleases [15] and TAL effector nucleases [16]. The authors of these studies reported elevated toxicity as a potential cause of these failures.

Cas9 is an RNA-programmable nuclease that functions in nature as a component of the bacterial adaptive immune system by recognizing and cleaving foreign DNA (e.g., from viruses or plasmids) [17]. We and others have repurposed this system as a powerful mammalian genome engineering tool [18, 19].

In this chapter, we describe a methodology for the complete inactivation of all PERVs from the porcine genome using a CRISPR-Cas9-based strategy.

2 Materials

2.1 Small Guide RNAs and Digital PCR

1. Small guide RNAs (gRNAs) (*see* Table 1).
2. DNeasy Blood & Tissue Kit (Qiagen).
3. MseI (New England Biolabs).
4. 2× BioRad dPCR master mix (BioRad) or equivalent.
5. Primers and probes (*see* Table 2).
6. Microfuge tubes, micropipettes, and tips.

2.2 Cell Line Transfection and Cas9 Activation

1. PK15 cells (ATCC CCL-33, ATCC).
2. Dulbecco's modified Eagle's medium (DMEM) with high glucose supplemented with 10% fetal bovine serum and 1% penicillin/streptomycin.

Table 1
gRNA sequences designed

ID	Sequence
gRNA 1	GGTGACCCTCCTCCAGTACGTGG
gRNA 2	GGTCATCCACGTACTGGAGGAGG

Table 2
dPCR primers and probes

ID	Sequence
POL1-FW	CGACTGCCCCAAGGGTTCAA
POL2-FW	CCGACTGCCCCAAGAGTTCAA
POL-RV	TCTCTCCTGCAAATCTGGGCC
POL PROBE	/56FAM/CACGTACTGGAGGAGGGTCACCTG/ BH/
ACTIN-FW	TAACCGATCCTTTCAAGCATTT
ACTIN-RV	TGGTTTCAAAGCTTGCATCATA
ACTIN PROBE	/5HEX/CGTGGGGATGCTTCCTGAGAAAG/BH/
GAPDH-FW	CCGCGATCTAATGTTCTCTTTC
GAPDH-RV	TTCACTCCGACCTTCACCAT
GAPDH PROBE	/5HEX/CAGCCGCGTCCCTGAGACAC/BH/

3. Cloned PiggyBac-Cas9/2gRNAs and Super PiggyBac Transposase plasmid (System Biosciences).
4. Lipofectamine 2000 (Invitrogen, Thermo Fisher Scientific).
5. Puromycin (Sigma).
6. Doxycycline (StemCell Technologies).
7. Tissue culture plastic: flasks, pipettes, and tips.
8. Equipment: humidified incubator at 37 °C and 5% CO_2, pipettors, and micropipettes.

2.3 Isolation of Clones

1. Selected PK15 cells.
2. TrypLE (Thermo Fisher Scientific).
3. ToPro-3 viability dye (Invitrogen, Thermo Fisher Scientific).
4. Tissue culture plastic: flasks, pipettes, and tips.
5. 96-well tissue culture plates.
6. Equipment: humidified incubator at 37 °C and 5% CO_2, pipettors, and micropipettes.
7. Other equipment: sorter such as BD FACSAria II SORP UV (BD Biosciences).

2.4 Genotyping

1. KAPA express extract (KAPA Biosystems).
2. KAPA 2G fast (KAPA Biosystems).
3. Primers (*see* Table 3).

Table 3
MISEQ Illumina primers

ID	Sequence
Miseq_pol1-FW	ACACTCTTTCCCTACACGACGCTCTTCC GATCTCGACTGCCCCAAGGGTTCAA
Miseq_pol2-FW	ACACTCTTTCCCTACACGACGCTCTTCC GATCTCCGACTGCCCCAAGAGTTCAA
Miseq_pol-RV	GTGACTGGAGTTCAGACGTGTGCTCTTC CGATCTTCTCTCCTGCAAATCTGGGCC

4. QIAquick Gel Extraction Kit (QIAGEN).
5. Microfuge tubes, micropipettes, and tips.
6. Equipment: electrophoresis system, blue-light transilluminator, and MiSeq Personal Sequencer (Illumina).
7. E-Gel EX Agarose gels (2%), which contain a proprietary fluorescent nucleic acid stain in the gel allowing detection down to 1 ng/band of DNA when visualized by blue-light transilluminator (Invitrogen, Thermo Fisher Scientific).

2.5 Reverse Transcriptase Assay

1. 0.45 μM Millex-HV Syringe Filter (EMD Millipore Corporation).
2. Amicon Ultra-15 Centrifugal Filter Unit (EMD Millipore Corporation).
3. RT reaction (*see* Table 4).
4. FW primer: 5′ ACCTTTGGCCCTGTTCATTT 3′.
5. E-Gel EX Agarose gels (2%) (Invitrogen, Thermo Fisher Scientific).
6. Microfuge tubes, micropipettes, and tips.
7. Equipment: water bath, PCR machine, electrophoresis system, blue-light transilluminator, and ultracentrifuge.

2.6 HEK293T-GFP Line Construction for Infectivity Assay

1. A Lenti-GFP construct [20] obtained from the plasmid pLVX-IRES-ZsGreen1 (Clontech).
2. ViraPower Lentiviral Packaging Mix (Invitrogen, Thermo Fisher Scientific).
3. Lipofectamine 2000 (Invitrogen, Thermo Fisher Scientific).
4. Lenti-X GoStix (Takara Clontech).
5. Puromycin (Sigma).
6. Equipment: humidified incubator at 37 °C and 5% CO_2, pipettors, and micropipettes.

Table 4
Reverse transcriptase reaction

Total volume: 20 µL		
1×	1× RT buffer	
0.5 mM	dNTPs	
0.5 µM	Influenza reverse primer	5′ CTGCATGACCAGGGTTTATG 3′
100 U	RNAseOUT (Thermo)	
100 U	SuperRnase Inhibitor (Thermo)	
5 µL	Sample lysis	
40 ng	Influenza RNA template	rA*rA*rC*rA*rU*rGrGrArArCrCrUrUrUrGrGrCrCr CrUrGrUrUrCrArUrUrUrUrArGrArArArUrCrArArGr UrCrArArGrArUrArCrGrCrArGrArArGrArGrUrArGrAr CrArUrArArArCrCrCrUrGrGrUrCrArUrGrCrArGrArCrCr U*rC*rA*rG*rU*rG 3′ (*phosphodiester bond)

3 Methods

3.1 Inactivation of PERVs in a PK15 Cell Line Using CRISPR-Cas9 (62 Copies)

3.1.1 Copy Number Estimation and gRNA Design

The PERV viral genome is composed of three genes: *gag* (core), *pol* (reverse transcriptase), and *env* (envelope). Given that the gene product of *pol* carries out reverse transcription and is thus essential for viral replication and infection, it is an adequate target to inactivate the PERVs. In our initial study [20], we used the sequences of 60 publicly available PERV genomes and 195 other endogenous retroviruses found in pigs. We identified a distinct clade of PERV elements; these comprise the only known family of active endogenous retroviruses in pigs [5].

1. Design a pair of gRNAs (such as those shown in Table 1) targeting the catalytic center [21] of the *pol* gene which is conserved in all PERV subgroups.
2. Perform multiple-genome alignment of PERVs using MUSCLE [22].
3. Build a phylogenetic tree of the sequences, and identify a clade that includes the PERVs.
4. Identify a functional site (catalytic site of pol gene).
5. Design specific and comprehensive gRNAs (*see* **Notes 1** and **2**).

6. Estimate copy number. Based on the identified PERV sequences (see above), design specific primers and probes (such as in Table 2).
7. Select multiple sequence alignments containing PERV pol. Design specific PCR primers and dPCR probe that target all PERVs but no other endogenous retroviral sequences. We use the R library DECIPHER for this purpose.
8. Estimate the copy number of the PERVs using droplet digital PCR. We have determined there are 62 copies of PERVs of all types in the PK15 cell line based on three independent benchmark loci.
9. Purify genomic DNA using the DNeasy Blood & Tissue Kit following the instructions of the manufacturer.
10. Digest 50 ng of genomic DNA with *Mse*I (10 U) at 37 °C for 1 h.
11. Prepare reaction mixture containing 10 μL of 2× BioRad dPCR master mix, 1 μL of 18 μM target primers and 5 μM target probe, 1 μL of 18 μM reference primers and 5 μM reference probe (FAM), 5 ng digested DNA, and water to bring to a total volume of 20 μL.
12. Perform PCR droplet generation, PCR amplification, and droplet analysis following BioRad dPCR guidelines.

PERVs copy number, and specific composition varies in different pig cell lines [6]. For some cell lines, better results could be achieved by cell-specific design. This can be performed by targeted sequencing of the PERV loci of your cell line of interest (see below) followed by design of customized gRNAs.

3.1.2 Cell line Preparation and Treatment

PiggyBac vector [23] can be used to deliver a payload containing a doxycycline-inducible Cas9 and the two gRNAs previously described, into the genome of a pig cell line. Continued induction of Cas9 leads to increased targeting frequency of the PERV. The maximum PERV targeting frequency in PK15 cell line has been observed between 2 and 3 weeks (~40% or 23 copies of PERV). Longer treatments do not lead to a higher number of PERV-inactivated populations probably due to toxicity of the treatment. In primary cell lines (such as fetal fibroblasts), we have observed higher editing efficiencies when adding to the medium PFT-alpha and bFGF [24]. A cocktail of these two substances increases the capacity of survival of highly edited cells.

The piggyBac-Cas9/2gRNAs construct we use is derived from a plasmid previously reported by Wang et al. [25]. The IDT gBlocks is an adequate format to synthesize DNA fragments encoding U6-gRNA1-U6-gRNA2 (Genewiz) and incorporated into a PiggyBac-Cas9 construct.

The following steps are conducted for the cell line transfection and Cas9 activation:

1. Grow PK15 cells in DMEM/10% FBS (*see* **Note 3**) in a humidified incubator at 37 °C and 5% CO_2.
2. Transfect 50,000 PK15 cells with 4 μg of cloned PiggyBac-Cas9/2gRNAs and 1 μg Super PiggyBac Transposase plasmid using Lipofectamine 2000 following the instructions of the manufacturer.
3. Select transfected cells using puromycin. We use 2 μg/mL of puromycin and 3 days of selection. Nevertheless, the length of time and exact concentration may need some adjustment using a positive and a negative control.
4. Induce Cas9 expression with 2 μg/mL of doxycycline.

3.1.3 Isolation of Clones

1. For isolation of clones with maximal PERV inactivation, dissociate PK15 cultures using TrypLE, and resuspend in medium with the viability dye ToPro-3 at a concentration of $1–2 \times 10^5$ cells/mL.
2. Conduct cell sorting of PK15 cells under sterile conditions using a sorter such as BD FACSAria II SORP UV with 100 mm nozzle. We use as settings SSC-H versus SSC-W and FSC-H versus FSC-W doublet discrimination gates and a stringent "0/32/16 single-cell" sorting mask to ensure that one and only one cell is sorted per well.
3. Sort cells into 96-well plates with each well containing 100 μL of PK15 medium.
4. After sorting, plates are centrifuged at $70 \times g$ for 3 min.
5. Grow the sorted PK15 cells for 2 weeks.

3.2 Genotyping of PK15 Cells After Genomic Edition

For genotyping of the clones, a fragment around the edited part of the *pol* gene is amplified by PCR and sequenced using Illumina MiSeq using version 2 kit PE250 or version 3 PE300. To this end:

3.2.1 PCR Amplification and Sequencing of Targeted Region

1. Prepare the two-week-old cell pellets by trypsinization and centrifugation, and resuspend in mixture containing 0.5 μL of 10× KAPA express extract buffer, 0.1 μL of 1 U/μL KAPA Express Extract Enzyme, and 4.6 μL of water.
2. Incubate the lysis reaction at 75 °C for 15 min, and inactivate the reaction at 95 °C for 5 min.
3. Add then all reactions to 25 μL of PCR reactions containing 12.5 μL of 2X KAPA 2G fast, 100 nM PERV Illumina primers, and 7.5 μL of water.
4. Incubate reactions at 95 °C for 3 min followed by 25 cycles of 95 °C, 10 s; 65 °C, 20 s; and 72 °C, 20 s.

5. To add the Illumina sequence adaptors, add 5 μL of reaction products to 20 μL of PCR mix containing 12.5 mL of 2 KAPA HiFi HotStart ReadyMix, 100 nM primers carrying Illumina sequence adaptors (Table 3), and 7 μL of water.
6. Incubate reactions at 95 °C for 5 min followed by 15–25 cycles of 98 °C, 20 s; 65 °C, 20 s; and 72 °C, 20 s.
7. Check PCR products on EX 2% gels following standard procedures in a blue-light transilluminator.
8. Followed by the recovery of 300–400 bp products from the gel.
9. After extraction from gel, purified bands are mixed to equal DNA amount.
10. Sequence with MiSeq Personal Sequencer following the instructions of the equipment or through service.

3.2.2 Computational Analysis of PERV Inactivation with CRISPR Genome Analyzer (CRISPR-GA) Algorithm (See Note 4) [5]

1. Combine the overlapping reads produced by PE250 or PE300 using PEAR [13].
2. Map the overlapping reads to the reference region using BLAT.
3. The mapping output is parsed to call the different insertions and deletions, as described by Güell et al. [12].
4. For edited clones only (after single-cell sorting), aligned reads can be grouped into sets containing specific combinations of haplotypes and indel types. Discard read sets with representation lower than 0.5% of the total number of mapped reads (*see* **Note 5**).

3.3 Functional Characterization

3.3.1 Reverse Transcriptase (RT) Assay

The RT assay is used to measure the presence of RT activity in the clones. Absence of activity confirms complete inactivation:

1. Start a culture of 50,000 cells in T75 cm^2 flasks for the test.
2. Collect supernatants after 4 days.
3. Filter the supernatant by using a 0.45 μM Millex-HV Syringe Filter.
4. Concentrate the filtered supernatant at 4000 × *g* for 30 min using Amicon Ultra-15 Centrifugal Filter Unit.
5. Spin the concentrated supernatant in ultracentrifuge at 24,000 × *g* for 60 min, and remove carefully the supernatant.
6. Lyse the virus pellet with 20 μL of 10% NP40 at 37 °C for 60 min.
7. Conduct the RT reaction following the specifications (Table 4) for 60 min at 37 °C.
8. Perform a PCR to detect the generated influenza cDNA following a standard PCR procedure with the indicated primers.
9. Inspect the PCR product in a 2% agarose gel in a blue-light transilluminator. A band of 72 bp indicates a positive RT assay result.

3.3.2 HEK293T-GFP Line Construction

This cell line is used for the infectivity assay described in Subheading 3.3.3:

1. To generate the lentivirus carrying Lenti-GFP, transfect ~5.10^6 293FT HEK cells with 3 μg of pVX-ZsGreen plasmid and 12 μg of ViraPower Lentiviral Packaging Mix using Lipofectamine 2000.
2. Collect lentiviral particles 72 h after transfection, and measure the viral titer using Lenti-X GoStix following the instructions of the manufacturer.
3. Transduce ~10^6 HEK293 cells with ~10^5 lentivirus particles, and select with puromycin to enrich the transduced cells during 5 days of culture.
4. Maintain the 293-GFP-Lenti cell lines with 0.5 μg/mL of puromycin.

3.3.3 Infectivity Assay

This assay allows the measurement of pig-to-human PERV transmission:

1. Co-culture 10^5 HEK293T-GFP cells with 10^5 PK15 wild-type cells in a 10 cm^2 plate for 1 week.
2. Conduct sequential cell sorting with a cell sorter to obtain GFP-positive HEK293T-GFP cells with high purity.
3. Culture HEK293T-GFP cells for 125 days.
4. Isolate genomic DNAs from the cells at different time points (day 76, day 80, and day 125).
5. Use dPCR to quantify PERV copy number as described in Subheading 3.1.1.
6. Optional: perform single-cell sorting from the purified infected HEK293T-GFP cells, and use dPCR to quantify PERV infection of the clonal cell lines.

In summary, we can successfully inactivate PERV in immortalized (PK15, 62 PERV copies) and primary cell lines (fetal fibroblasts, 25 PERV copies). Primary cells can be used for somatic cell nuclear transfer [26] and generate healthy animals free of active PERVs [24].

4 Notes

1. To minimize off-target activity, the gRNAs must be comprehensive and specific: targeting conserved sequences present in all PERVs but no other endogenous retrovirus elements or other parts in the pig genome. We used the R library DECIPHER to design specific gRNAs that target all PERVs but no other endogenous retroviral sequences (Table 1).

2. It may improve the gene-editing design process to know your exact PERV type. It may be beneficial to perform targeted sequencing of your cell line of interest. Specific haplotypes can be used to generate a more accurate design.
3. For primary cell lines, supplementing the media with apoptosis inhibitors such as PFT-alpha or growth factors such as bFGF may increase the resulting gene-editing efficiencies.
4. For large numbers of samples, current versions of CRISPR-GA may not be adequate for analysis of gene-editing data (future versions to be released on 2019 will support multiple samples). CRISPRESSO (http://crispresso.rocks) may be a good alternative.
5. This allows filtration of noise and to obtain a higher-quality genotyping assessment. Filtration threshold can be adjusted depending on copy number. We would always recommend a number lower than 1 (PERV copy number).

References

1. Ekser B, Rigotti P, Gridelli B, Coope DKC (2009) Xenotransplantation of solid organs in the pig-to-primate model. Transpl Immunol 21:87–92. https://doi.org/10.1016/j.trim.2008.10.005
2. ILAR J (1997) Xenotransplantation: science, ethics, and public policy. ILAR J 38:49–51
3. Schuurman H-J (2009) The International Xenotransplantation Association consensus statement on conditions for undertaking clinical trials of porcine islet products in type 1 diabetes—chapter 2: source pigs. Xenotransplantation 16:215–222. https://doi.org/10.1111/j.1399-3089.2009.00541.x
4. Weiss RA (2013) On the concept and elucidation of endogenous retroviruses. Philos Trans R Soc Lond Ser B Biol Sci 368:20120494. https://doi.org/10.1098/rstb.2012.0494
5. Patience C, Takeuchi Y, Weiss RA (1997) Infection of human cells by an endogenous retrovirus of pigs. Nat Med 3:282–286. https://doi.org/10.1038/nm0397-282
6. Lee D, Lee J, Yoon J-K et al (2011) Rapid determination of perv copy number from porcine genomic DNA by real-time polymerase chain reaction. Anim Biotechnol 22:175–180. https://doi.org/10.1080/10495398.2011.595294
7. Heneine W, Tibell A, Switzer WM et al (1998) No evidence of infection with porcine endogenous retrovirus in recipients of porcine islet-cell xenografts. Lancet 352:695–699. https://doi.org/10.1016/S0140-6736(98)07145-1
8. Dinsmore JH, Manhart C, Raineri R, Jacoby DB, Moore A (2000) No evidence for infection of human cells with porcine endogenous retrovirus (PERV) after exposure to porcine fetal neuronal cells. Transplantation 70:1382–1389
9. Takeuchi Y, Patience C, Magre S et al (1998) Host range and interference studies of three classes of pig endogenous retrovirus. J Virol 72:9986–9991
10. Wood JC, Quinn G, Suling KM et al (2004) Identification of exogenous forms of human-tropic porcine endogenous retrovirus in miniature Swine. J Virol 78:2494–2501
11. Ramsoondar J, Vaught T, Ball S et al (2009) Production of transgenic pigs that express porcine endogenous retrovirus small interfering RNAs. Xenotransplantation 16:164–180. https://doi.org/10.1111/j.1399-3089.2009.00525.x
12. Güell M, Yang L, Church GM (2014) Genome editing assessment using CRISPR Genome Analyzer (CRISPR-GA). Bioinformatics. 30 (20):2968–2970. https://doi.org/10.1093/bioinformatics/btu427
13. Zhang J, Kobert K, Flouri T, Stamatakis A (2014) PEAR: a fast and accurate Illumina Paired-End reAd mergeR. Bioinformatics. 30 (5):614–620. https://doi.org/10.1093/bioinformatics/btt593
14. Zamore PD, Tuschl T, Sharp PA, Bartel DP (2000) RNAi: double-stranded RNA directs the ATP-dependent cleavage of mRNA at 21 to 23 nucleotide intervals. Cell

101:25–33. https://doi.org/10.1016/S0092-8674(00)80620-0

15. Semaan M, Ivanusic D, Denner J (2015) Cytotoxic effects during knock out of multiple porcine endogenous retrovirus (PERV) sequences in the pig genome by zinc finger nucleases (ZFN). PLoS One 10:e0122059. https://doi.org/10.1371/journal.pone.0122059
16. Dunn D, DaCosta M, Harris M, Idriss R, O'Brien A (2015) Genetic modification of porcine endogenous retrovirus (PERV) sequences in cultured pig cells as a model for decreasing infectious risk in xenotransplantation. FASEB J 29:LB761
17. Jinek M, Chylinski K, Fonfara I, Hauer M, Doudna JA, Charpentier E (2012) A programmable dual-RNA-guided DNA endonuclease in adaptive bacterial immunity. Science 337:816–821. https://doi.org/10.1126/science.1225829
18. Mali P, Yang L, Esvelt KM et al (2013) RNA-guided human genome engineering via Cas9. Science 339:823–826. https://doi.org/10.1126/science.1232033
19. Cong L, Ran FA, Cox D et al (2013) Multiplex genome engineering using CRISPR/Cas systems. Science 339:819–823. https://doi.org/10.1126/science.1231143
20. Yang L, Güell M, Niu D et al (2015) Genome-wide inactivation of porcine endogenous retroviruses (PERVs). Science 350:1101–1104. https://doi.org/10.1126/science.aad1191
21. Sarafianos SG, Marchand B, Das K et al (2009) Structure and function of HIV-1 reverse transcriptase: molecular mechanisms of polymerization and inhibition. J Mol Biol 385:693–713. https://doi.org/10.1016/j.jmb.2008.10.071
22. Edgar RC (2004) MUSCLE: multiple sequence alignment with high accuracy and high throughput. Nucleic Acids Res 32:1792–1797. https://doi.org/10.1093/nar/gkh340
23. Wilson MH, Coates CJ, George AL (2007) PiggyBac transposon-mediated gene transfer in human cells. Mol Ther 15:139–145
24. Niu D, Wei H-J, Lin L et al (2017) Inactivation of porcine endogenous retrovirus in pigs using CRISPR-Cas9. Science 357:1303–1307. https://doi.org/10.1126/science.aan4187
25. Wang G, McCain ML, Yang L et al (2014) Modeling the mitochondrial cardiomyopathy of Barth syndrome with induced pluripotent stem cell and heart-on-chip technologies. Nat Med 20:616–623. https://doi.org/10.1038/nm.3545
26. Giraldo AM, Ball S, Bondioli KR (2012) Production of transgenic and knockout pigs by somatic cell nuclear transfer. Methods Mol Biol 885:105–123. https://doi.org/10.1007/978-1-61779-845-0_8

Chapter 11

Co-transplantation of Vascularized Thymic Graft with Kidney in Pig-to-Nonhuman Primates for the Induction of Tolerance Across Xenogeneic Barriers

Kazuhiko Yamada, Yuichi Ariyoshi, Thomas Pomposelli, and Mitsuhiro Sekijima

Abstract

Using advanced gene editing technologies, xenotransplantation from multi-transgenic alpha-1,3-galactosyltransferase knockout pigs has demonstrated marked prolongation of renal xenograft survival, ranging from days to greater than several months for life-supporting kidneys and >2 years in a heterotopic non-life-supporting cardiac xenograft model. However, continuous administration of multiple immunosuppressive drugs continues to be required, and attempts to taper immunosuppression have been unsuccessful. These data are consistent with previous reports indicating that the human-anti-porcine T cell response is similar or stronger than that across allogeneic barriers. Due to the strength of both the innate and adaptive immune responses in xenotransplantation, the level of continuous immunosuppression needed to control these responses and prolong xenograft survival has been associated with prohibitive morbidity and mortality. These facts provide compelling rationale to pursue a clinically applicable strategy for the induction of tolerance.

Mixed chimerism and thymic tissue transplantation have both achieved xenogeneic tolerance in pig-to-mouse models, and both have recently been extended to pig-to-baboon models. Although these strategies are promising in small animal models, neither direct intravenous injection of porcine bone marrow cells nor direct fetal thymic tissue transplantation into recipients was able to achieve >2 days chimerism following BM Tx or the engraftment of thymic tissues across xenogeneic barriers in pig-to-nonhuman primate models. Several innovative procedures have been largely developed by Kazuhiko Yamada to overcome these failures. These include vascularized thymic transplantation, combined with either thymokidney (TK) or vascularized thymic lobe (VTL) transplantation. Utilizing the strategy of transplanting vascularized thymic grafts with kidney from the same GalT-KO donor without further gene modification, we have achieved longer than 6 months survival of life-supporting kidneys in a baboon. Notably, the recipient became donor specific unresponsive and developed new thymic emigrants. In this chapter, we introduce a brief summary of our achievements to date toward the successful induction of tolerance by utilizing our novel strategy of vascularized thymic transplantation (including thymokidney transplantation), as well as describe the step-by-step methodology of surgical and in vitro procedures which are required for this experiment.

Key words Xenotransplantation, Transplantation, Kidney, Vascularized thymic transplantation, Tolerance

Cristina Costa (ed.), *Xenotransplantation: Methods and Protocols*, Methods in Molecular Biology, vol. 2110,
https://doi.org/10.1007/978-1-0716-0255-3_11,

1 Introduction

1.1 Xenogeneic T Cell Responses

Due to the intensity of antibody-derived rejection across xenogeneic barriers, researchers have not paid much attention to xenogeneic anti-donor T cell responses to vascularized donor organs. Early investigations into T cell responses exploited the mouse antihuman or anti-pig models that revealed a defective mouse CD4 T cell anti-stimulator MHC class II molecular interaction, effectively eliminating T cell activation via the direct pathway [1, 2], and suggested that if humoral mechanisms were overcome, cell-mediated rejection would prove to be a minor obstacle. However, we have clearly demonstrated in the early 1990s [3, 4] that (1) human T cells respond at least as well to xeno-MHC antigens (Ags) as they do to allo-MHC Ags in MLR. In addition, human-anti-pig T cell responses appear to share similar antigen-presenting cell (APC) requirements for stimulator (direct pathway) or responder (indirect pathway), and (2) the majority of the primary human-anti-pig xeno-response is directed toward porcine MHC class II Ags and involves interactions with human CD4 accessory molecules. These data taken together indicate that the human-anti-porcine T cell response is similar in strength and specificity to the allogeneic response [4].

1.2 The Importance of Inducing T Cell Tolerance for Successful Life-Supporting Xenogeneic Renal Transplantation

Efforts to avoid humoral rejection have included the use of transgenic donors expressing complement-inhibitory proteins, such as human decay-accelerating factor (hDAF, hCD55) [5], CD59 (membrane attack complex inhibitor) [6], and human membrane cofactor protein (hMCP, hCD46) [7] as well as the elimination of Gal expression on pig cells. Despite the early success of these efforts to control hyperacute rejection (HAR), both severe delayed xenograft rejection (DXR) and acute cellular xenograft rejection (ACXR) have remained obstacles to reach long-term survival of solid organ xenografts. We have achieved life-supporting hDAF-to-baboon renal survival for up to 29 days using chronic immunosuppression [8]. We hoped that GalT-KO pigs [9] would further prolong the survival of life-supporting renal xenografts; however, GalT-KO pigs only led to minimal improvement in survival (34 days) using immunosuppressive regimens that included ATG and anti-CD154 mAb [10]. Additionally, Zhong and his colleagues reported maximum survivals of only 16 days [11] using GalT-KO kidneys. Even with the use of heavy immunosuppression in these protocols, rejected GalT-KO grafts showed severe cellular and humoral rejection. The glomeruli and peritubular capillaries of these rejected xenografts stained brightly for IgG. Additionally, investigators observed consistent aberrations in the complement cascade.

The development of cellular rejection in this model despite heavy chronic immunosuppression suggests that small numbers of T cells can initiate both cellular and humoral responses. These data are consistent with reports by Korsgren and colleagues which indicate that even very small numbers of T cells suffice to initiate rejection of porcine islets by macrophages in T cell-deficient rodents [12, 13]. The level of immunosuppression needed to control these T cell responses and prolong xenograft survival has been associated with prohibitive morbidity and mortality. More recently, Shin et al. reported that pig islets, engrafted >500 days in nonhuman primates, were fully rejected by activated immune cells, particularly CD4+ and CD8+ T cells, when immunosuppressive maintenance drugs were discontinued [14, 15]. Therefore, strategies directed at inhibition of the direct and indirect pathways must be included for successful xenotransplantation between pigs and primates.

1.3 Yamada's Strategy to Overcome Xenogeneic T Cell Responses

Thymic tissue, as opposed to vascularized thymic grafts, was first demonstrated to induce xenograft tolerance in a pig-to-mouse transplant model by Sykes et al. Fetal thymic tissue was transplanted directly under the renal capsule in T cell-depleted thymectomized mice [16]. Their results in the pig-to-mouse model showed that murine recipients of porcine thymic tissue develop donor-specific unresponsiveness in mixed lymphocyte reaction (MLR) assays and subsequently accept skin grafts from the thymic donors [17, 18]. In order to extend these results to our pig-to-primate model, it was first necessary to demonstrate that thymic transplantation was possible in the pig. However, our initial attempts to implant minced thymic tissue as in the mouse model failed, with evidence of graft rejection within 15–30 days in an allogeneic pig model [19]. We hypothesized that ischemic damage to the transplanted thymus during revascularization led to an active immunologic response and that thymic tissue should be transplanted as a vascularized organ. Furthermore, during the tenuous revascularization period, unless T cell depletion was complete (which is very difficult in large animals), thymic tissue was rejected before it had the ability to induce tolerance.

In order to eliminate the revascularization period following allogeneic or xenogeneic transplantation, Yamada developed two procedures for vascularized thymic transplantation in miniature swine. One involves the preparation of composite "thymokidneys" in which autologous thymic tissue is allowed to engraft for 1–2 months under the donor's own kidney capsule before allogeneic transplantation [20–22] (*see* Fig. 1a) The other is the transplantation of an isolated vascularized thymic lobe (VTL) graft, or the so-called Yamada VTL procedure [23] (*see* Fig. 1b). Using these novel techniques, we demonstrated that vascularized thymic tissue induces tolerance and supports thymopoiesis across fully

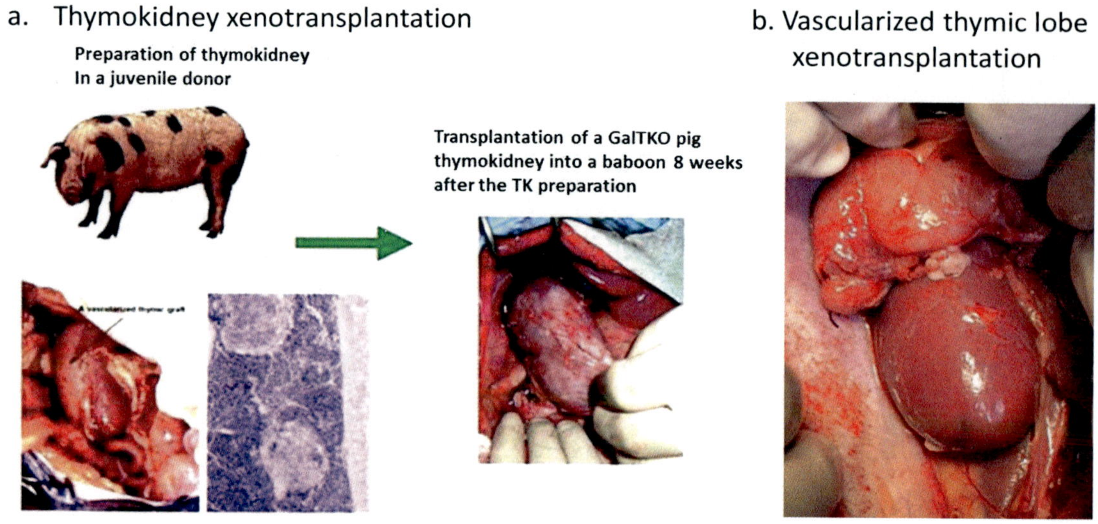

Fig. 1 Schematic diagram of preparation of the thymokidney in a donor pig and the pig thymokidney xenotransplantation in a baboon (**a**). HE specimen and a gross thymokidney (*left*) at 12 weeks after preparation in a donor pig (*left*). Gross GalT-KO thymokidney transplanted in a baboon (*right*). Schematic diagram of preparation of the vascularized thymic lobe xenotransplantation (**b**)

allogeneic barriers in MGH miniature swine [24]. In other experiments, when kidneys and non-vascularized thymic tissue from the same donor were transplanted as separate grafts, recipients rejected the non-vascularized allogeneic thymic grafts within 4 weeks, and kidney grafts were rejected within the second postoperative month. In contrast, recipients of thymus with simultaneous kidney (thymokidney), heart (heart plus VTL), or kidney following VTL transplant survived long term and demonstrated stable graft function and in vitro donor-specific unresponsiveness [21, 22, 24, 25]. Thus, the VTL-based approach can be used in combination with the described procedures for heart transplantation [26].

Based on these encouraging results, our team has extended this vascularized thymic transplant strategy to a pig-to-baboon model of renal xenotransplantation. In initial studies, recipients of GalT-KO kidneys and vascularized thymic grafts from the same donors maintained normal creatinine levels for up to 83 days, while GalT-KO kidneys without thymic grafts were rejected by 34 days [10], even with potent immunosuppression. In the past 5 years, further modification of the induction regimen facilitated survival of life-supporting thymokidneys for over 6 months [27, 28]. Notably, multiple recipients bearing functional thymokidneys for over 3 months developed T cells with the phenotype of new thymic emigrants (CD4+/CD31+/CD45+) as well as donor-specific unresponsiveness at the T and B cell levels. Because recipients were completely thymectomized prior to transplantation, these thymic emigrants most likely developed in the vascularized thymic grafts of the composite thymokidneys.

More recently, we have used hCD47/hCD55 transgenic GalT-KO pigs [29] as donors of VTL grafts. By expressing hCD47 on porcine thymic emigrants (which minimizes macrophage phagocytosis via the CD47-SIRP-alpha pathway), we found a CD4/CD8 double-positive population with a small number of CD4 or CD8 single positive cells in porcine VTL grafts in baboons, indicating active thymopoiesis as early as day 50 following VTL transplantation (Yamada K et al. manuscript in preparation).

The following section describes our method for kidney transplantation co-transplanted with a vascularized thymic graft in a pig-to-baboon model, in addition to helpful tips and pitfalls to avoid for this procedure.

2 Materials

2.1 Catheter Placement

1. A lightweight spandex/lycra jacket (Lomir Biomedical Inc., Malone, NY).
2. A rigid swivel that allows the animal to move freely in the cage without disturbing line connections.
3. A 1.0–1.5-m aluminum tether that attaches the swivel to the jacket.
4. Catheters, generally 1.5–2.0 m flexible, clear Tygon microbore tubing (Norton Performance Plastics, Akron, OH).

2.2 Thymectomy (Baboons) and VTL Procurement (Pigs)

1. Surgical telescopes (preferably $\times 3.5$ or greater) for the primary assistant and the primary surgeon.
2. Surgical draping and instruments including scalpel, bone saw and/or heavy bone scissors, Weitlaner retractors, electrocautery, hemoclips, and bone wax as well as suture materials (PDS, Prolene, and silk ties).
3. Cotton swab applicators for gentle separation of the thymic tissue from the pericardium.
4. Anesthesia and mechanical ventilator.
5. Euro-Collins solution or UW solution.

2.3 Thymokidney, VTL, and Kidney Transplantations (Pig to Baboon)

1. Surgical draping and surgical instruments are similar to those in clinical pediatric kidney transplantation, including scalpel, electrocautery, fine-tip/Castro needle drivers, $\times 2.5$–4.5-magnification surgical loupes, 6-0 to 8-0 nonabsorbable surgical suture, Euro-Collins or UW solution, Balfour retractors, and sterile ice.
2. Anesthesia and mechanical ventilator.

2.4 Flow Cytometry to Assess T Cell and B Cell Depletion and Host Thymopoiesis in Porcine Thymic Grafts

1. Antihuman (cross-reactive to baboons) antibodies for CD3 (polyclonal rabbit antihuman CD3, A0452; Dako North America, Inc., Carpinteria, California, USA), CD4 (mouse antihuman CD4 mAb, 1F6; Invitrogen, Carlsbad, California, USA), CD8 (mouse antihuman CD8; BD Biosciences, Franklin Lakes, New Jersey, USA), CD20 (mouse antihuman CD20; BD Biosciences), and CD25 (mouse antihuman, BD Biosciences). All conjugated to a fluorochrome.
2. Anti-pig monoclonal antibodies (mAb, available in our research center): Anti-porcine CD1, CD2, CD3, CD4, CD8, CD25, CD45RA, Class I, Class II, PAA (produced internally). All conjugated to fluorochrome.
3. Conjugated anti-mouse IgG1 and anti-mouse IgG2a to be used as controls.
4. Cold FACS media: 1 L of Hanks' balanced salt solution, 1 g of bovine serum albumin (BSA), 1 g of sodium azide.
5. FACS tubes.

2.5 Baboon Mixed Lymphocyte Reaction

1. Reagents for peripheral blood monocytes (PBMC) isolation: Hanks' balanced salt solution (Life Technologies, Grand Island, NY, USA), Histopaque 1077 (Sigma-Aldrich, St. Louis, MO, USA), ACK lysing buffer (BioWhittaker, Maryland, MD, USA).
2. MLR medium: AIMV media (*see* Subheading 2 and **Note 1**).
3. MLR plate: 96-well flat-bottom plate (Costar, Cambridge, MA, USA).
4. Cell Proliferation Dye eFluor 670 (Invitrogen, Carlsbad, California, USA).
5. Antihuman (cross-reactive to baboons) antibodies, the same as was previously described (*see* Subheading 2.4).
6. FACS tubes.
7. Cold FACS media: 1 L of Hanks' balanced salt solution, 1 g of BSA, 1 g of sodium azide.

2.6 Antibody/Complement-Mediated Cytotoxicity Assay

1. Target pig PBMC.
2. Medium 199 with 2% fetal bovine serum (FBS).
3. Decomplemented serum samples (56 °C for 30 min).
4. 7-AAD (Sigma) diluted to 5 μg/mL in FACS media.
5. Rabbit complement (at a predetermined working dilution).
6. 96-Well U-bottom plates and lids, non-sterile.
7. Combitips, combi-pipettor, micropipette tips, multichannel pipettor, Pasteur pipettes, micropipettor, truncated pipette tip, plastic trough.

8. 37 °C, CO_2 incubator.
9. Centrifuge.
10. FACS tubes.
11. Cold FACS media: 1 L of Hanks' balanced salt solution, 1 g of BSA, 1 g of sodium azide.
12. Ice and ice bucket.

3 Methods

Though significant academic progress has been made, xenotransplantation has not yet become a clinical reality. Below is a brief introduction to the surgical and immunologic procedures compiled from the most successful pig-to-nonhuman primate renal transplantation protocol developed by Yamada.

3.1 Catheter Placement

At the authors' institution, kidney xenotransplantation is conducted in life-supporting models, and a bilateral native nephrectomy is typically performed at the time of transplantation. Thus, daily laboratory values are required to monitor the animal's health and graft function. For these reasons, the placement of central venous and arterial catheters or "lines" is required for animal care during the induction, peri-transplantation, and post-transplantation periods. Lines provide the caregiver with the ability to both deliver drug therapy and draw blood for diagnostic testing. This is of particular importance in baboons or monkeys because it is difficult to perform a physical examination beyond simple observation. Although central catheters confer a risk of infection, there is greater risk associated with sedation for daily or twice-daily blood draws and drug-administration, especially in potentially fragile post-xenotransplantation recipients.

The lines are utilized for (1) continuous blood pressure monitoring, (2) blood samples for CBC, blood chemistry, and drug levels, and (3) the number of drugs administered (continuous and/or intermittent) and use of different lines due to drug compatibility. We used to place three lines (two venous, one arterial) 1 week prior to transplantation. The arterial line is used primarily for blood draws, and the two venous lines are used for drug delivery. However, we have recently started placing only two venous lines with success in order to lessen the risk of infection by minimizing indwelling lines. We have found that continuous blood pressure monitoring is generally not needed once the animal is recovered from anesthesia, and blood draws can be performed from one of the two venous lines. Importantly, soluble MMF must be placed on its own dedicated D5%W carrier line and not mixed with normal saline.

The following is our procedure for placing central lines in the jugular veins as well as the carotid.

1. Generally, central lines are placed in the great vessels of the neck, on the left side.
2. Make a 3-cm transverse incision sharply, one fingerbreadth above the clavicle, starting from the midline and extending across the anterior border of the sternocleidomastoid (SCM).
3. Electrocautery and blunt dissection should be used to transect the platysma and isolate the external jugular (EJ) vein, the internal jugular (IJ) vein, and the carotid artery.
4. Make a small incision in the animal's back (between shoulder blades), and tunnel the lines through the subcutaneous tissue and into the open neck wound.
5. Dissect the sternocleidomastoid and retract medially or laterally for access to the carotid sheath. The easier approach is to retract laterally.
6. Open the carotid sheath and isolate the carotid, and then tie off distally.
7. Place a small bulldog clamp on the vessel, 1 cm proximally.
8. Create an arteriotomy using tenotomy scissors and then cannulate the vessel.
9. Release clamps and advance the arterial catheter (*see* **Note 3**).
10. A similar procedure is performed for the IJ followed by the EJ (*see* **Note 4**).
11. Close the skin with buried subcuticular sutures.
12. Place a protective line jacket on the animal in standard fashion (Fig. 2. *See* **Note 5**).

3.2 Host Thymectomy and VTL Procurement from Donor Pigs (See Notes 6–9)

Our previous data in pig-to-mouse models have demonstrated that recipient thymectomy is required for the induction of xenogeneic tolerance [18, 24]. Three weeks prior to transplantation and 2 weeks prior to line placement, animals on the tolerance-induction (thymokidney) protocol undergo thymectomy.

3.2.1 Host Thymectomy

1. After intubating the donor pig, the animal is placed supine on the operating table.
2. First, expose the median pectoral groove using electrocautery, blunt dissection, and finger fracture.
3. Dissect the manubrium, exposing the jugular notch, and then perform a sternotomy (*see* **Note 7**).
4. Dissect and remove the thymic gland (*see* **Note 8**) with gentle traction using cotton swab applicators.

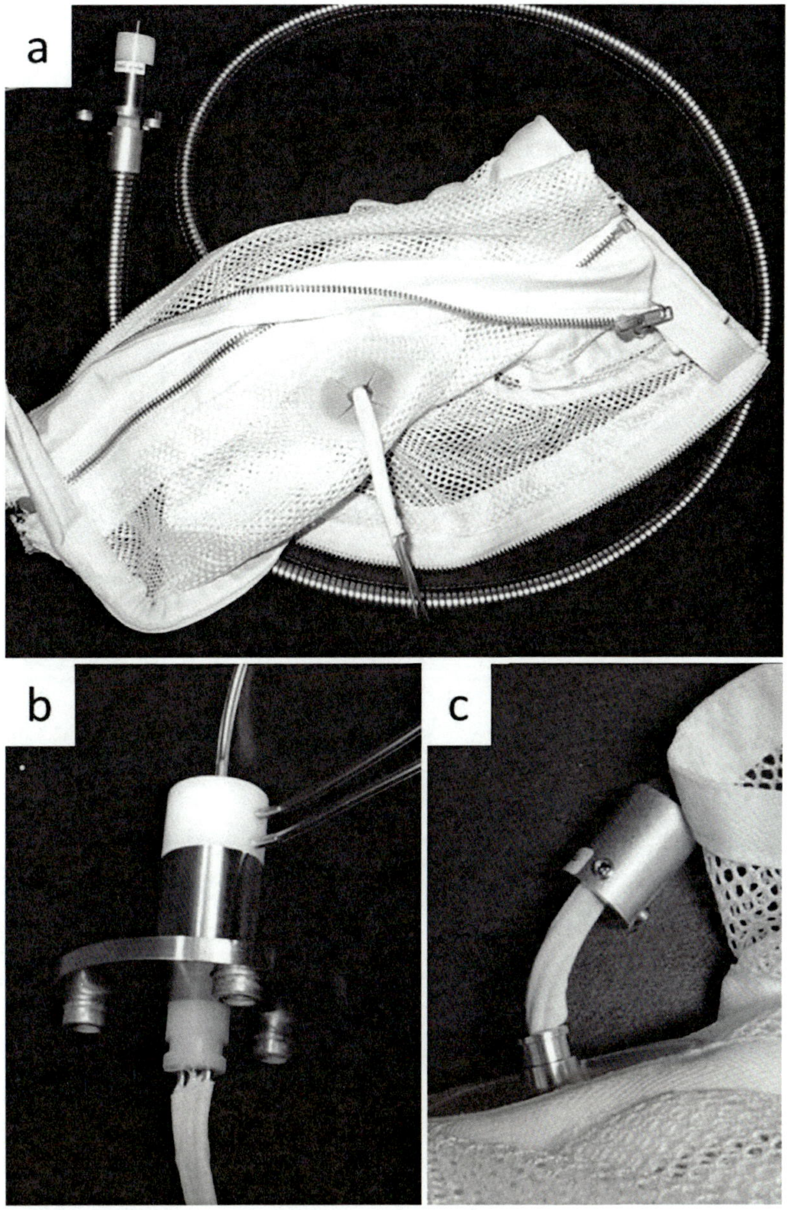

Fig. 2 A picture of the line system used for baboon recipients of xenografting procedures. The mesh jacket is rigidly attached to an aluminum tether through which 1–3 catheters are placed. The tethers are 4–6 ft. in length. (**b**) A close-up view of the swivel shown in (**a**), which has been unattached from the aluminum tether. As many as three catheters are securely affixed to each of three ports on the swivel. This allows the animal to move within the cage without twisting the lines. (**c**) A close-up view of the connection between the jacket and the tether. In this image, the jacket has been unattached from the tether, showing the tape-wrapped catheters traveling through the jacket and into the tethering system

5. Weigh the excised thymus and take several pieces for histology and immunologic assays.
6. Test for air leak with approximately 5 mL of normal saline poured into the thorax. If no leak is observed, go to the next step. If an air leak is observed, repair with an absorbable suture.
7. Following meticulous hemostasis, close chest with interrupted 0-PDS sutures (*see* **Note 9**).
8. Before completely closing the sternum, ask the anesthesiologist to provide a sustained breath to fully expand the lungs. Again, assess for air leak with 3–5 mL of normal saline on the sternum to determine if a chest tube is required. If thymectomy is appropriately performed, it should be done in 60 min, and no chest tube is typically required.

3.2.2 VTL Procurement from Donor Pigs (Yamada's Procedure) [23, 24]

The VTL harvest and VTL transplantation procedures are technically challenging. To improve the likelihood of success, the surgeon should have extensive experience with porcine thymectomy [30] as well as vascular surgical techniques.

1. To procure thymic lobe(s) from the donors, make a longitudinal cervical incision to expose the porcine cervical thymus [30].
2. Expose the pre-tracheal muscles and the connective tissue sheath overlying the thymus to provide access to the left and right lobes [23].
3. Carefully dissect the cervical thymus away from the connective tissue and ligate vessels that are not feeding or drainage vessels of the thymic lobes (*see* **Note 10**).
4. Excise thymic lobe with well-preserved feeding arteries and drainage veins and perfuse the excised thymic lobe with Euro-Collins solution.

3.3 Thymokidney, VTL, and Kidney Transplantations

3.3.1 Thymokidney Preparation in GalT-KO Pig Donors (Yamada's Thymokidney Procedure) [20, 21]

The thymokidney model of xenotransplantation is the most successful life-supporting xenograft protocol for large animals [10]. In this model, donor-specific T cell tolerance to the donor graft is induced by co-transplantation of donor thymic tissue. The procedure for the preparation of a composite thymokidney was developed because transplantation of prevascularized thymic tissue helps to prevent revascularization-associated ischemia. For this reason, thymic tissue in this model is placed beneath the autologous native renal capsule in the donor (7–12-week-old juvenile miniature swine) to create a composite thymokidney 6–8 weeks prior to composite transplantation (*see* **Note 11**). During this time, the thymic tissue becomes vascularized by the autologous kidney. The prevascularized thymokidney graft is then transplanted as a composite kidney graft using the kidney transplantation procedure described below [20–22] (*see* Fig. 1a).

1. Excise cervical thymic lobe(s) (*see* **Note 12**) through a midline or 1-cm lateral neck incision [21].
2. Excise a cervical thymic lobe and mince the excised thymic tissue with Metzenbaum scissors or a scalpel on the back table (*see* **Note 13**), and then close cervical incision.
3. Expose the kidney using a retroperitoneal flank incision (right and left thymokidney preparation should be performed through two separate flank incisions to prevent abdominal adhesions as both kidneys will be procured in several weeks) and then place thymic tissue beneath the autologous renal capsule (Fig. 1a left. *See* **Note 14**).
4. Close the edge of the capsular window using electrocautery or with a single 7-0 Prolene suture.
5. Close flank incision in layers with 3-0 or 2-0 vicryl. Nylon can be used for pig skin.

3.3.2 Thymokidney Procurement from GalT-KO Pig Donors

The thymokidney procurement for pig-to-baboon xenotransplantation is similar to human kidney procurement. However, since the donor kidney contains thymus underneath the kidney capsule, it is necessary to preserve the donor's perinephric fat. This is because local tissues may be adherent resulting from the thymokidney preparation. Careful manipulation and prudent use of electrocautery are required to avoid further damage to the fragile thymic tissue vascularized by the kidney.

1. A midline incision from xiphoid to pubis is used to provide access to the abdomen (*see* **Note 15**).
2. Expose thymokidneys (*see above*), and isolate the renal vein (RV) and artery (RA), transecting each vessel at the level of the aorta (Aa) (*see* **Note 16**).
3. Remove thymokidney(s) and flush via the RA on the back table using Euro-Collins or UW solution and immediately place on ice.

3.3.3 GalT-KO Thymokidney Transplantation to Baboons

Surgical thymokidney transplantation is generally performed using methods similar to those of pediatric kidney transplantation. Although primary kidney transplantation in pig-to-nonhuman primates may be described by other groups, the unique and complex nature of the procedures described in this chapter will help readers perform successful xenogeneic kidney transplantation. In addition, while many groups have experimented with xenotransplantation, the procedures developed by Dr. Yamada have resulted in both clinical (minimal acute tubular damages) and immunologic (extended graft survival) successes for our group. Because porcine kidneys are predisposed to vasoconstriction, surgeons should minimally manipulate the graft as described below (*see* **Note 17**). Each vascular anastomosis (renal vein and renal artery) should be done within 12–15 min, ideally under 10 min.

1. Make a midline abdominal incision from xiphoid to pubis.
2. Gain exposure using a Balfour retractor and retract the animal's bowels to the left, allowing access to the right retroperitoneum.
3. For nephrectomy of the right kidney, focus inferiorly and isolate the ureter. Ligate the ureter such that at least 50% of the native length remains attached to the bladder (this is done so that a future uretero-ureteral anastomosis can be performed if a technical complication arises).
4. Skeletonize the host renal vessels to the level of inferior vena cava (IVC) and the aorta (Aa). Then, ligate the RA using 3-0 silk ties followed by ligation of the RV.
5. Place hemoclips proximally.
6. For kidney transplantation (*see* **Note 16**), dissect the Aa and IVC such that the surgeon has sufficient access to allow placement of the vascular clamp and anastomosis (care is taken to avoid injuring the inferior mesenteric artery).
7. Heparinize the recipient with 70–100 IU/kg at least 2 min prior to beginning the transplant.
8. First, cross clamp recipient IVC with a Satinsky clamp (*see* **Note 16**). The venous anastomosis is done first because the RA will over lie the IVC following anastomosis (*see* **Note 17**).
9. Create venotomy with an 11-blade or Potts' scissors.
10. Typically, place two stay sutures (one superiorly and one inferiorly on the cava) and sew first the back wall, followed by the front wall with 7-0 Prolene (8-0 if baboons are <5 kg). Senior surgeons may prefer one running suture method. The venous anastomosis should be completed under 12 min.
11. Place a bulldog on the graft RV and gently release the Satinsky clamp from the recipient IVC, allowing venous blood to backfill the RV (*see* **Note 18**, important).
12. After clamping the Aa with a Satinsky clamp, use an 11-blade and Potts' scissors to create an arteriotomy.
13. Perform the arterial anastomosis using 7-0 running Prolene (8-0 if baboons are <5 kg). The author highly recommends using a parachute (3 o'clock single knot) technique performed at least under 15 min as this minimizes manipulation of the graft and ischemic time (*see* **Note 19**).
14. Release the aortic Satinsky clamp. Obtain hemostasis using interrupted 7-0 Prolene sutures as needed if bleeding is encountered. In our experience with this technique, the kidney becomes pink and warm and produces urine within minutes of revascularization.
15. After confirming the graft is producing urine, perform a left-sided nephrectomy (*see* **Note 20**).

16. Perform the neo-ureterocystostomy using 2 or 3 running 7-0 PDS or Prolene sutures. Placement of a 3F or 4F stent is recommended (the stent can be removed after day 30).
17. Cystopexy can be performed if necessary (such as a short donor ureter).
18. Close the abdomen in 3–4 layers. Our practice is to close the peritoneum with absorbable suture, followed by closure of the fascia with a running permanent suture, and then closing the superficial fascia and skin with a running absorbable suture.
19. Wean off the ventilator and replace the jacket. The animal should be perched within 2 h of extubation. *See* **Note 21** for animal care recommendations.

3.3.4 VTL Transplantation (Yamada's VTL Procedure)

Like the VTL harvest, the VTL transplantation is technically challenging, and extensive experience with vascular surgical techniques will increase the likelihood of success. The team should practice with a pig-to-pig allogeneic model before applying these techniques to a pig-to-nonhuman primate model. The advantage of using the thymokidney is that the prevascularized thymus is transplanted simultaneously with the kidney. However, the VTL procedure can accompany the transplantation of any organ (including the kidney). In contrast to the prevascularized thymokidney, this procedure is more complex as the actual thymic lobe must be anastomosed to the recipient vessels. In theory, a co-transplanted thymic graft should confer tolerance to the recipient regardless of the transplanted organ (i.e., liver, heart, etc.). In an allogeneic model of VTL and simultaneous heart transplantation [25], the recipient becomes tolerant in a donor-specific manner. Although attempts to prepare thymohearts have been undertaken, these results have not been encouraging. One reason for failure of the thymoheart may relate to the anatomic challenges (limited space) beneath the pericardium. Thymoliver grafts have, as of yet, not been attempted. The VTL transplantation from the GalT-KO pig into baboons is typically completed using end-to-side anastomoses for the thymic artery and vein to the recipient animal Aa or IVC above the renal vessel anastomosis for simultaneous VTL and kidney transplantation (*see* Fig. 1b) or carotid artery and IJ for isolated VTL transplant, respectively [23, 24]. This is done in a similar manner to that of kidney transplantation (*see* above and Subheading 4). Either 7-0 or 8-0 Prolene should be used, and ×3.5 or greater (greater than ×4.5 for end-to-end) surgical loupes are recommended. *See* **Note 21** for animal care recommendations.

3.4 Flow Cytometry to Assess for T Cell and B Cell Depletion and Host Thymopoiesis in Porcine Thymic Grafts

In order to assess T cell or B cell levels as well as host T cell development in the transplanted thymus (thymopoiesis), antihuman monoclonal antibodies (cross-reactive to baboon molecules) are used. Anti-porcine mAb are used to assess pig thymocytes in the thymic grafts after xenotransplantation. Some anti-porcine antibodies may not be commercially available.

1. Resuspend 1×10^6 PBMC or thymocytes in 100 μL of Hanks' balanced salt solution containing 0.1% BSA and 0.05% sodium azide.
2. Incubate for 30 min at 4 °C with conjugated antibodies. Use conjugated anti-mouse IgG1 and anti-mouse IgG2a as controls for nonspecific binding.
3. After three washes, analyze cells by flow cytometry using the FSC-H/FSC-A Area Scaling to discriminate doublets.

3.5 Baboon Mixed Lymphocyte Reaction

The MLR assesses immunologic responses to the donor (mainly CD4 T cell responses against donor antigens on MHC class II) as compared to that against a third party to determine whether (1) the recipient is immunocompetent and (2) the recipient has developed a donor-specific response. Here, we describe bulk MLR assays using PBMC.

1. Isolate and resuspend responders and stimulators at 4×10^6 cells/mL in AIMV media (*see* **Note 1**).
2. For eFluor 670 labelling, prepare a 5 μM solution of eFluor 670 in room-temperature HBSS by adding 10 μL of 2.5 mM eFluor 670 stock to 5 mL of HBSS. This will be mixed 1:1 with the responder cell suspension.
3. After vortexing the cells, incubate for 10 min at 37 °C in the dark.
4. Stop the labeling by adding 1 mL of FBS and five times volume of cold AIMV media.
5. Incubate on ice for 5 min.
6. Wash twice with AIMV media.
7. Count the cells and resuspend labelling responder cells at 4×10^6 cells/mL in AIMV media.
8. Plate labelling responder cells at 4×10^5 cells/well with an equal volume of stimulators in a 96-well flat-bottom plate. All samples should be tested in triplicate.
9. Incubate in 5% CO for 5 days at 37 °C.
10. After 5 days of incubation, remove cells from wells by pipetting up and down into FACS tube.
11. Wash wells with HBSS and add to appropriate FACS tube.
12. Wash twice with FACS media.

13. Add conjugated mAb and incubate at 4 °C for 30 min.
14. Wash twice as previously described.
15. Resuspend cells in 400 μL of FACS media.
16. Acquire cells using a flow cytometer.

3.6 Antibody/Complement-Mediated Cytotoxicity Assay

The 7-AAD-dye-exclusion FACS procedure is designed for the titration of cytotoxic antibodies to cell surface antigens, using an antibody/complement reaction, followed by a dye exclusion assay and FACS acquisition. A similar procedure is used for SLA genotype testing of swine peripheral blood lymphocytes.

1. Process blood and count PBMC.
2. Prepare cell solutions of 5×10^6 cells/mL. The dilutions are made in Medium 199 (2% FBS, i.e., 10 mL of FBS and 490 mL of Medium 199). Keep the diluted suspensions on ice.
3. Identify the test panel and determine the appropriate dilutions for experimental serum samples. Also, decide which pig PBMC will be used as target cells. Typical setup involves serial dilutions of selected serum samples beginning at 1:2.
4. For sera serial dilutions, add 25 μL of Medium 199/2% FBS to each well of the 96-well round-bottom plate and 25 μL of each serum to the first well of each row, diluting the serum samples across the row using a multichannel pipette up to 1:256 dilution. Mix the dilutions of the samples by pipetting up and down five times.
5. You also need to prepare three control wells per target cell containing either media alone, the diluted rabbit complement, or both. These will act as the viability and complement background controls.
6. Add 25 μL of the appropriate cell suspension (pig target cells) to each well, including the complement and medium control wells.
7. Incubate the plate in the CO_2 incubator for 15 min at 37 °C.
8. After incubation, add 125 μL of Medium 199/2% FBS to each well in the plate. Centrifuge at 260 × *g* for 5 min at 4 °C (brake on).
9. Dilute the complement to an appropriate (predetermined) dilution in Medium 199 (1:8, i.e., 0.5 mL of rabbit complement and 3.5 mL of M199).
10. Dispense 25 μL of the diluted complement into the appropriate wells.
11. Seal the plate with a plate sealer and shake the plate on the tray shaker for 30 s.
12. Incubate the plate in the CO_2 incubator at 37 °C for 30 min.

13. Put the plate on ice.
14. Add 200 μL cold FACS media to all wells containing cells.
15. Transfer the contents of each well to the FACS tubes.
16. Add 5 μL of 7-AAD antibody to all tubes.
17. Place tubes either on ice or at 4 °C for 30 min.
18. Acquire cells using FACS machine looking at FSC vs. FL-3. The dead cell population will be stained with 7-AAD and be positive in FL-3.

4 Notes

1. For the MLR assay, use an AIM-V media which is serum-free. This is important because serum-based media are associated with higher levels of background stimulation, especially across a xenogeneic barrier.
2. The EJ should be tagged with a vessel loop. This is done because the vein is very small and frequently experiences vasospasm. To prevent visual obstruction, the EJ catheter is placed after the IJ vein and arterial catheters.
3. The arterial catheter should not be advanced more than 6 cm in a 6-kg baboon.
4. The IJ and EJ are low pressure vessels. After distal ligation, a proximal clamp is typically unnecessary prior to venotomy. Because anatomically, the EJ communicates with the subclavian vein at an acute angle, the EJ line is typically only advanced 1–2 cm. The catheters are tacked to the SCM, deep to the reapproximated platysma. This may help to avoid contamination if a wound infection (rare) occurs.
5. Catheter-related issues: If line infection occurs or the lines thrombose, they must be removed. In order to do so, the animal is taken to the OR for line removal through the same incision used for placement. Expect dense scar tissue around the lines, and extreme care must be taken to avoid puncture or damaging the lines during dissection, as this would introduce a potentially fatal air embolus or hemorrhage. If additional lines are required after removal, the femoral vessels can be used. In doing so, the femoral artery will need to be tied distally, which is tolerated remarkably well in baboons. Both ipsilateral femoral vessels can be used during a single line placement. In some cases, it may be necessary to subcutaneously tunnel a second venous line across the animal's anterior pelvis. Two and sometimes three lines can be tunneled safely through the subcutaneous tissues over the animal's back (at the level of L1). The longitudinal groin incisions, despite not being covered by the

jacket, are rarely a problem. If needed, all four vessels (two femoral arteries and two femoral veins) can be ligated without causing clinically significant morbidity.

6. Thymectomy and VTL procurement are technically challenging and should be performed by an experienced surgeon.
7. A small finger (typically the fifth digit) is used to develop the plane between the sternum and the mediastinum, caudally. The same is then done from the xiphoid process, rostral. Once the plane from the manubrium to the xiphoid has been fully developed, electrocautery, a bone saw, and heavy scissors are used to perform a sternotomy. A heavy Weitlaner retractor is used to provide exposure.
8. In juvenile animals, the intact thymus is easily distinguished from fat by its texture and lighter color. Taking care to avoid damaging the pleura and pericardium, the thymus is gently dissected off the heart using cotton-swabbed applicators. Small but major vessels to the thymus are present at the superior aspect of the mediastinal thymus, which should be identified and clipped or ligated with 3-0 or 4-0 silks.
9. Insight following thymectomy: Prior to tying the last two sutures, the anesthesiologist is instructed to provide a sustained breath. If a lung injury is present, air bubbles will rise from the pooled mediastinal saline. Once no more bubbles are encountered, the chest closure is completed, and the animal is weaned from the ventilator as quickly as possible. This is important in case poor respiratory drive is encountered due to technical issues, allowing it to be dealt with prior to leaving the operating room. The chest is closed in three layers, and the skin is closed using an absorbable subcuticular suture.
10. The blood supply to the thymic lobes is variable through branches that typically arise from the subclavian artery opposite from the internal mammary artery. The thymic veins primarily drain into the native internal jugular vein. At the time of thymic lobe procurement, both the feeding arteries (subclavian or thymic) and the internal jugular vein are taken en bloc. In addition, the rostral thymus may have collateral blood flow from the carotid. In this case, the carotid artery is procured en bloc as well.
11. Fetal porcine thymic tissue has been shown to proliferate under mice renal capsules and subsequently induce tolerance. However, in our experience, when only several pieces of autologous porcine thymic tissue were placed subcapsularly, the tissue did not grow well. Because of poor fetal thymic availability, we chose to use juvenile pigs (7–12 weeks) for thymokidney preparation. In addition, we have observed poor growth of thymi from aged swine (>12 months old) when placed beneath the renal capsule. For this reason, juvenile swine (6–10 kg) are preferred [21].

12. Miniature swine, unlike baboons or humans, have a cervical thymic lobe in addition to the mediastinal thymus. The cervical thymus is ideal for thymokidney preparation or VTL transplantation because it is easily accessed using a longitudinal off-midline cervical incision. A single, fully excised cervical thymic lobe is typically required for the preparation of two thymokidneys. Thus, only half of a thymic lobe is sufficient for a single thymokidney.
13. Upon removal of the thymic tissue, the gland is minced prior to implantation. However, we have observed poor results when the thymus is minced too finely or when the thymus is injected subcapsularly using an angiocatheter. This is likely due to increased cell lysis and inflammatory changes.
14. The kidney is reached using a flank incision, and a capsular window is created using an 11-blade or angiocatheter. The tissue must be placed carefully under the renal capsule, taking care not to tear the capsule. The optimal amount of thymus to procure from the juvenile is typically related to the size of the kidney itself. Unpublished data from our laboratory has indicated that when only a few small pieces of thymus were implanted, subsequent growth of the transplanted thymus was insufficient. We have observed improved results when 40–60% of the superior surface of the kidney was covered with thymus [20].
15. If the recipient has a second kidney (or thymokidney) to be used at a later date, the animal must be recovered. Thus, in transecting the RA, care must be taken not to take much of a Carrel patch. Typically, 1–2 mm of aortic patch is sufficient for the recipient transplant. A small Carrel patch allows the surgeon to perform a primary longitudinal closure of the donor Aa using 6 or 7-0 Prolene suture. If the Carrel patch is larger, it will be necessary to close the Aa using a Dacron or Gore-Tex patch. Thereafter, the abdomen is closed in the standard three-layer fashion and the animal is recovered.
16. Transplantation should be performed by a skilled surgeon, and the vascular anastomosis should be completed within ideally 20 min, at most 30 min. If a young surgeon is performing the xenokidney transplant, an attending surgeon should be mindful of the anastomosis time and ensure that the graft is minimally manipulated.
17. It is important to note the blood pressure before and after clamping the IVC as hemodynamics may vary dramatically with clamping and unclamping the IVC. If blood pressure decreases, the distal Aa can be clamped using a bulldog to improve circulation.

18. Clamping the Aa will diagnose bleeding from the venous anastomosis. To avoid narrowing the venous anastomosis, sutures should not be tied prior to unclamping the vein. This will allow the RV to engorge prior to tying the final knot. If the surgeon is concerned about bleeding, sutures can be tied loosely prior to unclamping. The venous anastomosis should be completed quickly (under 15 min) and preferably within 10 min.
19. As porcine kidneys have a tendency to suffer vasoconstriction due to manipulation, the authors recommend an anastomotic approach that minimizes touching the xenograft. To do so, the surgeon should stand on the animal's left side and be able to perform both the arterial and venous anastomoses quickly and skillfully. A large Carrel patch for the donor vessel is not recommended.
20. In our standard model of xenotransplantation, we generally remove both kidneys prior to xenograft transplantation. However, in some circumstances, if investigators are expecting an increased risk of graft rejection, the renal graft (pig) may be transplanted before or after the left kidney is removed. In some cases, such as the presence of high levels of preexisting anti-non-Gal antibody or if the transplant is expected to undergo early rejection, it may be beneficial to leave the contralateral native kidney and ligate the ureter. Removing the ligature will save the life of the animal to follow immunologic assessment even after removal of the rejected graft.
21. Animal care: Following line placement and prior to transplantation, the animals undergo a brief, but stringent induction using T cell and B cell depleting agents. These are generally monoclonal or polyclonal antibodies that are commercially available. Administration of any cellular depleting antibody should be preempted by flow-cytometric T and B cell subset phenotyping. T cell-depleting agents have been tested at multiple doses and at multiple time points; however, the general goal is to achieve near-complete T cell depletion (50–150 cells/μL) for the first 2 weeks which protects the transplanted thymic grafts in the induction period (Yamada, K. et al., manuscript in preparation). Because immunosuppression is extensive during the first 2 weeks, recipients are started on antibiotics to avoid opportunistic infection. Our standard regimen includes cefazolin, Levaquin, and ganciclovir. The animal's WBC typically falls below 1000 cells/μL for the first week. While the animal is leukopenic, platelet counts are a good surrogate for WBC concerning infection. Clinically, we have observed that platelets fall in the setting of bacteremia, despite a very low WBC, in a potentially normal appearing animal. Like humans, immunosuppressed baboons may display no signs of sickness while immunosuppressed, despite active

infection. For this reason, any clinically concerning signs (decreased appetite, activity, etc.) should prompt blood cultures and empiric widening of antibiotic coverage. Typically, coverage is increased to vancomycin and cefazolin is stopped. If there is concern for gram-negative infection, piperacillin tazobactam can be started. In these situations, it is best to consult an infectious disease expert.

22. Antibody/complement-mediated cytotoxicity assay: Rabbit complement is kept on ice to avoid deactivation. Appropriate dilution should be determined by using control naive porcine serum. To avoid background staining, 7-AAD should be stopped within 60 min.

Acknowledgments

We thank Ms. Haruna Shimizu for her editorial assistance. This research was supported by NIH grant (NIAID 5P01AI045897). All procedures and animal care were performed in accordance with the Principles of Laboratory Animal Care formulated by the National Society for Medical Research and the Guide for the Care and Use of Laboratory Animals prepared by Columbia University Medical Center.

References

1. Moses RD, Pierson RN 3rd, Winn HJ, Auchincloss H Jr (1990) Xenogeneic proliferation and lymphokine production are dependent on CD4 + helper T cells and self antigen-presenting cells in the mouse. J Exp Med 172(2):567–575
2. Deschamps JY, Roux FA, Sai P, Gouin E (2005) History of xenotransplantation. Xenotransplantation 12(2):91–109. https://doi.org/10.1111/j.1399-3089.2004.00199.x
3. Murray AG, Khodadoust MM, Pober JS, Bothwell AL (1994) Porcine aortic endothelial cells activate human T cells: direct presentation of MHC antigens and costimulation by ligands for human CD2 and CD28. Immunity 1 (1):57–63
4. Yamada K, Sachs DH, DerSimonian H (1995) Human anti-porcine xenogeneic T cell response. Evidence for allelic specificity of mixed leukocyte reaction and for both direct and indirect pathways of recognition. J Immunol 155(11):5249–5256
5. Cozzi E, White DJ (1995) The generation of transgenic pigs as potential organ donors for humans. Nat Med 1(9):964–966
6. Mollnes TE, Fiane AE (2003) Perspectives on complement in xenotransplantation. Mol Immunol 40(2–4):135–143
7. Adams DH, Kadner A, Chen RH, Farivar RS (2001) Human membrane cofactor protein (MCP, CD 46) protects transgenic pig hearts from hyperacute rejection in primates. Xenotransplantation 8(1):36–40
8. Buhler L, Yamada K, Kitamura H et al (2001) Pig kidney transplantation in baboons: anti-Gal (alpha)1-3Gal IgM alone is associated with acute humoral xenograft rejection and disseminated intravascular coagulation. Transplantation 72(11):1743–1752
9. Kolber-Simonds D, Lai L, Watt SR et al (2004) Production of alpha-1,3-galactosyltransferase null pigs by means of nuclear transfer with fibroblasts bearing loss of heterozygosity mutations. Proc Natl Acad Sci U S A 101 (19):7335–7340. https://doi.org/10.1073/pnas.0307819101
10. Yamada K, Yazawa K, Shimizu A et al (2005) Marked prolongation of porcine renal xenograft survival in baboons through the use of alpha1,3-galactosyltransferase gene-knockout donors and the cotransplantation of vascularized thymic tissue. Nat Med 11(1):32–34. https://doi.org/10.1038/nm1172
11. Chen G, Qian H, Starzl T et al (2005) Acute rejection is associated with antibodies to

non-Gal antigens in baboons using Gal-knockout pig kidneys. Nat Med 11 (12):1295–1298. https://doi.org/10.1038/nm1330

12. Benda B, Karlsson-Parra A, Ridderstad A, Korsgren O (1996) Xenograft rejection of porcine islet-like cell clusters in immunoglobulin- or Fc-receptor gamma-deficient mice. Transplantation 62(9):1207–1211
13. Sandberg JO, Benda B, Lycke N, Korsgren O (1997) Xenograft rejection of porcine islet-like cell clusters in normal, interferon-gamma, and interferon-gamma receptor deficient mice. Transplantation 63(10):1446–1452
14. Shin JS, Kim JM, Kim JS et al (2015) Long-term control of diabetes in immunosuppressed nonhuman primates (NHP) by the transplantation of adult porcine islets. Am J Transplant 15 (11):2837–2850. https://doi.org/10.1111/ajt.13345
15. Shin JS, Min BH, Kim JM et al (2016) Failure of transplantation tolerance induction by autologous regulatory T cells in the pig-to-non-human primate islet xenotransplantation model. Xenotransplantation 23(4):300–309. https://doi.org/10.1111/xen.12246
16. Zhao Y, Swenson K, Sergio JJ, Arn JS, Sachs DH, Sykes M (1996) Skin graft tolerance across a discordant xenogeneic barrier. Nat Med 2(11):1211–1216
17. Nikolic B, Gardner JP, Scadden DT, Arn JS, Sachs DH, Sykes M (1999) Normal development in porcine thymus grafts and specific tolerance of human T cells to porcine donor MHC. J Immunol 162(6):3402–3407
18. Rodriguez-Barbosa JI, Zhao Y, Barth R et al (2001) Enhanced CD4 reconstitution by grafting neonatal porcine tissue in alternative locations is associated with donor-specific tolerance and suppression of preexisting xenoreactive T cells. Transplantation 72(7):1223–1231
19. Haller GW, Esnaola N, Yamada K et al (1999) Thymic transplantation across an MHC class I barrier in swine. J Immunol 163 (7):3785–3792
20. Yamada K, Shimizu A, Ierino FL et al (1999) Thymic transplantation in miniature swine. I. Development and function of the "thymokidney". Transplantation 68(11):1684–1692
21. Yamada K, Shimizu A, Utsugi R et al (2000) Thymic transplantation in miniature swine. II. Induction of tolerance by transplantation of composite thymokidneys to thymectomized recipients. J Immunol 164(6):3079–3086
22. Yamada K, Vagefi PA, Utsugi R et al (2003) Thymic transplantation in miniature swine: III. Induction of tolerance by transplantation of composite thymokidneys across fully major histocompatibility complex-mismatched barriers. Transplantation 76(3):530–536. https://doi.org/10.1097/01.Tp.0000080608.42480.E8
23. LaMattina JC, Kumagai N, Barth RN et al (2002) Vascularized thymic lobe transplantation in miniature swine: I. Vascularized thymic lobe allografts support thymopoiesis. Transplantation 73(5):826–831
24. Kamano C, Vagefi PA, Kumagai N et al (2004) Vascularized thymic lobe transplantation in miniature swine: thymopoiesis and tolerance induction across fully MHC-mismatched barriers. Proc Natl Acad Sci U S A 101 (11):3827–3832. https://doi.org/10.1073/pnas.0306666101
25. Nobori S, Shimizu A, Okumi M et al (2006) Thymic rejuvenation and the induction of tolerance by adult thymic grafts. Proc Natl Acad Sci U S A 103(50):19081–19086. https://doi.org/10.1073/pnas.0605159103
26. Postrach J, Bauer A, Schmoeckel M, Reichart B, Brenner P (2012) Heart xenotransplantation in primate models. Methods Mol Biol 885:155–168. https://doi.org/10.1007/978-1-61779-845-0_10
27. Tanabe T, Watanabe H, Shah JA et al (2017) Role of intrinsic (graft) versus extrinsic (host) factors in the growth of transplanted organs following allogeneic and xenogeneic transplantation. Am J Transplant Off J Am Soc Transplant Am Soc Transplant Surg 17 (7):1778–1790. https://doi.org/10.1111/ajt.14210
28. Rivard CJ, Tanabe T, Lanaspa MA et al (2018) Upregulation of CD80 on glomerular podocytes plays an important role in development of proteinuria following pig-to-baboon xeno-renal transplantation—an experimental study. Transpl Int 31(10):1164–1177. https://doi.org/10.1111/tri.13273
29. Tena AA, Sachs DH, Mallard C et al (2017) Prolonged survival of pig skin on baboons after administration of pig cells expressing human CD47. Transplantation 101(2):316–321. https://doi.org/10.1097/TP.0000000000001267
30. Yamada K, Gianello PR, Ierino FL et al (1997) Role of the thymus in transplantation tolerance in miniature swine. I. Requirement of the thymus for rapid and stable induction of tolerance to class I-mismatched renal allografts. J Exp Med 186(4):497–506

Chapter 12

Xenogeneic Lung Transplantation Models

Lars Burdorf, Agnes M. Azimzadeh, and Richard N. Pierson III

Abstract

Study of lung xenografts has proven useful to understand the remaining barriers to successful transplantation of other organ xenografts. In this chapter, the history and current status of lung xenotransplantation will be briefly reviewed, and two different experimental models, the ex vivo porcine-to-human lung perfusion and the in vivo xenogeneic lung transplantation, will be presented. We will focus on the technical details of these lung xenograft models in sufficient detail, list the needed materials, and mention analysis techniques to allow others to adopt them with minimal learning curve.

Key words Xenotransplantation, Xenogeneic, Pulmonary xenograft, Lung transplantation, Lung perfusion

1 Introduction

1.1 Perspective

Solid organ xenograft survival and function have improved over the last decade in association with successful generation of genetically modified pigs [1]. Development of α1,3-galactosyltransferase knockout (GalTKO) pigs has helped to overcome the preformed-antibody-mediated hyperacute rejection [2] that porcine organs exhibit when transplanted into nonhuman primates [3]. Expression of human complement pathway regulatory proteins (hCPRP), such as hCD46, hCD55, or CD59, controlled the activation of complement which occurs usually within a very short time with wild-type or GalTKO organs and significantly delays the organ injury [4]. Similarly, human transgenes controlling the coagulation cascade activation and the inflammation response have further improved the outcome in xenogeneic perfusion and transplantation [5]. Despite these genetic modifications, pig lungs activate and sequester platelets and neutrophils and activate the coagulation cascade, when perfused ex vivo with human blood or transplanted into baboons. These phenomena illustrate that significant major hurdles remain to be overcome before reaching clinical application of lung xenotransplantation. Further, we suggest that the processes

Cristina Costa (ed.), *Xenotransplantation: Methods and Protocols*, Methods in Molecular Biology, vol. 2110,
https://doi.org/10.1007/978-1-0716-0255-3_12,

causing lung xenograft injury may be important to control in order to obtain improved results with other organ xenografts.

1.2 Historical and Scientific Background

The first experience with xenogeneic pulmonary organs was reported in 1968 by Bryant et al. [6]. In these experiments, an ex vivo apparatus integrating porcine lungs was used to oxygenate human blood, with the idea that they might be used for patients undergoing cardiac surgical procedures. The lungs rapidly exhibited increased pulmonary vascular resistance and pulmonary edema. Since those early days, many groups have conducted multiple studies in cardiac, kidney, and islet xenotransplantation. In contrast, only a few groups have evaluated pulmonary xenotransplantation (historical groups; Michler, Machiarini, Davis, Fodor, Pierson/Azimzadeh). Whether as a cause or a consequence, lung xenograft outcomes remain relatively poor.

One unusual feature of the porcine lung xenograft is the presence of pulmonary intravascular macrophages (PIMs), which we [7, 8] and others [9] have shown to elaborate thromboxane and vWF multimers and are associated with severe constriction of the pulmonary vasculature. In addition, lung xenografts are associated with rapid disappearance of platelets and neutrophils from the circulation even when anti-pig antibody and complement activation are potently inhibited. We believe that both of these phenomena are simply more readily apparent for the lung than for other organs due to the relatively high number of PIMs and the very large and highly vasoreactive endothelial surface area in the lung. Therefore, although lungs have played a relatively minor role in the field of xenotransplantation to date, in our estimation, lung xenograft studies offer a unique, valuable opportunity to contribute to future progress in the field.

1.3 Animal Models

To study the outcome of pulmonary xenografts or drugs and methods that might improve the function of those organs, different animal models have been developed.

1.3.1 Rodent Models

The use of rodents as organ donor and recipient offers the possibility to conduct studies in a relatively rapid, high-throughput, and cost-effective fashion. Another advantage lies in the reduced amount of pharmacologic or molecular reagents needed to perform mechanistically informative studies when compared to larger animals and thus lower costs to perform definitive experiments.

Schroeder et al. [10] published a study in 2003 in which lungs from male wild-type mice were perfused ex vivo with human blood. After the removal of the heart-lung bloc, the main pulmonary artery was cannulated with an intravenous catheter. For the experiment, the lungs were ventilated with room air. During the first 15 min, lungs were flushed with 5% human albumin in saline at a continuous flow rate of 0.2–0.5 mL per min. Thereafter, the albumin solution was replaced by the human blood perfusate at unchanged flow conditions. As a control, lungs were also perfused with autologous blood.In vivo orthotopic left lung xenotransplantation models from hamster or guinea pig to rat have also been performed [11–14]. Orthotopic mouse lung allograft models have more recently been developed that can take advantage of the multiple genetic modifications to the mouse, including GalTKO mice that could be used as recipients for theoretically "discordant" Gal-expressing wild-type lungs [15]. To the best of our knowledge, this model has not been developed for the study of lung xenotransplantation.

However, the relevance of these rodent models to the clinically relevant pig-to-human combination is doubtful, in particular because many approaches (genetic modification to the pig) and reagents (e.g., antihuman or anti-pig monoclonal antibodies) that are developed for clinical use do not cross-react with rodents.

1.3.2 Concordant Large Animal Models

Fox-to-dog [16] as well as macaque monkey-to-baboon [17] orthotopic lung transplantations have been performed in the past, using either no immunosuppression or FK506. Another series of concordant transplantations was conducted by Sadeghi et al. [18] who transplanted cynomolgus monkey heart-lung blocs into baboons. Recipient animals received cyclosporine and steroid-based immunosuppression.

1.3.3 Large Animal Models of Discordant Lung Xenotransplantation

In 1998, a cross-circulation lung perfusion model was described as a method to deplete anti-GalT antibodies from the blood of the "recipient" animal [19]. The baboon underwent thoracotomy and anticoagulation, and an "outflow" cannula was placed in the inferior vena cava. Via a roller pump, the blood was perfused through the main pulmonary artery of the porcine lung, drained by gravity into a heated reservoir, and returned to the baboon via a venous cannula in the right atrial appendage (Fig. 1).

With the same aim to effect immunoabsorption of anti-GalT antibodies, Macchiarini et al. [20] performed experiments in which the recipient animal (goat) was used to cross-circulate the right lung from a GalT-expressing pig prior to receiving the left lung of the same donor animal in an orthotopic transplantation.

Pigs have been identified as the most feasible source of organs for use in humans. Thus, pig-to-human and pig-to-primate models are now preferred because they closely simulate the intended application in the clinic. Importantly, these models permit evaluation of reagents that interact with pivotal pig proteins or with key human and nonhuman primate molecules. In particular, pig-to-primate orthotopic left lung xenotransplantation and ex vivo perfusion of porcine lungs with fresh human blood are currently the most widely used to study discordant pulmonary xenograft injury [21, 22].

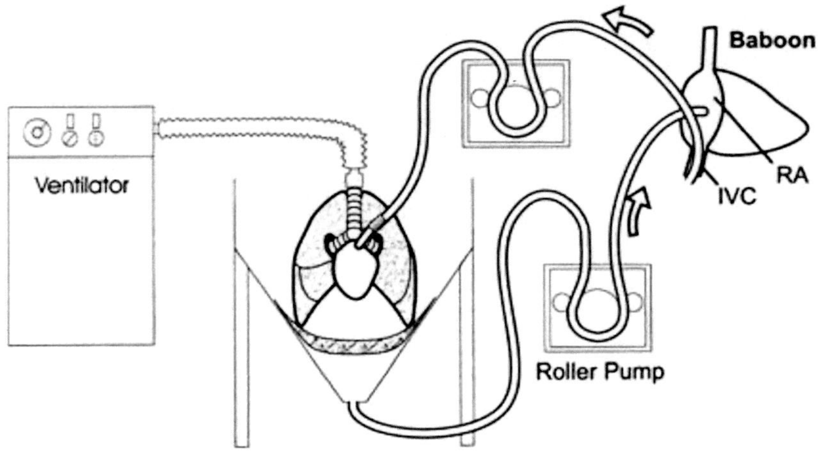

Fig. 1 Diagram of extracorporeal lung perfusion circuit used by Daggett et al. [19]. Blood is removed from the baboon inferior vena cava and delivered to the swine main pulmonary artery with a roller pump. Blood is drained into a reservoir and returned to the baboon with a second roller pump

In the following paragraphs, these two models will be reviewed and explained in detail.

2 Materials

2.1 Drugs and Medications

1. Calcium chloride, 100 mg/mL.
2. Dopamine HCL inj., 40 mg/mL.
3. Heparin, 1000 IU/mL.
4. Human albumin 5%, 250 mL.
5. Isoflurane.
6. Ketamine HCI inj. 100 mg/mL.
7. Noradrenalin (norepinephrine), 1 mg/mL.
8. Perfadex (Xvivo Perfusion AB, Gothenburg, Sweden).
9. Propofol (Abbott Laboratories, North Chicago, IL 60064, USA).
10. Vasopressin inj. 10 U/0.5 mL.
11. Xylazine (AnaSed Injection), 20 mg/mL.
12. 8.4% Sodium bicarbonate inj.1 mEq/mL.
13. 0.9% Sodium chloride.
14. 1-Benzylimidazole (1-BIA)
15. Famotidine, Benadryl.

2.2 Ex Vivo Lung Perfusion Circuit and Accessories

1. Glass beaker (Ace Glass Incorporated, Vineland, NJ 08360, USA).
2. Blood reservoir 1000 mL (Ace Glass Incorporated, Vineland, NJ 08360, USA).
3. Latex tubing 3/8″.
4. Y-3/8″ × 3/8″ × 3/8″ blood tubing connector.
5. Straight metal tubing-PA adapter (custom-made).
6. Metal rig.
7. Blood pump (Polyston A/S, Vaerlose, Denmark).
8. Stirrer (Thermolyne Nuova II, Dubuque, ID 52001, USA).
9. Reservoir-water heater (American Pharmaseal Company RK 300, Valencia, CA 91355, USA).
10. Biological atmosphere mixture gas (Airgas Inc., Salem, NH 03079, USA).
11. Human blood (freshly collected) in blood collection bags containing CPDA-1 (Terumo, Somerset, NJ 08873, USA).
12. Fresh-frozen human plasma (two bags per lung side).

2.3 Monitoring and Anesthesia

1. Isoflurane vaporizer.
2. Endotracheal tube (ETT) and mask.
3. Ventilator.
4. Veterinary anesthesia ventilator.
5. Flowmeter.
6. Pressure monitoring.
7. Pressure transducer.
8. Monitoring line.
9. Data acquisition program (LabChart Pro, ADInstruments, Colorado Springs CO 80906, USA).
10. Syringe pumps.
11. Biological atmosphere mixture gas in tank.
12. 95% oxygen, 5% carbon dioxide in tank.

2.4 Sutures, Catheters, and Disposables

1. Prolene 4-0, 5-0, 6-0, Silk 0 and #1 ties (Ethicon, Somerville, NJ 08876, USA).
2. Vessel loops VL-205 (Key Surgical Inc., Hannover, Germany).
3. Pediatric venous cannula 12–20 Fr (Medtronic Perfusion Systems, Brooklyn Park, MN 55428, USA).
4. Swan-Ganz 7.5F catheter (Edwards Lifesciences, Irvine, CA 92614, USA).
5. Arrow-flex polyurethane sheath catheter (Arrow, Reading, PA 19605, USA).

6. Angiocatheter 18-22G (Jelco, Medex Medical Ltd., Rossendale, Lancashire, UK).
7. Tracheal tubes 3.5–7.0 mm.
8. BD Vacutainer (EDTA, Serum, CTAD) (Becton Dickinson, Franklin Lakes, NJ 07417, USA).
9. Ethyl alcohol 190 Proof, 95%.

2.5 Equipment and Reagents for Analyses and Assays

1. Computer linked to the detectors (Transonic Systems Inc., Model T206, Ithaca, NY, USA) to record flow and pressure continuously.
2. Automated blood cell analyzer.
3. Flow cytometer and antibodies for P-selectin and CD41 for measuring platelet activation at the cellular level.
4. Asserachrom-βTG enzyme-linked immunosorbent assay (ELISA) from Diagnostica Stago (Parsippany, NJ, USA) to assess platelet activation by beta-thromboglobulin (βTG) level formation.
5. Enzygnost micro F1+2 ELISA (Dade Behring, Marburg, Germany) for measuring thrombin formation.
6. C3a and/or C5a ELISA (Quidel, San Diego, CA, USA).
7. Cytokine Luminex arrays (Luminex Corporation, Austin, TX 78727, USA) for measuring cytokines, chemokines, and other molecules of interest.

3 Methods (*See* Note 1)

3.1 Ex Vivo Porcine-to-Human Lung Perfusion Model

Ex vivo xenogeneic perfusion of a porcine lung using human donor blood is the most clinically relevant and thus important model than can be used to assess the impact of interventions on short-term outcomes (within 6–8 h). It closely replicates the kinetics, physiology, and immunohistochemical findings seen during in vivo pig lung transplantation in a nonhuman primate. All changes that occur during in vivo hyperacute rejection (HAR), such as increased pulmonary vascular resistance (PVR), loss of transpulmonary blood flow, failure of oxygenation, and vascular barrier function injury, leading to tracheal or lung edema, are observed in both models. Only perfusion of pig lungs with whole human blood offers the possibility to perform mechanistic studies in the clinically relevant species combination and to evaluate interaction of formed blood elements across species. Functional parameters as well as blood and tissue samples can be collected at pre-specified intervals over a specified time range to study the mechanisms of rejection (*see* **Note 2**).

It is possible to perfuse one lung, both lungs en bloc, or to divide the pulmonary bloc into left and right lungs and perfuse the "pair" separately. We primarily utilize the paired model because the impact of modifications to one circuit can be assessed in the context of a contemporaneous contralateral reference ("control") experiment where only the intervention under study differs between the two circuits. This approach allows biologically significant differences to become more readily apparent, especially given biologic variability among pigs, and since uncontrolled variables (such as anti-pig antibody titer) could otherwise exert confounding influences on experimental outcome and complicate study interpretation (*see* **Note 3**).

3.1.1 Donor Organ Procurement

1. Premedicate the pig with intramuscular ketamine and xylazine and isoflurane by mask (*see* **Note 4**).
2. Intubate through a midline surgical tracheostomy, and insert an endotracheal tube (ETT) of appropriate size secured with multiple heavy ties (*see* **Note 5**).
3. Ventilate and anesthetize the pig with inhaled isoflurane through the ETT.
4. Open the chest with a median sternotomy contiguous with the tracheostomy incision.
5. Obtain vascular access peripherally or by cannulating a vessel in the surgical field with IV tubing (*see* **Note 6**).
6. After heparinization (*see* **Note 7**), introduce a large angiocatheter (>14 gauge) or an angulated cannula (aortic or venous cannula, 12–20 Fr) into the pulmonary trunk through a Prolene purse string to introduce the flush solution.
7. Administer premedication (e.g., prostacyclin).
8. Ligate the right superior and inferior venae cavae and left superior vena cava, and cross-clamp the ascending aorta, before administration of the flush solution (*see* **Note 8**).
9. Make an incision in the left atrium, or amputate the left ventricular apex, to allow free drainage of the pulmonary vein effluent.
10. After administration of the flush solution, inflate the lungs, place a clamp on the trachea or endotracheal tube, and explant the lungs en bloc with the heart, removing the trachea, which for convenience can be done including the tracheostomy site with the ETT in place.
11. If the experimental design calls for perfusion of both lungs at the same time, secure a return cannula into the left ventricle (it can be, but need not be, introduced across the mitral valve into left atrium), with umbilical tape gently but securely tightened around the LV apex.

12. Establish pulmonary artery (PA) perfusion via the angulated cannula in the pulmonary trunk through which the flush solution is administered, with care to preserve competence of the pulmonic valve during subsequent perfusion.
13. Alternatively (and more commonly) introduce a flanged metal or plastic cannula through an incision in the right ventricular outflow tract, through the pulmonary valve, and secure with a heavy tie around the main pulmonary artery distal to the pulmonoplegia site. Cannula sizes and dimensions take into account the size of the vessels to be cannulated (which depend on pig size) and the size of tubing used for the perfusion circuit (typically ¼″ (6.4 mm) or 3/8″ (9.5 mm)).
14. Insert an endotracheal tube into the trachea at a convenient cervical location (safely above the high take-off of the right upper lobe "pig" bronchus), and attach securely with umbilical tapes or a strong silk ligature if the tracheostomy tube from the procurement is not transferred to the perfusion apparatus.
15. Put the gently inflated lungs (to avoid barotraumas) with the ETT clamped on ice until the blood for the perfusion has been prepared and the experiment can be started.
16. For the "paired" perfusion system (side-by-side individual right and left lungs), separate surgically the lungs as follows. Because the right main pulmonary artery is relatively short, cannulate it before dividing the left pulmonary artery, securing the right PA cannula with a heavy tie just distal to the main PA bifurcation (*see* **Note 9**).
17. Then, divide the left pulmonary artery just past the main PA bifurcation, usually leaving sufficient length to secure a cannula (e.g., stainless-steel cannula) with a simple heavy silk suture (*see* **Note 10**). The PA cannulas will later be connected to the inflow perfusion tubing.
18. Insert a size-matched balloon-cuffed ETT into the cervical trachea as described above. The trachea must be left with the right lung due to origination of the right upper lobe ("pig bronchus") from the trachea.
19. Divide the left main bronchus flush with the trachea, and close the resulting tracheal defect with a continuous 4-0 or 5-0 Prolene suture.
20. Cannulate the left main bronchus directly, securing the endotracheal tube (3.0–6.0 mm) with its tip just proximal to the left main bronchial bifurcation with a 4-0 Prolene suture, taking multiple partial-thickness bites in the outer airway, tieing the suture tightly, and then securing the same ligature over the ET tube (we prefer to use an ET tube with a balloon, to assure secure fixation of the ETT). Thus, inflate the tube balloon gently.

21. To prevent the ETT from being dislodged, extend the bronchial Prolene suture proximally along the ETT, and additionally tie the suture around the ETT just proximal to the proximal end of the balloon.
22. For storage, inflate each lung gently (avoiding barotraumas), clamp the ETT, place the lung in a plastic bag, and cool on water or saline ice.

3.1.2 Human Blood Preparation

The preparation of the human blood used for the perfusion of the porcine lungs can be done in various ways. It is essential to use some freshly collected blood from healthy donors in order that the platelet and neutrophil function simulate in vivo biologic phenomena (*see* **Note 11**).

1. Use regular clinical blood banking collection bags containing CPDA-1 to collect the fresh blood (~450 mL/one blood unit). Depending on the size of the pigs and thus the size of the lungs, it needs to be determined how many individual donors are needed (*see* **Note 12**).
2. In the "paired" lung model, apportion the blood between the circuits in approximate proportion to relative pig lung weight (60% right and 40% left can be assumed, in our experience).
3. To prevent clotting of the blood in the circuit while reconstituting calcium-dependent physiologic mechanisms (the coagulation and complement pathways are both critically dependent on calcium), add heparin (3 IU/mL blood-plasma mixture) prior to adding calcium chloride (1.6 mg/mL blood-plasma mixture) to neutralize the CPDA chelating agent. The activated clotting time (ACT) should be above 500 s. If it is below, more heparin should be added.
4. If necessary, add sodium bicarbonate to obtain a target pH of 7.4.

3.1.3 Organ Perfusion

The system described permits perfusion of single lungs or lungs en bloc in a simple, safe, highly reproducible manner (Fig. 2). The perfusion circuit can of course be modified for other purposes (e.g., to study ischemia-reperfusion injury) or for ex vivo lung allograft experiments.

1. Clean the tubing system and any other system components that will have contact with the blood, rinse extensively with clean water, disinfect (ethyl alcohol 95%), and dry overnight, before blood is introduced into the perfusion circuit.
2. Coat all blood-contacting circuit surfaces with human albumin 5% in 0.9% "normal" saline to minimize blood activation by contact with recycled glass, plastic, and silicon circuit components (*see* **Note 13**).

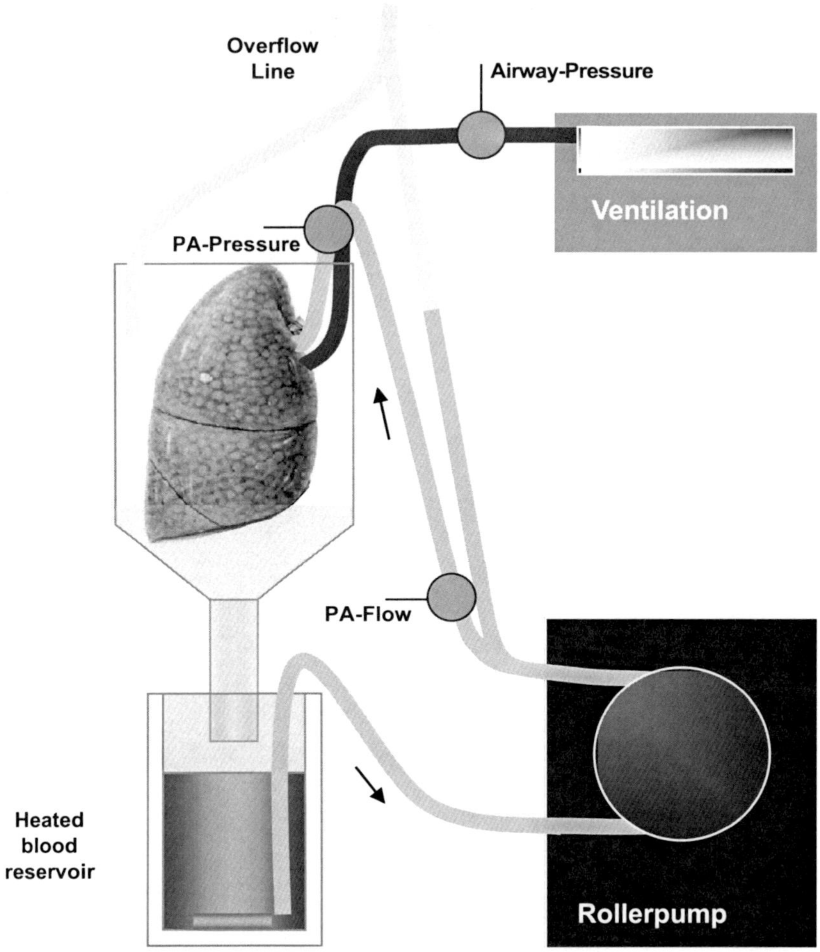

Fig. 2 Diagram of the ex vivo lung perfusion circuit. Blood is pumped via silicon tubing into the pulmonary artery by a roller pump. After the perfusion of the lung, the blood is drained into a heated blood reservoir from which it is recirculated to the lung

3. Suspend the pig lung in a funnel over the jacketed glass beaker that will collect the blood issuing from the pulmonary veins.
4. Warm the reservoir beaker with a circulating water pump to 37 °C.
5. Pump blood from the reservoir via the tubing system, and return to the lung via the PA (*see* **Note 14**).
6. Connect an appropriately adjusted volume-controlled ventilator to the endotracheal tube and ventilate the lung with a custom-ordered gas mixture of 21% oxygen and 10% carbon dioxide balanced with nitrogen. This concentration of inhaled carbon dioxide maintains the blood pCO_2 at 35–45 mmHg and prevents the profound alkalosis that would otherwise occur since the lung is not a significant source of CO_2, which is derived mainly from tissue metabolism.

7. To evacuate air from the blood-containing tubing, pour the prepared human blood into the reservoir and start the roller pump.
8. De-air the circuit and connect the inflow tubing to the pulmonary artery. The flow setting of the pump should initially be set to 150 mL/min and adjusted subsequently (*see* **Note 15**).
9. For side-by-side bilateral lung perfusion, devote 55–60% of blood volume and flow to the right lung and 40–45% to the left side, based on the normal proportion of relative lung weight.
10. Ventilate the lungs at 7 mL/kg BW and adjust ventilator tidal volume according to the lung compliance, seeking to achieve peak airway pressure of 12–15 cm H_2O (~8–11 mmHg). It is advisable to use a PEEP valve with a setting of 5 cmH_2O.
11. To record physiologic lung function parameters, use pressure transducers to record PA and airway pressures and a flow probe to continuously monitor the blood flow into the lung in the inflow tubing just proximal to the pulmonary artery. Calculate the PVR by dividing mean PA pressure by blood flow. To facilitate the recording of the parameters, a PC-based data acquisition program can be used.
12. Collect blood samples directly from the pulmonary vein effluent.
13. Excise lung biopsies behind tissue clamps; leave the tissue clamps in place for the duration of the experiment, suspended from the scaffolding to prevent distortion of the lung anatomy by the weight of the clamp (tissue stapling devices could be used, but add considerable expense).
14. Trisect each lung biopsy and process for later histological, molecular, and immunohistological analyses.
15. Assess lung function by blood gas analyses, measuring pO_2, pCO_2, and pH at scheduled intervals over time under consistent circumstances. By intermittently switching from the ventilator gas mixture (21% O_2) to 100% oxygen for 3–5 min, measure the capacity of lung oxygenation. If preferred, an oxygenator swept with 8% CO_2 and 10–14% O_2 balanced with nitrogen can be integrated into the circuit (*see* **Note 16**). This approach simplifies assessment of lung oxygen transfer capacity by step-up of pO_2 during stable ventilation with room air, but it adds the cost of the oxygenator and its potentially confounding effects on blood.

3.2 In Vivo Porcine Lung Transplantation Model

As for other xenogeneic organ transplants, transplantation of a porcine pulmonary graft into a nonhuman primate represents the gold standard to test lung xenograft function, to observe blood and lung tissue parameter changes, and to determine survival of the graft and recipient in a clinically relevant "translational" xenogeneic model. Even though the ex vivo perfusion model facilitates informative mechanistic studies, it will never offer the possibility to discover perhaps unexpected phenomena that may occur during attempted life support in a living animal and allow monitoring of perturbations in the physiology of the recipient. Importantly, the in vivo model is not perturbed by the artificial elements of the ex vivo circuit (e.g., blood tubing and blood-air interactions) or by unappreciated or inconsistent effects of "pooling" multiple human donors.

Although studies in cynomolgus monkeys have been reported [23], most in vivo lung xenotransplant work has been done in baboons, where adult animals are of appropriate size (>10 and preferably >15 kg) to accept left lung grafts from 6 to 14-kg pigs [24–26] (*see* **Note 17**).

3.2.1 Donor Organ Procurement

1. Perform the anesthesia of the donor pig and surgical procurement of the heart-lung bloc as described in the ex vivo lung perfusion section.
2. For the in vivo lung transplantation, separate the left porcine lung surgically from the heart.
3. Excise the pulmonary veins as an island from the left atrium.
4. Because there is often a large distance between the upper and lower left pulmonary veins, imbricate the intervening left atrial wall to reduce the donor vein cuff size to more closely approximate the smaller baboon pulmonary vein cuff.
5. Divide the PA just proximal to the left main pulmonary artery bifurcation (*see* **Note 18**).
6. Finally, clamp the left main bronchus to ensure that the lung stays inflated and is cut close to the bifurcation of the trachea.
7. Store the lung on ice until the implantation of the organ.

3.2.2 Preparation of the Recipient

1. Sedate baboon recipients with ketamine (10 mg/kg BW, i.m.).
2. After the intubation, maintain general anesthesia with isoflurane (1–3%, i.h.).
3. For initial intravenous access, place an angiocatheter in a superficial extremity vein.
4. Place an arterial line in the femoral artery for monitoring of hemodynamic parameters (*see* **Note 19**).
5. If desired, prepare the femoral vein and cannulate directly with an inducer sheath for a later Swan-Ganz catheter positioning.

6. To prevent blood clotting at the catheters, give heparin as a bolus (70 IU/kg BW, i.v.) and infuse at 200 IU/h continuously. Flush lines with heparinized saline until use.
7. Open the chest of the animal with a transverse "clamshell" incision at the fourth or fifth intercostal space (*see* **Note 20**).
8. Open both the pericardium and the pleura with longitudinal incisions.
9. A flow-directed PA balloon catheter can be introduced into the femoral vein, and manually guide across the tricuspid valve (*see* **Note 21**) and out the pulmonary outflow tract to the proximal PA.
10. Position a snared vessel loop around the right pulmonary artery, working between the aorta and SVC. Securing this loop allows reversible selective perfusion of the transplanted left lung and thus intermittent or continuous rigorous assessment of left lung function.
11. A flow probe (Transonic Systems Inc., Ithaca, NY) can be placed on the ascending aorta to allow measurement of the cardiac output (*see* **Note 22**). To quantify the proportion of cardiac output transiting the xenograft, a second flow probe can be placed around the right pulmonary artery between the aorta and the superior vena cava.

3.2.3 Explantation of the Left Native Baboon Lung

1. For the explantation of the native lung, displace the heart from the pericardium and into the right chest, avoiding hemodynamic embarrassment by incising the right inferior pericardium along the diaphragm. To this end, retract a suture tied on the left atrial appendage toward the right, and protect it with a rubber tube (Rummell snare) from rubbing against the contracting heart. Sutures through the pericardium just anterior to the left hilum, similarly protected with rubber tubing and retracted anteriorly and to the right, aid in exposure with minimal hemodynamic compromise.
2. Working within the pericardium (particularly helpful!) and outside of it, identify and dissect the left hilar structures.
3. Ligate left upper and lower pulmonary veins centrally and peripherally with heavy suture (Silk 0 or #1) and divide.
4. Use a straight vascular clamp to occlude the left PA as close to the mediastinum as possible, and divide the LPA distally just proximal to the left main PA bifurcation. By cutting the recipient PA on a bevel, the anticipated size discrepancy (larger donor PA) can be more easily accommodated.
5. After dividing the inferior pulmonary ligament, divide the left main bronchus at the left main bronchial hilum, and control bronchial and mediastinal vessels with suture ligatures. For

non-survival procedures, where vagus function and durable bleeding control are less essential, cautery may be substituted.

6. Restore ventilation to the right lung by placing a straight clamp on the left main bronchus (*see* **Note 23**).

3.2.4 Implantation of the Xenogeneic Organ

1. Tailor the donor bronchus, pulmonary artery, and vein to accommodate size discrepancies and avoid redundant vascular tissue, which may otherwise kink.
2. Place the cold xenogeneic lung in the left pleura of the recipient baboon, preferably atop an insulating layer.
3. First, anastomose donor and recipient bronchus with a continuous suture (4/0 or 5/0 Prolene). An interrupted 4/0 PDS or other monofilament suture is more time-consuming, but it may be preferred if long-term survival with bronchial healing is expected.
4. The clamp on the left main bronchus can be removed without significant interference in completing the vascular anastomoses because the pig lung intrinsically has reduced compliance relative to the native baboon lung.
5. For the end-to-end anastomosis of the pulmonary artery, use a continuous Prolene 6/0 suture, taking care not to "purse-string" this anastomosis.
6. For later de-airing, keep the anastomosis open (not yet tied) (*see* **Note 24**).
7. To simplify the connection of the pulmonary veins, free up the recipient pulmonary vein and left atrium from the pericardium posteriorly to allow placement of a clamp (Satinsky/C-clamp) proximally on the left atrium.
8. After placing a broad-based vascular clamp across the left atrium, excise the ligatures on the pulmonary veins, preserving maximal recipient cuff diameter. Incise the atrial tissue between the upper and lower recipient vein orifices to obtain one large lumen.
9. Anastomose the donor and recipient left atrial cuffs with a 4/0 or 5/0 Prolene, leaving the completed suture line untied until after de-airing.
10. De-air the transplanted lung by first opening the clamp on the pulmonary veins; tie the venous sutures and assure hemostasis of the venous anastomosis.
11. Open the proximal clamp on the pulmonary artery slowly, to first evacuate air from the PA and then allow controlled low-pressure, low-flow reperfusion of the xenograft.
12. Tie the PA suture after venting, fully establishing blood flow to the xenograft.

13. Assure hemostasis throughout the surgical field and close the thoracotomy.

3.2.5 Longitudinal Graft Assessment

After transplantation of the xenogeneic organ, it is essential to control the hemodynamic parameters and stabilize the recipient. Inotropic medication is almost always required during graft implantation (*see* **Note 25**). In 2007, Nguyen et al. [24] used the described in vivo model to perfuse a porcine lung in a baboon for up to 255 min. At the same laboratory, recent experiments have shown that it is possible to keep the transplanted recipient animal alive under general anesthesia for 30 h (unpublished data). Others have done so for up to 3–5 days [26]. During this time, it is possible to collect not only tissue samples from the xenogeneic lung but to also record ventilation parameters and systemic, pulmonary arterial, and central venous pressures. Flow probes on the ascending aorta and on one pulmonary artery (we prefer the right, to avoid potentially compromising flow to the xenograft) allow monitoring cardiac output and flow to the transplanted lung. By pulling gently on the snared vessel loop on the right pulmonary artery, it is possible to partially or completely occlude this vessel and thereby adjust recipient dependence on the pulmonary xenograft (*see* **Note 26**). During single left lung perfusion, cardiac output, systemic arterial pressure, and systemic arterial oxygen saturation indicate whether the pulmonary vascular resistance and gas transfer of the transplanted left lung are sufficient to support life.

3.2.6 Adapting into a Lung Survival Model

Recently, improved gene editing techniques have significantly accelerated generation of genetically engineered pigs with multiple porcine gene knockouts and expression of various human transgene combinations [27, 28]. The use of these new xenogeneic donor lungs in conjunction with refined immunosuppressive and supportive drug regimens [29, 30] has been associated with life-supportive organ function in most of our recent experiments. A "survival" model, in which the chest is closed and animals are awakened after the procedure, enables study of immunologic responses and survival of the recipient animals [31, 32]. Chest drains with a Heimlich valve placed under the recipient's jacket prevent pneumothorax. Chest closure in multiple layers simulates clinical practice, with primary rib approximation (interrupted or figure-of-eight sutures with heavy Vicryl 1 or similar), and muscle, subcutaneous tissue, and skin are approximated anatomically with absorbable suture. After resuscitation and stabilization, the lung recipient is extubated and monitored in its cage. We use an oxygen-enriched environment (FiO_2 40–50%, e.g., with Plexiglas insulated recovery cage) to assure physiologic O_2 saturation of the blood. To date, the longest survival after pig-to-baboon pulmonary xenotransplantation has been 31 days, achieved by the Pierson laboratory (manuscript in preparation).

3.3 Analyses and Assays

To compare the function associated with different genetics using in vivo or ex vivo xenogeneic lung models, a traditional parameter is the time after transplant or reperfusion until the organ loses its function. For lung xenografts, loss of blood flow through the lung due to increased vascular resistance, failure of blood oxygenation, and occurrence of tracheal edema occur in both models and with similar kinetics when in vivo and ex vivo experiments are performed using lungs from one lung donor; ex vivo, the loss of perfusate volume within the lung xenograft may occur before another "lung failure" criterion is met.

To understand the biology of pig lungs in a preclinical xenogeneic system, we analyze functional and hemodynamic results, as well as hematologic, immunologic, and immunohistologic changes. The described models provide access to these parameters by recording pressures and flows and collecting tissue and blood samples in the run of the experiment.

3.3.1 Functional and Hemodynamic Analyses

1. Measure the function of the organ by the pulmonary vascular resistance (PVR) (*see* **Note 27**) and by the ability of the lung to oxygenate the blood. Especially within the first 30 min of perfusion and also before lung demise, an increase in PVR can be observed. If possible, record those parameters continuously, or in representative intervals during this time. Our preferred approach is to record flow and pressure continuously on a computer linked to the detectors using proprietary software.
2. Measure the lung's ability to transport oxygen into the blood by analyzing the pulmonary vein blood gases after a short time (3–5 min) on 100% FiO_2, record as the amount of step-up in pO_2 relative to pulmonary vein pO_2 on 21% FiO_2, and compare the result to reservoir blood measurements. If a deoxygenator is incorporated into the circuit, this maneuver and calculation of measuring lung gas transport function intermittently during the experiment becomes unnecessary.
3. In the in vivo model, transplant function can intraoperatively be assessed by temporarily occluding the blood flow to the right lung (e.g., with a vessel loop on the right PA). Life-supportive function of the xenolung is confirmed if systemic blood pressure and O_2 saturation in the peripheral blood remain stable during the occlusion. After chest closure, blood flow to the transplanted graft can be confirmed via telemetric flow measurement (Transonic PhysioGear or EndoGear, Ithaca, NY 14850, USA). During the later postoperative time, chest radiographs (X-ray or CT scan), intermittent bronchoscopies, and lung biopsies can provide further information on the xenograft condition (Fig. 3).

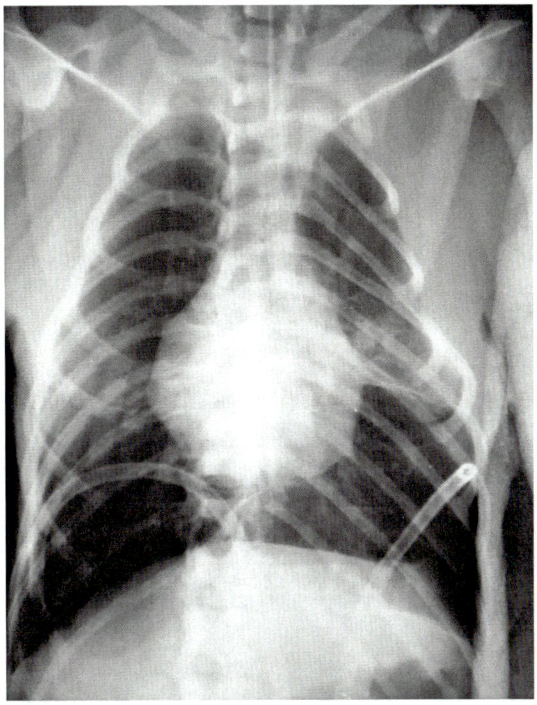

Fig. 3 Routine chest X-ray, taken on postoperative day 2 after left-sided xenogeneic lung transplantation. The radiograph shows a well-ventilated graft without significant interstitial edema, consolidation, or pleural effusion. The recipient baboon was respiratorily and hemodynamically stable. Bilateral chest drains were removed after the X-ray was taken

3.3.2 Hematologic Analyses

With the xenogeneic perfusion of an organ, multiple hematological changes can be observed in the blood perfusate.

1. After establishing blood flow through the organ, collect blood samples. We collect them at 5-, 15-, and 30-min intervals, as sequestration and activation of cells occurs immediately. Thereafter, predetermined intervals of 60 min and then 60–120-min intervals thereafter are sufficient to track trends over time and compare results between groups treated in various ways.
2. For cell counts, enumerate white blood cells, including neutrophils, monocytes, and platelets by standard automated techniques in perfusate/blood samples (*see* **Note 28**).
3. For measuring platelet activation at the cellular level, a very useful parameter is the cell surface expression of P-selectin (CD62P) determined by flow cytometry. To measure CD62P expression on platelets, stain blood samples by monoclonal antibodies specific for CD41 (as a marker for platelets) and CD62P (expressed by activated platelets) according to the instructions of the manufacturer. Identify platelets and platelet aggregates by size and by the presence of CD41 staining, and

analyze later for expression of CD62P. Express results as the percentage of CD62P-positive cells among CD41-positive cells (*see* **Note 29**).

4. To measure platelet activation and coagulation cascade in archived plasma samples, assess platelet activation by beta-thromboglobulin (βTG) level formation using a commercial ELISA (Asserachrom-βTG). Measure thrombin formation as the production of prothrombin fragments 1+2 (F1+2) generated by the cleavage of prothrombin in thrombin by ELISA (*see* **Note 30**).
5. Determine the activation of the complement cascade by measuring C3a or C5a levels in plasma samples collected in EDTA blood tubes and stored in EDTA at −70 °C (C3a and C5a ELISA; Quidel, San Diego, CA).

3.3.3 Cytokine Measurements

As cytokines are signaling molecules, secreted by numerous cells of the immune system, the measurement and level of cytokines can be used to analyze the strength of immune activation and immune response. The analyses may include interleukin 1β (IL-1β), IL-6, IL-8, interferon gamma (IFNγ), TNF, chemokines, soluble CD40L and sCD62P (platelet activation), and other pro- and anti-inflammatory cytokines and biomarkers. New high-throughput technologies (Luminex, arrays) efficiently measure multiple targets simultaneously in a small volume of plasma [33].

3.3.4 Histology and Immunohistology

At various time points throughout the experiments, tissue samples from the xenogeneic organ should be collected to prepare histological slides. A standard staining to visualize cell infiltration of the tissue and vessel thrombosis is the hematoxylin and eosin stain (H&E). Figure 4 shows histologic lung tissue, collected from a transgenic lung on postoperative day 4 after xenogeneic lung transplantation. It demonstrates findings consistent with acute xenolung rejection, illustrating loss of integrity of alveolar endothelial wall structures and massive intra-alveolar hemorrhage. Immunohistochemistry (IHC) is typically performed on snap-frozen lung tissue insufflated with OCT, diluted with saline (~1:1), into the airways through a small-gauge angiocatheter (22–24G) and syringe, to reduce edge artifact. IHC staining includes visualization of human monocytes (CD14), neutrophils (myeloperoxidase), T cells (CD3, CD4, CD8) or B cells (CD20), total (CD41) and activated (P-selectin) platelets, tissue factor, IgG and IgM, complement pathway products (C4d, C5b-9), and vWF (to identify endothelium)

3.3.5 PCR/Gene Expression Analyses

The rejection of an organ does not only affect its function and tissue structure, it also modulates genes in both pig lung and human or NHP blood cells [34]. Snap-frozen tissue samples collected at

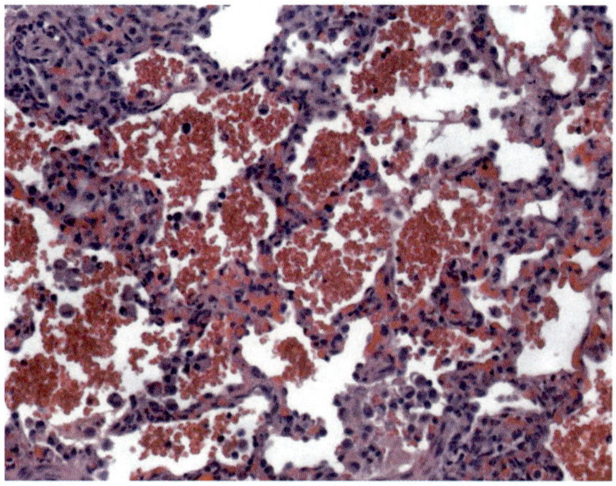

Fig. 4 Example of acutely rejected lung tissue (H&E, ×200). The slide shows massive intra-alveolar hemorrhage and tissue consolidation, correlating with the clinical symptoms of dyspnea and severe tracheal edema

various time points during the experiment can be used to measure gene expression by real-time RT-PCR or cDNA microarray analysis [35]. By designing PCR primers and probes that recognize pig or human sequences only, it is possible to monitor species-specific changes in gene expression within lung tissue parenchymal cells or within human infiltrating cells.

4 Notes

1. All transplant experiments require surgical and anesthesia instruments, equipment, and expertise for the procurement of the organ and, if performed, for the performance and monitoring of the transplant. Involvement of a surgical team trained in transplantation and preferably cardiothoracic surgical techniques is essential to successfully conduct lung xenotransplant research.
2. It is advisable to have short intervals after initiation of blood flow into the lung xenograft (as little as 1–5 min for some parameters) for blood and tissue sampling and to continuously record functional parameters from the beginning of the perfusion. Especially during the first few min, blood (cell sequestration) and functional changes (increase in PVR) are frequently observed.
3. The perfusion system (e.g., flow rate and cannula sizes) and other logistics (e.g., the human blood volume) can be adapted to perfuse lungs of pigs with a bodyweight of 2–50 kg or more.

4. We use a combination of ketamine (10–15 mg/kg i.m.) plus xylazine (0.5–1 mg/kg i.m.) for the anesthesia induction. Anesthesia is maintained with isoflurane (0.5–4.0% inhaled).
5. Ethicon silk 1-0 ties (or equal) can be used. It is possible to suspend the lungs from those ties.
6. Especially in small pigs, it can be difficult to get i.v. access on a peripheral vessel (e.g., ear vein). In that case, it is possible to place an angiocatheter in the internal jugular vein through the midline surgical tracheostomy.
7. We give a heparin dose of 500 IU/kg.
8. We use Perfadex (50 mL/kg BW).
9. If the right PA is not long enough to insert a cannula to allow perfusion of the upper right lobe, it is instead possible to secure the cannula to the main pulmonary artery and then close the hole where the left PA was dissected with suture. It is essential to assure that the closure suture does not obstruct the vessel as this would significantly reduce the blood flow to the lung and increase the measured PA pressure. We use a Prolene 5-0 double layer mattress/simple continuous suture.
10. As mentioned for the ties that secure the ETT to the airway, we recommend using the ties on the PA catheter/connector as well to suspend the lungs from the aluminum rig framework. This reduces the weight of the lung on any single anatomic structure and prevents PA vessel kinking.
11. Examples in the literature using non-diluted blood [33, 36] or blood diluted with blood-type-matched human plasma [37] for xenogeneic lung perfusions have been conducted. Unpublished work shows that dilution of blood with electrolyte solutions or use of refrigerated or overnight-stored blood significantly attenuates the pace of hyperacute heart or lung rejection.
12. For pigs <20 kg, we utilize two units (two individual donors, each contributing 450 mL) of blood expanded with four 240-mL individual donor "units" of thawed "recovered" or "fresh-frozen" plasma. One fresh human donor and two plasma "units" are usually sufficient for an "en bloc" experiment.
13. While some activation of complement and adhesion of formed blood elements is demonstrable in our circuit, this effect is readily controlled for by measuring these parameters after a consistent (20 min) interval of perfusate circulation through the circuit. Blood perfused for longer intervals or in the context of iso- or allo-lung perfusion exhibits little additional alteration over relevant study intervals (unpublished observations).

14. Our circuit design incorporates an overflow line that allows the blood to continue to mix and circulate while bypassing the lung when high pulmonary vascular resistances occur, to avoid PA pressures above 45 cmH_2O (equivalent to 4.4 kP or 33 mmHg) and avoid stasis in the tubing. This peak PA pressure is chosen to simulate the maximum pressure that we presume a normal right ventricle could overcome for a sustained period in vivo.
15. In our experiments, we have seen that the PA flow can show a wide range of values. Certainly, the genetics of the donor pig (e.g., wild-type, GalTKO, or multi-transgenic pig) and also other parameters (ischemia-reperfusion) have influence on the blood flow and contribute to the variability in pulmonary vascular resistances in different experiments.
16. An oxygenator adds substantial expense, and it is a source of significant blood activation.
17. It is important to match for size donor and recipient not only for physiologic but also surgical reasons. The main bronchus of the pig has usually a wider lumen than the primate bronchus. To facilitate the bronchus anastomosis, it is thus important to use a donor pig which has a 10–20% lower bodyweight than the recipient baboon.
18. A pig PA 50% larger than the baboon left main PA should be anticipated and managed using standard surgical tailoring approaches.
19. Arterial lines can be inserted into both femoral arteries. Having two lines in place allows the collection of arterial blood (e.g., for arterial blood gas measurements) without losing the real-time invasive blood pressure measurement for the time of the drawing which can be crucial in animals with unstable or critical condition.
20. Alternatively, a lateral thoracotomy can be used for the lung transplantation procedure. This access is surgically less invasive but does however not allow to test the life-supporting function of the transplanted graft by temporarily manually occluding the right PA.
21. If utilized, we suggest advancing the PA balloon-directed catheter after opening the chest, as placement of the catheter often requires manual guidance across the tricuspid and pulmonic valves.
22. Obtaining a stable ascending aorta flow signal using the flow probe can be difficult; ultrasound gel improves the signal in most cases.
23. Although, only the right lung is ventilated during the implantation of the left lung. As in clinical practice, it is usually not

necessary to change the settings of the ventilation machine. During the single-lung ventilation, arterial blood gases should be collected to assure an appropriate ventilation of the recipient.

24. We typically place a clamp on the donor PA to facilitate alignment, position it adjacent to the recipient LPA, and avoid tension on the anastomosis during its performance and thus prevent stricture.
25. In our experiments, we use dopamine (2–20 μg/kg/min), noradrenalin (2–20 μg/kg/min), and vasopressin (0.2–1.0 IU/min). Additionally, we also routinely use inhaled nitric oxide to minimize pulmonary vascular resistance during graft implantation and subsequent assessment; the utility of this approach has not been rigorously assessed.
26. We recommend to perform the life-supporting xenogeneic lung maneuver only after 30–45 min of reperfusion time and to occlude the right PA only partly at first. The transplanted lung initially shows a higher vascular resistance and a lower compliance. By completely occluding the right PA shortly after the implantation, this could lead to a cardiac decompensation with acute right heart failure.
27. We estimate the PVR by dividing the pulmonary arterial pressure by the pulmonary flow.
28. Samples are collected in ethylenediaminetetraacetic acid (EDTA) anti-coagulated blood tubes.
29. The blood can be fixed immediately after collection to avoid nonspecific platelet activation during sample processing and to allow simultaneously antibody staining of all time points at the same time. Additional markers (such as glycoprotein V or activation of GPIIb/IIIa receptor) can also be analyzed in a similar way.
30. Importantly, plasma samples used in coagulation studies must be collected in CTAD tubes and processed according to the manufacturer's instructions.

References

1. Pierson RN 3rd, Dorling A, Ayares D et al (2009) Current status of xenotransplantation and prospects for clinical application. Xenotransplantation 16:263–280
2. Schroeder C, Allan JS, Nguyen BN et al (2005) Hyperacute rejection is attenuated in GalT knockout swine lungs perfused ex vivo with human blood. Transplant Proc 37:512–513
3. Hisashi Y, Yamada K, Kuwaki K et al (2008) Rejection of cardiac xenografts transplanted from alpha1,3-galactosyltransferase gene-knockout (GalT-KO) pigs to baboons. Am J Transplant 8:2516–2526
4. Burdorf L, Stoddard T, Zhang T et al (2014) GalTKO.hCD46 lung xenograft injury is associated with thrombin generation and platelet activation. Am J Transplant 14:1084–1095
5. Burdorf L, Rybak E, Zhang T et al (2013) Human EPCR expression in GalTKO.hCD46 lungs extends survival time and lowers PVR in a

xenogenic lung perfusion model. J Heart Lung Transplant 32:S137

6. Bryant L, Eiseman B, Avery M (1968) Studies of the porcine lung as an oxygenator for human blood. J Thorac Cardiovasc Surg 55:255–263
7. Collins BJ, Blum MG, Parker RE et al (2001) Thromboxane mediates pulmonary hypertension and lung inflammation during hyperacute lung rejection. J Appl Physiol 90:2257–2268
8. Burdorf L, Harris D, Dahi S et al (2018) Thromboxane and histamine mediate PVR elevation during xenogeneic pig lung perfusion with human blood. Xenotransplantation 3: e12458
9. Cantu E, Balsara KR, Li B et al (2007) Prolonged function of macrophage, von Willebrand factor-deficient porcine pulmonary xenografts. Am J Transplant 7:66–75
10. Schroeder C, Wu GS, Price E et al (2003) Hyperacute rejection of mouse lung by human blood: characterization of the model and the role of complement. Transplantation 76:755–760
11. Komatsu K, Youm W, Konishi H et al (1996) Prolonged survival of hamster-to-rat pulmonary xenografts by tacrolimus (FK506) and cyclophosphamide. J Heart Lung Transplant 15:722–727
12. Miyata Y, Ohdan H, Yoshioka S et al (1998) Relationship of xenogeneic microchimerism to graft outcome in hamster-to-rat lung xenotransplantation. J Heart Lung Transplant 17:233–240
13. Nonaka M, Kadokura M, Kataoka D et al (2000) Hyperacute xenorejection of Guinea pig-to-rat lung transplantation can be attenuated by blood which has perfused another xenograft. Ann Thorac Cardiovasc Surg 6:146–150
14. De Perrot M, Keshavjee S, Tabata T et al (2002) A simplified model for en bloc double lung xenotransplantation from hamster to rat. J Heart Lung Transplant 21:286–289
15. Yang YG, deGoma E, Ohdan H et al (1998) Tolerization of anti-Galalpha1-3Gal natural antibody-forming B cells by induction of mixed chimerism. J Exp Med 187:1335–1342
16. Veith FJ, Richards KU, Hagstrom JWC, Montefusco CM (1981) Intrafamilial lung xenograft from fox to dog. J Thorac Cardiovasc Surg 81:546–552
17. Takeda M, Kawauchi M, Nakajima J, Furuse A (1995) Methotrexate in rescue therapy for xenotransplanted lung in primes. In: (Abstract po-46) Third International Congress of Xenotransplantation, Boston, 1995
18. Sadeghi AM, Laks H, Drinkwater DC et al (1991) Heart-lung xenotransplantation in primates. J Heart Lung Transplant 10:442–447
19. Daggett CW, Yeatman M, Lodge AJ et al (1998) Total respiratory support from swine lungs in primate recipients. J Thorac Cardiovasc Surg 115:19–27
20. Macchiarini P, Oriol R, Azimzadeh A et al (1999) Characterization of a pig-to-goat orthotopic lung xenotransplantation model to study beyond hyperacute rejection. J Thorac Cardiovasc Surg 118:805–814
21. Sahara H, Watanabe H, Pomposelli T et al (2017) Lung xenotransplantation. Curr Opin Organ Transplant 22:541–548
22. Burdorf L, Azimzadeh AM, Pierson RN III (2018) Progress and challenges in lung xenotransplantation: an update. Curr Opin Organ Transplant 23:621–627
23. Sahara H, Sekijima M, Ariyoshi Y et al (2018) Effects of carbon monoxide on early dysfunction and microangiopathy following GalT-KO porcine pulmonary xenotransplantation in cynomolgus monkeys. Xenotransplantation 25 (1). https://doi.org/10.1111/xen.12359
24. Nguyen BN, Azimzadeh AM, Zhang T et al (2007) Life-supporting function of genetically modified swine lungs in baboons. J Thorac Cardiovasc Surg 133:1354–1363
25. Yeatman M, Daggett CW, Parker W et al (1998) Complement-mediated pulmonary xenograft injury - studies in swine-to-primate orthotopic single lung transplant models. Transplantation 65:1084–1093
26. Cantu E, Parker W, Platt JL, Duane Davis R (2004) Pulmonary xenotransplantation: rapidly progressing into the unknown. Am J Transplant 4(Suppl 6):25–35
27. Harris DG, Quinn KJ, French BM et al (2015) Meta-analysis of the independent and cumulative effects of multiple genetic modifications on pig lung xenograft performance during ex vivo perfusion with human blood. Xenotransplantation 22:102–111
28. Laird CT, Burdorf L, French BM et al (2017) Transgenic expression of human leukocyte antigen-E attenuates GalKO.hCD46 porcine lung xenograft injury. Xenotransplantation 24 (2). https://doi.org/10.1111/xen.12294
29. French BM, Sendil S, Sepuru KM et al (2018) Interleukin-8 mediates neutrophil-endothelial interactions in pig-to-human xenogeneic models. Xenotransplantation 25:e12385
30. Burdorf L, Riner A, Rybak E et al (2016) Platelet sequestration and activation during GalTKO.hCD46 pig lung perfusion by human blood is primarily mediated by GPIb,

GPIIb/IIIa, and von Willebrand Factor. Xenotransplantation 23:222–236

31. Burdorf L, Laird C, Sendil S et al (2018) Progress in xenogeneic lung transplantation using multi-transgenic donor pigs and targeted supportive drug treatments. Transplantation 102: pS106
32. Watanabe H, Sahara H, Nomura S et al (2018) GalT-KO pig lungs are highly susceptible to acute vascular rejection in baboons, which may be mitigated by transgenic expression of hCD47 on porcine blood vessels. Xenotransplantation 25:e12391
33. Nguyen BN, Azimzadeh AM, Schroeder C et al (2011) Absence of gal epitope prolongs survival of swine lungs in an ex vivo model of hyperacute rejection. Xenotransplantation 18:94–107
34. Azimzadeh AM, Dawson H, Nguyen BN et al (2005) aGal antigen modulates the expression of pro-inflammatory genes in hyperacute lung rejection. In: 8th International Xenotransplantation Congress, Goteborg, Sweden, September 2005
35. Knosalla C, Yazawa K, Behdad A et al (2009) Renal and cardiac endothelial heterogeneity impact acute vascular rejection in pig-to-baboon xenotransplantation. Am J Transplant 9:1006–1016
36. Kim HK, Kim JE, Wi HC et al (2008) Aurintricarboxylic acid inhibits endothelial activation, complement activation, and von Willebrand factor secretion in vitro and attenuates hyperacute rejection in an ex vivo model of pig-to-human pulmonary xenotransplantation. Xenotransplantation 15:246–256
37. Schroeder C, Allan J, Nguyen BN et al (2004) Xenogenic ex vivo perfusion of lungs from galt K/O pigs: initial results. Transplantation 78 (Suppl 2):20

Chapter 13

Liver Xenotransplantation in a Nonhuman Primate Model

Nalú Navarro-Alvarez and Parsia A. Vagefi

Abstract

Xenotransplantation using pigs is now considered a potential option to tackle the prevalent scarcity of liver grafts if the immunological and coagulation barriers can be overcome. Significant improvements have been made in overcoming graft loss due to hyperacute rejection with the development of genetically engineered α1,3-galactosyltransferase KO (GalT-KO) pigs. However, survival after liver xenotransplantation (LXT) has remained short, mainly due to the severe thrombocytopenia seen right after graft reperfusion, resulting in profound bleeding complications, intense graft thrombotic microangiopathy (TMA), and the subsequent graft loss. We have recently proven that thrombocytopenia and TMA can be overcome with exogenous administration of human coagulation factors, and thus, survival has improved. We here describe the technical procedure of our pig-to-baboon liver xenotransplantation model using exogenous coagulation factors and give detailed information on peri- and postoperative care of the transplanted animals.

Key words Xenotransplantation, Pig, Baboon, Orthotopic, Liver transplantation, Coagulation factors

1 Introduction

The current major limitation of clinical liver transplantation is the shortage of suitable donor organs. According to UNOS reports, more than 13,726 patients were waiting for a liver transplantation as of December 31, 2016, and only 7841 patients had undergone liver transplantation in 2016 [1].

Despite the numerous innovative approaches, such as the use of split livers [2], living donors [3], and extended criteria donors [4, 5] to address this critical shortage, the gap between transplants performed and patients on the waiting list is increasing every year [1, 6].

The field of xenotransplantation is steadily progressing, with the development of genetically engineered pigs which has to some extent overcome the main immunological barriers [7–11]. However, some physiological incompatibilities remain, such as coagulation dysregulation caused presumably because of molecular incompatibilities between pigs and primates that promote or fail

Cristina Costa (ed.), *Xenotransplantation: Methods and Protocols*, Methods in Molecular Biology, vol. 2110, https://doi.org/10.1007/978-1-0716-0255-3_13,

to regulate pathological clotting, that have yet to be overcome [12–14].

Following liver xenotransplantation, profound thrombocytopenia has been well documented as a major complication, with the deadly consequences of bleeding and severe systemic consumptive coagulopathy that eventually leads to graft failure [15–17]. A number of preventive therapies such as platelet transfusion and administration of antifibrinolytic agents have been attempted in order to address recipient thrombocytopenia with some success [15, 18, 19]; however, these approaches have failed to further extend animal survival beyond 9 days [15, 20].

More recently, our group has demonstrated the efficacy of exogenous coagulation factor administration [11], not only in maintaining stable platelet levels and decreased red blood cell transfusions, but more importantly showing the absence of TMA in the liver xenografts [11]. While exogenous coagulation administration alone did not result in an improved survival, there was successful prolonged survival of porcine liver xenotransplant recipients for up to nearly 1 month when combined with co-stimulatory blockade, completely avoiding coagulation dysregulation and organ rejection [10, 21].

In view of these promising results, interest in the field of xenotransplantation has significantly risen given the potential of using pig liver xenotransplantation as a potential therapeutic option to bridge human patients while waiting for a liver transplant [10, 21, 22]. In this chapter, we give detailed description of the technical procedure of our pig-to-baboon liver xenotransplantation model with the administration of exogenous coagulation factors and provide thorough information on peri- and postoperative care of the transplanted animals.

2 Materials

2.1 Donor Animals

In order to overcome the hyperacute rejection, it is recommended to use genetically modified donor pigs. For this specific protocol, we used the following:

1. GalT-KO miniature swine, which were derived through breeding at our own facility via homologous recombination and nuclear transfer, as previously described [7] (*see* **Note 1**).

2.2 Donor Anesthesia and Liver Procurement

1. Drugs for sedation: Glycopyrrolate, xylazine, and telazol for intramuscular administration.
2. Drugs for anesthesia: Isoflurane for induction and maintenance of anesthesia.
3. Drugs for analgesia: Fentanyl transdermal patch (Alvogen Inc. (12 μg/h)).

4. Standard monitorization devices (pulse oximetry, EKG, noninvasive blood pressure device (SurgiVet 3 Parameter Advisor Vital Signs Monitor from Smiths Medical)) for continuous monitorization of heart rate, blood pressure, and oxygen saturation.
5. Matrx model 3000 Veterinary anesthesia ventilator, an electronically controlled, time-cycled, pressure-limited ventilator for mechanical ventilation after endotracheal intubation.
6. Endotracheal tube (*see* **Note 2**).
7. I.v. lines × 2 (arm/leg).
8. I.v. infusion pumps (Q Core Medical, Sapphire multi-therapy infusion system).
9. Intra-arterial line to be inserted into the femoral artery.
10. ABG machine to measure blood gases (i-STAT®1).
11. Bair Hugger (Thermacare convection warming system).
12. Standard instruments for liver surgery including atraumatic dissection forceps, Metzenbaum dissecting scissors, microsurgical scissors and needle holder, clamps (mosquito), fine right angle (big and small), fine Jake Schnidt, fine Castros, and forceps (Adson).
13. Sutures: Ethibond Excel 2 (Ethicon, Norderstedt, Germany) and double-armed polypropylene sutures (Prolene 5-0 RB-2, Ethicon).
14. UW solution (Barr Pharmaceuticals, Pomona, NY, USA) cooled to 4 °C.
15. 16-Fr cannula connected to a Y-cystoscopy tubing system (Baxter Healthcare).
16. Ringer's lactate solution (Hospira) cooled to 4 °C.
17. Normal saline solution (Hospira) frozen (*see* **Note 3**).
18. Metal bowl and metal tray.
19. Scale.
20. Tissue-Tek (O.C.T. Compound, Sakura Finetek).

2.3 Line Insertion and Preoperative Care

1. Baboon recipients (*Papio hamadryas*) (female and male) between 6 and 8 kg (we purchase baboons from the Mannheimer Foundation (Homestead, FL, USA) to be used as liver recipients (*see* **Note 1**)).
2. Jacket and tether (Lomir, Malone, NY, USA).
3. Angiocath 20G (Becton Dickinson).
4. Tunneled intravenous catheter using 18-gauge Tygon tubing (Saint-Gobain Corp, Valley Forge, PA) to be placed in the internal and external jugular veins for continuous venous access.

5. Drugs for sedation: Atropine and ketamine for intramuscular administration.
6. Same reagents and equipment described for anesthesia, monitorization, and ventilation of the donor animal.
7. Continuous infusion system (Large Animal Infusion Set, Lomir Biomedical Inc., Quebec, Canada). This infusion system provides maximum procedural flexibility. The infusion set consists of a tether (metal tube of approximately 2–3 m) that is attached to a special baboon jacket. The tether is connected to the cage by a 3-lumen swivel and to the jacket by an end plate (*see* **Note 4**).
8. Immunosuppressants: Antithymocyte globulin (ATG), methylprednisolone, cobra venom factor (CVF), and FK506. To be given as follows: Thymoglobulin induction (10 mg/kg), methylprednisolone (20 mg/kg), and cobra venom factor (100 μg/kg) (*see* **Note 5**).

2.4 Xenogeneic, Orthotopic Liver Transplantation

1. Baboon recipients (as described in Subheading 2.3).
2. Standard and microsurgical instruments set, including vascular instruments.
3. Endotracheal tube (*see* **Note 2**).
4. Continuous autotransfusion system (CATS) (*see* **Note 10**).
5. Retractors including Bookwalter and/or Balfour abdominal retractor and self-retaining (Weitlaner) retractor.
6. Cardiovascular forceps (for the clamping of abdominal vessels).
7. Sutures (Ethicon, Somerville, NJ): 5-0 and 4-0 silk ties, Silk 3/0 (for hand ties), double-armed polypropylene, Prolene 6-0, 7-0, and 8-0, for vascular and bile duct anastomoses, Vicryl 1-0 and 3-0 for peritoneum and subcutaneous closure, respectively, Monocryl 3-0 for subcuticular skin closure.
8. Plaster spray to cover the wound and Silvadene.

2.5 Coagulation Factor Administration

1. Coagulation factors: Octaplex®, a second-generation human prothrombin complex concentrate (hPCC) containing the coagulation factors II, VII, IX, and X and the inhibitors of coagulation, proteins C and S (generously provided by Octapharma, Lachen, CH).
2. NovoSeven®, recombinant activated factor VIIa (Novo Nordisk, Plainsboro, NJ, USA).

2.6 Animal Follow-Up and Postoperative Care

1. Maintenance immunosuppression: Continuous FK506 infusion (0.2 mg/kg/day) (target serum levels 10–20 ng/mL) and methylprednisolone (*see* **Note 6**).

2. Rejection therapy: Methylprednisolone and FK506 (*see* **Note 7**).
3. Analgesics: Fentanyl patch (1–4 μg/kg/h TD) and buprenorphine PNR.
4. Antimicrobials and antiviral drugs: Cefazolin (Ancef generic), ampicillin/sulbactam (Unasyn generic), sulfamethoxazole/trimethoprim (Bactrim generic), and ganciclovir (Valcyte generic) (*see* **Note 8**).
5. Miscellaneous drugs: Proton pump inhibitor (GI) esomeprazole (Nexium) (*see* **Note 8**).

3 Methods

3.1 Donor Animals

A common issue frequently observed following liver transplantation is the development of small- or large-for-size syndrome; therefore, caution must be taken in the selection of the donor and recipient (*see* **Note 1**).

In addition, any animal undergoing any sort of surgery or experimental procedure must complete a period of acclimatization to prevent stress-related problems. Therefore, pigs are transferred to the animal facility at least 1 week before donation.

3.2 Donor Anesthesia and Liver Procurement

1. Animals will be kept without food but with free access to water for 12 h prior to the scheduled operation in order to prevent aspiration and surgical complications.
2. Approximately 1 h prior to the operation, donor pigs are sedated with Telazol at 2 mg/kg i.m. or i.v., single dose, onset of action 15 min, duration of effect 90 min; glycopyrrolate at 0.01 mg/kg i.m. or i.v., single dose, onset of action 15 min, duration of effect 60 min; and xylazine at 2 mg/kg i.m. or i.v., single dose, onset of action 15 min, duration of effect 60 min.
3. Premedicate with atropine 0.04 mg/kg i.m. or i.v., single dose.
4. Bring the animal to the operating table and place him on supine position.
5. Start anesthesia with inhaled isoflurane 0.5–1.5% and proceed to perform endotracheal intubation (*see* **Note 2**).
6. Chest, abdomen, and posterior aspect of both calves are shaved.
7. Start an intravenous line in marginal ear vein using a #24 angiocath for the administration of 5% dextrose and half normal saline maintained at 40–50 mL per h.

8. Keep a continuous mechanical ventilation and monitoring (ECG, oxygen saturation, and noninvasive arterial pressure via a cuff) of the animal.
9. A Bovie cautery pad is placed around the leg.
10. The abdomen is then prepped with Betadine and sterilely draped for surgery.
11. Perform a midline incision to access the peritoneal cavity. Expose the abdominal aorta, the portal vein, and the inferior vena cava (IVC) as well as the bile duct.
12. Mobilize the liver by division of falciform and triangular ligaments, with particular care taken not to injure the phrenic vessels.
13. The fine areolar tissue between the IVC and the liver is carefully divided.
14. Before the administration of heparin i.v., take a blood sample and save it for later measurement of coagulation factors and CMV testing (*see* **Note 9**).
15. Insert a needle-vent cannula into the aortic root.
16. Cross-clamp the supraceliac descending aorta and start flushing with cold Ringer's lactate solution first followed by UW solution through the aortic root cannula. Open the IVC above and below the liver for venting.
17. Perform the excision of the donor liver as follows: First divide the bile duct, the portal vein, and then the celiac axis with a cuff of aorta, and then excise all the remaining attachments and remove the liver.
18. Submerge the liver in a bowl filled with ice-cold Ringer's solution and take it to the back table and continue with more flushing and with the preparation of the vessels for implantation and removal of the diaphragm. This step should be done with the liver on ice (*see* **Note 3**).
19. Prepare for implantation. A representation of the pig liver anatomy is displayed in Fig. 1.
20. Do not forget to weigh the liver before implantation into the recipient.
21. Before closing the animal, take a spleen sample for CMV testing.
22. Confirm the demise of the donor animal by checking vitals and heartbeat.
23. Close the baboon in layers.
24. Dispose the pig carcass according to regulations.

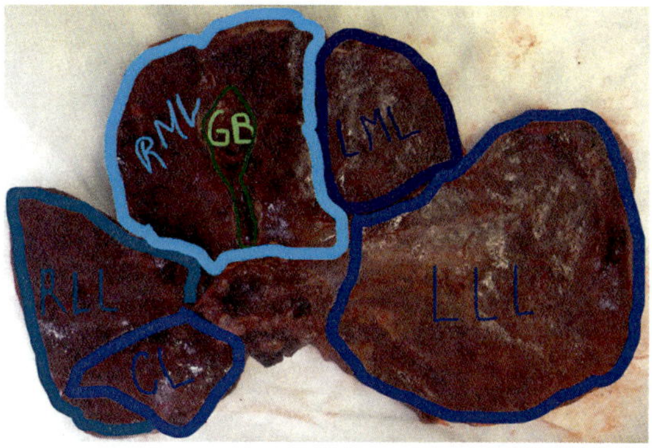

Fig. 1 Caudoventral view of the visceral aspect of a swine liver. GB, RLL, LLL, RML, and LML correspond to gall bladder, right lateral lobe, left lateral lobe, right median lobe, and left median lobe, respectively

3.3 Recipient Line Insertion and Preoperative Care (See Note 4)

1. Premedicate the baboon with i.m. atropine (0.01 mg/kg) and 15 min later sedate with i.m. ketamine (20 mg/kg). Use a squeeze back system in the cages for intramuscular injection.
2. Take the animal to the operating room and place him on a mask to start isoflurane at 3–5% for anesthesia induction, while you get intravenous access by placing a 20-g angiocath into the right or left lower peripheral vein.
3. 30 min prior to initiation of procedure, administer buprenorphine 0.01–0.03 mg/kg i.m. or i.v., single dose, onset of action 1–2 h, duration of action 8–12 h.
4. Proceed to intubation using an endotracheal tube (*see* **Note 2**).
5. Continue maintenance anesthesia with 1–1.5% isoflurane.
6. Prepare in right neck and back area and drape in a usual sterile fashion.
7. After localizing right carotid arterial pulse, perform a 4–5-cm incision in the right neck along the edge of trapezius muscle. Divide bluntly the platysma muscle and loose connective tissue around external jugular vein using electrocautery.
8. Visualize the external jugular vein, clean it, and proceed to tie it off at the proximal end.
9. Do the same with the internal jugular vein.
10. In the meantime, make a small incision slightly median to the lower end of right scapula.
11. Using a trocar, make a tunnel between the two incisions and pull through the two central lines which were previously connected to a primate jacket and a flexible metal tube (Large Animal Infusion System). This is a very important point.

12. Perform a small cut first into the wall of the right external jugular vein and insert one line, followed by the insertion of the other line in the right internal jugular vein. After adjusting the length, place a tie around the vein to fixate the lines in place.
13. Bring the muscle together using Vicryl 2-0. Close the incision using continuous subcuticular suture technique with Monocryl 3-0.
14. Close the incision on the back using 2-0 silk.
15. Check the patency of the lines and proceed to dress the wounds.
21. Postoperatively, the animal will receive a fentanyl transdermal patch (1–4 μg/kg/h transdermally (t.d.), applied on the day of surgery, onset of action 3–4 h, duration of action 72 h).
16. Place the jacket on the animal, extubate, and allow him/her to wake up before returning the animal to his/her cage.
17. Once in the cage, provide water and allow the baboon to take food whenever it feels ready.
18. Assess daily the general condition of the recipient.
19. Start the immunosuppression regimen as follows: For induction: start on day −1 with antithymocyte globulin (ATG) (10 mg/kg), methylprednisolone (20 mg/kg), cobra venom factor (CVF) (100 μg/kg) for complement factor (CH50) depletion given on day −1 and FK506 (*see* **Note 5**).
20. Continue with maintenance immunosuppression as follows: Continuous FK506 infusion (0.20 mg/kg/day) (target serum levels 10–20 ng/mL) starting on day −2 and a methylprednisolone taper starting on day 0 (*see* **Note 6**).

3.4 Xenogeneic, Orthotopic Liver Transplantation

1. Premedicate and sedate and administer analgesics to the animal in a similar fashion as for line insertion, with the difference that drugs will be administered i.v., since the animal has already an i.v. line in place.
2. Once the animal is sedated, remove the jacket so you can proceed to shave the abdomen before being transferred to the operating room.
3. Place the animal in supine position.
4. Obtain arterial access from the left or right femoral artery with a 20-G angiocath (Becton Dickinson).
5. Prepare and drape the abdomen in a sterile fashion.
6. Open the abdomen through a midline incision from the xiphoid and circumventing the umbilicus.

7. Divide and ligate the falciform ligament. Inspect the abdominal cavity and place the Bookwalter retractor (Bookwalter TM) for good exposure of the cavity and abdominal vessels.
8. Perform a splenectomy by mobilizing the organ from the peritoneum, ligating and dividing the splenic hilar vessels and short gastric vessels.
9. Turn your attention to the liver. Use the Balfour retractor for good exposure of the liver. Divide the left and right triangular ligaments and rotate the liver so that the IVC can be isolated above and below the liver.
10. Turn your attention to the liver hilum; proceed to divide the common bile duct and hepatic artery. Use the self-retaining Weitlaner retractor if needed.
11. Give heparin i.v. (*see* **Note 9**).
12. Place a straight vascular clamp on the portal vein and then Satinsky clamps on the infrahepatic IVC and then the suprahepatic IVC. Divide these vessels and remove the liver. Do not forget to record the weight (*see* **Note 10**).
13. Wash the abdomen with warm saline while allowing time for the donor liver to be prepared for transplantation.
14. Before starting the actual donor transplantation, administer 120 mg i.v. of Solu-Medrol.
15. Bring the donor graft into the field and proceed to do the orthotopic liver transplant using the caval replacement technique as follows.
16. The suprahepatic IVC anastomosis is completed with 5-0 Prolene, the posterior wall in an everting fashion and the anterior wall in an over-and-over fashion.
17. The infrahepatic IVC anastomosis is completed with 5-0 Prolene in an end-to-end fashion; do not tie down this suture.
18. The portal vein anastomosis is completed in an end-to-end fashion using 7-0 Prolene.
19. Open the portal vein clamp and flush blood through the infrahepatic IVC anastomosis (blood flush). Reapply the portal vein clamp and tie down the IVC suture.
20. After completion of these three anastomoses, the liver is reperfused by removing the vascular clamps, first suprahepatic vena cava, then infrahepatic vena cava, then portal.
21. At this point, if the anastomosis was adequate, the liver will turn red rapidly (Fig. 2).
22. The hepatic artery anastomosis may be completed in an end-to-end fashion with 7-0 or 8-0 Prolene (*see* **Note 10**).

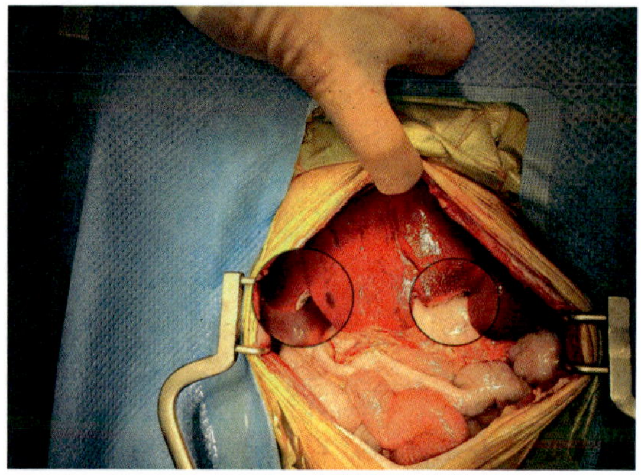

Fig. 2 Verification of liver appearance after anastomosis is completed and careful inspection for bleeding in the sites of biopsy and the cavity before abdomen closure

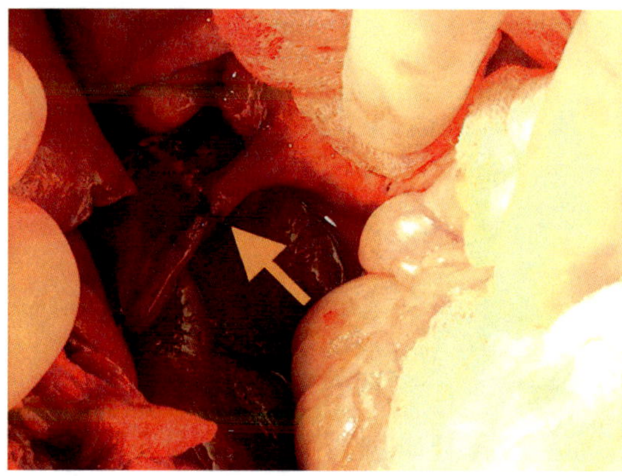

Fig. 3 Duct-to-duct anastomosis performed with 7-0 PDS

23. After arterialization of the liver and careful inspection for bleeding, a duct-to-duct anastomosis is done with 7-0 PDS (Fig. 3).
24. Cover all vascular anastomoses with Surgicel.
25. Take a liver biopsy at 30 and 60 min post-reperfusion (Fig. 2) and place it on Tissue-Tek.
26. Inspect the abdomen for bleeding until you are sure that adequate hemostasis has been achieved (Fig. 2).
27. The abdomen is closed in layers using running 0 PDS and 2-0 Vicryl.
28. The skin is closed with running 2-0 Nylon and finally apply Dermabond to the wound and dress it.
29. Remove the arterial catheter and apply pressure for about 2–3 min and place a stitch to promote hemostasis.

30. Turn off isoflurane and allow the animal to fully wake up before returning to the cage.
31. Postoperatively, the animal will receive a fentanyl transdermal patch (1–4 μg/kg/h t.d., applied on the day of surgery, onset of action 3–4 h, duration of action 72 h) and buprenorphine PNR given every 8 h.

3.5 Coagulation Factor Administration

In this part of this chapter, we describe the use of exogenous coagulation factors as one of the key features of our protocol to improve the significant challenges associated with large animal liver xenotransplantation (pig to primate), including uncontrolled coagulopathy, thrombocytopenia, and TMA. We have already proven that the administration of two types of compounds containing coagulation factors is effective in the prevention of the above related complications.

It is critical to remember that even though platelet sequestration and consumptive coagulopathy tend to occur within minutes of graft reperfusion, our protocol instructs to initiate these strategies (to alleviate this problem) 6 h after transplantation. This time point was established to account for the clearance of circulating baboon factors following reperfusion of the porcine xenoliver. With factor VII having the shortest half-life of 3–6 h, we begin exogenous coagulation factor supplementation at 6 h post-transplant in anticipation of supplementing the natural degradation. In addition, the time delay was also set to prevent any potential immediate post-transplant graft thrombosis.

1. Octaplex®: a second-generation human prothrombin complex concentrate (hPCC) containing the coagulation factors II, VII, IX, and X and the inhibitors of coagulation, proteins C and S (generously provided by Octapharma, Lachen, CH). If you choose to use this strategy, make sure you give it in a continuous infusion at 1 mL/h and further titrated to maintain an international normalized ratio (INR) of less than 2 (*see* **Note 11**).
2. NovoSeven®, recombinant activated factor VIIa, (Novo Nordisk, Plainsboro, NJ, USA). If you decide to use this strategy, make sure you give it at 1 μg/kg/h and further titrate the levels to maintain factor VII activity to >60% (*see* **Note 12**).

3.6 Animal Follow-Up and Postoperative Care

This section is one of the most critical parts in order to maintain an adequate animal care that could impact beneficially the survival of the animal. One must be vigilant of the following points and take immediate action.

1. Continue with the maintenance immunosuppression: continuous FK506 infusion (0.2 mg/kg/day) (target serum levels 10–20 ng/mL) and methylprednisolone (*see* **Note 6**).

2. Postoperatively, the animal will receive a fentanyl transdermal patch (1–4 μg/kg/h t.d., applied on the day of surgery, onset of action 3–4 h, duration of action 72 h) and buprenorphine PNR given every 8 h.
3. Antimicrobials and antiviral drugs: cefazolin (Ancef generic), ampicillin/sulbactam (Unasyn generic), sulfamethoxazole/trimethoprim (Bactrim generic), and ganciclovir (Valcyte generic) (*see* **Note 8**).
4. Miscellaneous drugs: proton pump inhibitor (GI) esomeprazole (Nexium) (*see* **Note 8**).
5. If during the course of the follow-up the baboon presents with signs of rejection, use the following rejection therapy regimen: methylprednisolone and FK506. At rejection, a high-dose methylprednisolone (10 mg/kg) is given for 3 days. ATG (1.5 mg/kg i.v.) for 5 days (*see* **Note** 7) could be considered for steroid-resistant rejection.

3.6.1 Immediate Postoperative Care

The animals should be monitored directly in the study cage following surgery until the animal is awake and moving purposefully around its cage. Careful attention must be paid to detect any signs of pain and distress. Additional analgesics, other medications, and i.v. fluids should be administered as needed.

After the immediate postoperative period, the animals should be observed at least three times daily. If the animal is known to have active infection, rejection, or declining liver function, it should be monitored more frequently (5–6 times daily) for signs of pain and distress. The signs of pain and distress include animal sitting quietly avoiding any movement, increased heart rate, increased respiration rate, lack of interest in food or water, and the animal's body hair "standing on end". In an animal with infection or a failing liver graft, some of these signs may be observed as well.

3.6.2 Blood Draw and Blood Support

This must be maintained at the absolute minimum needed as phlebotomy may lead to anemia. Blood products will be collected from donor baboons to be used to hemodynamically support xenograft recipients in case they suffer internal hemorrhage. During surgery, animals whose hematocrit goes below 20% should receive packed red blood cells with every 30-mL unit of blood expected to raise the hematocrit by 3% using the CATS (*see* **Note 10**).

3.6.3 Liver Function Test Monitoring

Liver function should be monitored by liver function tests and percutaneous biopsies scheduled at days 7, 14, and 30 and then monthly. However, it is imperative to mention that if the animal is doing well and the biopsy procedure is considered too risky, it is better to let the animal follow his/her normal course and do not intervene. We have observed that animals do deteriorate following a surgical procedure.

However, the observation of signs of rejection may lead to necessitate unscheduled percutaneous liver biopsy.

Rejection is diagnosed by a combination of abnormal liver function test and histology as obtained by liver biopsy. Rejection can be treated with high-dose steroids which would cause the liver function test to normalize. Clinically, the animal may be less active and have decrease appetite. Rejection does not usually cause pain, but the animal should be closely monitored for signs of pain and distress if a rejection crisis is suspected. If the animal develops ascites as part of the deterioration of liver function, it might be necessary to perform a paracentesis (only if the animal strictly requires it).

4 Notes

1. In order to avoid small-for-size or large-for-size syndrome, we recommend trying to match as much as possible donor and recipients' sizes. If that is not possible, try to adjust the donor size to no more than 10% ±the recipient size.
2. Use an endotracheal tube size 3–4 if the animal weighs <8 kg and size 4–4.5 if the animal is >8 kg.
3. It is really important to maintain the liver on ice, and this must be sterile; therefore, a frozen saline solution is a good alternative. Do not let the bag to be completely frozen; this should be more like slush instead of hard ice.
4. Lomir jacket and tether system: The baboons are habituated to the jackets and to the tether system, which are produced in different sizes (from 6 to 25 kg). The training is initiated 4 weeks before transplantation. Start acclimatizing the animal to the jacket only for about 1–2 weeks. Continue by attaching the tether to the jacket without connecting it to the cage; just drop it inside the cage so the animal gets use to it (1 week). Then proceed to connect it to the cage system (1 week), and once you are sure the animal is going to tolerate it, you can proceed to the first experimental procedure.
5. Before the administration of rATG, it is important to give a single bolus of diphenhydramine (Benadryl generic) at a dose of 5 mg/kg to decrease the adverse effects caused by the cytokine storm due to ATG treatment. This should be given at least 15 min before ATG. In addition, it is also necessary to give a bolus of Solu-Medrol at 20 mg/kg.
6. Initiate with a dose of methylprednisolone at 10 mg/kg and continue with postoperative dose taper as follows: 150, 120, 60, 40, 20, 10, 5 mg thereafter for as long as the animal is alive.

7. Rejection therapy: It will depend on the levels of LFTs, but a typical course of 3-day pulse of corticosteroids will suffice to prevent any signs of rejection. If you notice that rejection is not being controlled and therefore you are facing a steroid-resistant rejection episode, consider the use of ATG.
8. It is important to provide wide coverage for gram-positive and gram-negative bacteria from the time the animal undergoes line insertion and continue until the animal's demise. Right after line insertion, initiate with cefazolin (Ancef generic) at 25 mg/kg twice a day (BID) for just 3 days. It is also important to protect the animal from gastrointestinal bleeding due to stress; therefore, administer a proton pump inhibitor like esomeprazole (Nexium) at a dose of 10 mg/day. Once the animal undergoes liver transplantation, you need to widen your coverage; therefore, initiate with ampicillin/sulbactam (Unasyn generic) at 75 mg/kg kg four times a day (QID) and sulfamethoxazole/trimethoprim (Bactrim generic) at 10 mg/kg BID. In addition, there is a high likelihood of potential CMV infection resulting from the immunosuppression. Therefore, it is important to prevent it with the administration of ganciclovir (Valcyte generic) at 2.5 mg/kg BID.
9. In the case of porcine liver retrieval, give a bolus of 3000 UI i.v. at the time of clamping. In the case of the baboon liver transplantation, give a bolus of 800–1000 UI i.v. at the time of clamping (100 UI/kg).
10. It is possible that during surgery, the animal presents a drop on blood pressure, in which case you will need to administer PRBCs. Animals whose hematocrit goes below 20% will receive packed red blood cells with every 30-mL unit of blood expected to raise the hematocrit by 3%.
11. Make sure you prepare a fresh vial of Octaplex every time you need to replace it, as its activity will decrease if you prepare it in advance.
12. Given the short half-life of NovoSeven® (approximately 2 h), the dose needs to be empirically increased approximately every 18–24 h in an effort to maintain stable activity. Typical increases range from 0.5 to 1 μg/kg/h which should be established based on the factor VII activity.

References

1. OPTN (2016) Organ procurement and transplantation network. Available from: https://optn.transplant.hrsa.gov/data/
2. Dalal AR (2015) Split liver transplantation: what's unique? World J Transplant 5:89–94
3. Miller CM, Quintini C, Dhawan A et al (2017) The international liver transplantation society living donor liver transplant recipient guideline. Transplantation 101:938–944
4. Trapero-Marugán M, Little EC, Berenguer M (2018) Stretching the boundaries for liver transplant in the 21st century. Lancet Gastroenterol Hepatol 3:803–811
5. Gao Q, Mulvihill M, Scheuermann U et al (2019) Improvement in liver transplant outcomes from older donors: a US national analysis. Ann Surg 270(2):333–339
6. Charlton MR (2016) Roadmap for improving patient and graft survival in the next 10 years. Liver Transpl 22:71–78
7. Kolber-Simonds D, Lai L, Watt S et al (2004) Production of alpha-1,3-galactosyltransferase null pigs by means of nuclear transfer with fibroblasts bearing loss of heterozygosity mutations. Proc Natl Acad Sci U S A 101:7335–7340
8. Tseng Y-LL, Kuwaki K, Dor F et al (2005) alpha1,3-Galactosyltransferase gene-knockout pig heart transplantation in baboons with survival approaching 6 months. Transplantation 80:1493–1500
9. Liu F, de Ruyter E, Athorn R, et al (2019) Effects of L-citrulline supplementation on heat stress physiology, lactation performance and subsequent reproductive performance of sows in summer. J Anim Physiol Anim Nutr (Berl) 103(1):251-257
10. Shah JA, Navarro-Alvarez N, DeFazio M et al (2016) A bridge to somewhere: 25-day survival after pig-to-baboon liver xenotransplantation. Ann Surg 263:1069–1071
11. Navarro-Alvarez N, Shah J, Zhu A et al (2016) The effects of exogenous administration of human coagulation factors following pig-to-baboon liver xenotransplantation. Am J Transplant 16:1715–1725
12. Hammer C, Thein E (2002) Physiological aspects of xenotransplantation, 2001. Xenotransplantation 9:303–305
13. Lin CC, Cooper DK, Dorling A (2009) Coagulation dysregulation as a barrier to xenotransplantation in the primate. Transpl Immunol 21:75–80
14. Cowan PJ, Robson SC, d' Apice AJ (2011) Controlling coagulation dysregulation in xenotransplantation. Curr Opin Organ Transplant 16:214–221
15. Kim K, Schuetz C, Elias N et al (2012) Up to 9-day survival and control of thrombocytopenia following alpha1,3-galactosyl transferase knockout swine liver xenotransplantation in baboons. Xenotransplantation 19:256–264
16. Shimizu A, Yamada K, Yamamoto S et al (2005) Thrombotic microangiopathic glomerulopathy in human decay accelerating factor-transgenic swine-to-baboon kidney xenografts. J Am Soc Nephrol 16:2732–2745
17. Shimizu A, Hisashi Y, Kuwaki K et al (2008) Thrombotic microangiopathy associated with humoral rejection of cardiac xenografts from alpha1,3-galactosyltransferase gene-knockout pigs in baboons. Am J Pathol 172:1471–1481
18. Iwase H, Ezzelarab MB, Ekser B, Cooper DK (2014) The role of platelets in coagulation dysfunction in xenotransplantation, and therapeutic options. Xenotransplantation 21:201–220
19. Cowan PJ, d' Apice AJ (2008) The coagulation barrier in xenotransplantation: incompatibilities and strategies to overcome them. Curr Opin Organ Transplant 13:178–183
20. Ekser B, Long C, Echeverri GJ et al (2010) Impact of thrombocytopenia on survival of baboons with genetically modified pig liver transplants: clinical relevance. Am J Transplant 10:273–285
21. Shah JA, Patel MS, Elias N et al (2017) Prolonged survival following pig-to-primate liver xenotransplantation utilizing exogenous coagulation factors and costimulation blockade. Am J Transplant 17:2178–2185
22. Cooper DKC, Wijkstrom M, Hariharan S et al (2017) Selection of patients for initial clinical trials of solid organ xenotransplantation. Transplantation 101:1551–1558

Chapter 14

Intra-bone Bone Marrow Transplantation in Pig-to-Nonhuman Primates for the Induction of Tolerance Across Xenogeneic Barriers

Kazuhiko Yamada, Yuichi Ariyoshi, Thomas Pomposelli, and Kazuhiro Takeuchi

Abstract

Mixed chimerism and thymic tissue transplantation strategies have achieved xenogeneic tolerance in pig-to-mouse models, and both have been extended to pig-to-baboon models. A mixed chimerism strategy has shown promise toward inducing tolerance in allogeneic models in mice, pigs, nonhuman primates (NHP), humans, and a rat-to-mouse small animal xeno-model. However, even though α-1,3-galactosyltransferase gene knockout (GalTKO) pigs have been used as bone marrow (BM) donors, direct intravenous injection of porcine BM cells was detected for only up to 4 days (peripheral macro-chimerism) in one case, and the rest lost chimerism within 2 days.

Recent data in allogeneic models demonstrated that direct injection of donor BM cells into recipient BM spaces (intra-bone bone marrow transplantation: IBBMTx) produces rapid reconstitution and a higher survival rate compared to i.v. injection. In order to minimize the loss of injected porcine BM peripherally before reaching the BM space, Yamada developed a xeno-specific regimen including IBBMTx coated with a collagen gel matrix in a preclinical pig-to-baboon model (Yamada IBBMTx). This strategy aims to achieve improved, persistent macro-chimerism as well as engraftment of BM across a xenogeneic barrier. The initial study published in 2015 demonstrated that this IBBMTx strategy leads to markedly prolonged peripheral macro-chimerism detectable for up to 23 days. Furthermore, a more recent study using human CD47-transgenic (Tg) GalTKO pigs as xeno-donors achieved long-lasting macro-chimerism >60 days with evidence of reduction of anti-pig natural antibodies (nAb). This is the longest macro-chimerism that has ever been achieved in a preclinical large animal xenotransplant model to date. In this chapter, we introduce a brief summary of our achievements in regard to successful tolerance induction by utilizing our novel strategy of IBBMTx as well as describe the step-by-step methodology of surgical and in vitro procedures that are required for this project.

Key words Xenotransplantation , Intra-bone bone marrow transplantation, Human CD47 transgenic, Pig-to-baboon model, Tolerance

1 Introduction

Xenotransplantation is becoming a more realistic strategy for solving the organ shortage crisis. However, even with high-dose anti-CD40 or anti-CD154 mAb-based immunosuppression combined with

Cristina Costa (ed.), *Xenotransplantation: Methods and Protocols*, Methods in Molecular Biology, vol. 2110, https://doi.org/10.1007/978-1-0716-0255-3_14, © Springer Science+Business Media, LLC, part of Springer Nature 2020

multiple immunosuppressive drugs, recipients typically succumb to infection associated with chronic immunosuppression or graft rejection [1–8]. These data are consistent with previous reports [9, 10] indicating that the human-anti-porcine T-cell response is similar or stronger than the response seen across allogeneic barriers. Because of the strength of both innate and adaptive immunity in xenotransplantation, the level of continuous immunosuppression needed to control these immune responses and prolong xenograft survival has been associated with prohibitive morbidity and mortality [7, 11, 6]. These results provide compelling rationale to pursue a clinically applicable strategy for the induction of tolerance.

Despite the great immunologic differences between differing species, both mixed chimerism and thymic transplantation strategies have demonstrated efficacy in inducing tolerance of human T cells to porcine xenografts in mouse models [12–16]. However, these strategies require fundamental modifications when applied to pig-to-primate xenograft tolerance. Details of our strategy for thymic transplantation in a pig-to-baboon model are described in Chap. 11 [17–19].

Approaches toward achieving mixed chimerism by infusion of donor BM have successfully induced tolerance in clinical renal transplantation [20, 21]. However, because of the activation of T cells, which in turn activate B cells and NK cells sequentially [22–25], durable chimerism is likely required to induce stable T/B/NK-cell tolerance across xenogeneic barriers. In support of this, previous studies have shown that (1) mouse anti-rat natural antibodies, which have undefined specificities, disappear when non-myeloablative mixed chimerism is induced [25, 26] and (2) human NK cells are rendered tolerant by porcine mixed chimerism. This was reflected in some cases by specific unresponsiveness with otherwise preserved NK-cell function and, in others, global hyporesponsiveness in the pig-to-human mixed chimerism model in humanized mice. Even though specific humoral unresponsiveness was induced in some recipients [27], most porcine cells were cleared within 24–48 h post-GalTKO BM infusion [27–29]. These data indicate that in order to achieve durable chimerism across xenogeneic barriers for the induction of tolerance in pig-to-nonhuman primates or humans, further innovative strategies are required.

1.1 Our Strategy of IBBMTx Has Achieved >60 Days Durable Chimerism

1.1.1 IBBMTx Using GalTKO Pigs Facilitates the Prolongation of Macro-chimerism from Days to Up to 3 Weeks

A previous study performed in mice showed a smaller proportion of donor BM cells (1–2%) localized to the BM in recipients that were preconditioned by irradiation [30], indicating that the localization of donor BM cells at the host BM is a critical component of both BM engraftment and the reconstitution of host hematopoiesis. Therefore, we hypothesized that direct injection of collagen-coated porcine BM cells would not only minimize the loss of BM cells peripherally, but also the collagen coating would facilitate that porcine BM cells stay inside the host BM. This would result in the persistence of macro-chimerism and engraftment of porcine BM cells. Our results demonstrate that our new strategy involving IBBMTx (Yamada IBBMTx, detailed below) leads to (1) a high percentage of durable macro-chimerism for up to 23 days and (2) a high incidence (four of six animals) of BM engraftment with hyporesponsiveness across xenogeneic barriers in our pig-to-baboon model [31].

1.1.2 IBBMTx Using hCD47-Tg GalTKO Pigs Achieves Long-Lasting Chimerism >6 Months, Which Is the Longest Ever Attained in a Large Animal Xenotransplant Model

Recent work to circumvent the rapid clearance of BM cells has focused on innate immunity and macrophage-associated mechanisms. Specifically, attention is on SIRP-alpha, a transmembrane protein with intracellular tyrosine kinase activity that is present on macrophages, dendritic cells, and neutrophils. The ligand of this receptor is CD47, and recognition of CD47 by SIRP-alpha down-regulates phagocytosis by macrophages [32, 33]. In a small animal study in which human hematopoietic stem cells were transplanted into severely immunocompromised mice to allow for human hematopoiesis, the authors found significant improvement in hematopoiesis and engraftment in mice transgenic for human SIRP-alpha [34]. These data suggest that expression of human CD47 on porcine BM cells may minimize the loss of circulating porcine BM cells in nonhuman primates. Hawley et al. first attempted to infuse hCD47/hCD55 Tg GalTKO BM cells intravenously into baboons. Contrary to high expectations, injected hCD47/hCD55 Tg GalTKO BM cells disappeared within 7 days.

Durable mixed xenogeneic chimerism is required for the induction of tolerance of human T-cell-independent B cells producing anti-pig Abs [35], which potentially results in elimination or minimization of non-Gal nAb-related renal graft injury as well as innate or T-cell rejection. Therefore, we recently combined our strategy of IBBMTx with hCD47 transgenic (Tg) donors in order to achieve further durable mixed xenogeneic chimerism [36]. Briefly, we have achieved long-lasting macro-chimerism (longer than 8 weeks) following hCD47/hCD55-Tg GalTKO IBBMTx in baboons, while neither GalTKO alone or human complement regulatory genes without hCD47 Tg maintained macro-chimerism for longer than 5 weeks. In addition, preformed nAb markedly decreased in recipients of hCD47-Tg IBBMTx, especially after 2 months following IBBMTx. Our data has been partially presented at the TTS 2019, Madrid, Spain (Watanabe H), and our original article describing the details of our results is in preparation (Yamada K et al). The following section describes our methods for IBBMTx, in addition to helpful tips and pitfalls to avoid for this procedure.

2 Materials

2.1 Catheter Placement

1. A lightweight spandex/lycra jacket (Lomir Biomedical Inc., Malone, NY).
2. A rigid swivel that allows the animal to move freely in the cage without disturbing line connections.
3. A 1.0–1.5-m aluminum tether that attaches the swivel to the jacket.
4. Catheters, generally 1.5–2.0 m flexible, clear Tygon microbore tubing (Norton Performance Plastics, Akron, Ohio).

2.2 Splenectomy

1. Recipient baboons.
2. Surgical telescopes (preferably ×3.5 or greater) for the primary assistant and the primary surgeon.
3. Surgical draping and surgical instruments including Weitlaner retractors, electrocautery, as well as suture materials (silk ties).
4. Anesthesia and mechanical ventilator.

2.3 Irradiation for NHP

1. Small container for animal transport.
2. Irradiation machine.

2.4 Bone Marrow Transplantation

1. Collagen gel culturing kit (Wako Chemicals, USA).
2. Bone marrow injection gun (PerSys Medical, USA).
3. Surgical draping and infusion set.
4. 0.9% NaCl (NS).

2.5 Flow Cytometry to Assess T-Cell and B-Cell Depletion and Macro-chimerism

1. Antihuman (cross-reactive to baboons) antibodies for CD3 (polyclonal rabbit antihuman CD3, A0452; Dako North America, Inc., Carpinteria, California, USA), CD4 (mouse antihuman CD4 mAb, 1F6; Invitrogen, Carlsbad, California, USA), CD8 (mouse antihuman CD8; BD Biosciences, Franklin Lakes, New Jersey, USA), CD20 (mouse antihuman CD20; BD Biosciences), CD25 (mouse antihuman, BD Biosciences), and CD45 (mouse anti-NHP; BD Biosciences). All conjugated to a fluorochrome.
2. Anti-pig monoclonal antibodies (mAb, available in our research center): Anti-porcine CD1, CD2, CD3, CD4, CD8, CD25, CD45, Class I, Class II, Pan Pig (produced internally). All conjugated to fluorochrome.
3. Conjugated anti-mouse IgG1 and anti-mouse IgG2a to be used as controls.
4. Cold FACS media: 1 L of Hanks' balanced salt solution, 1 g of bovine serum albumin (BSA), 1 g of sodium azide.
5. FACS tubes.

2.6 CFU Assay

1. Reagents: fetal bovine serum (FBS), porcine stem cell factor, porcine GM-CSF, porcine IL-3, recombinant human erythropoietin (epoetin alfa), and IMDM MethoCult H4230 (STEMCELL Technologies) for CFU medium.
2. Cell-culture dish (Fisher Scientific).
3. 40-micron filter.
4. DNA extraction kit (Qiagen).

3 Methods

3.1 Catheter Placement

At the authors' institution, kidney xenotransplantation is conducted in life-supporting models, and a bilateral native nephrectomy is typically performed at the time of transplantation. Thus, daily laboratory values are required to monitor the animal's health and graft function. For these reasons, the placement of central venous and arterial catheters or "lines" is required for animal care during the induction, peri-transplantation, and post-transplantation periods. Lines provide the caregiver with the ability to both deliver drug therapy and draw blood for diagnostic testing. This is of particular importance in baboons or monkeys because it is difficult to perform a physical examination beyond simple observation. Although central catheters confer a risk of infection, there is greater risk associated with sedation for daily or twice-daily blood draws and drug-administration, especially in potentially fragile post-xenotransplantation recipients.

The lines are utilized for (1) continuous blood pressure monitoring; (2) blood samples for CBC, blood chemistry, and drug levels; and (3) the number of drugs administered (continuous and/or intermittent) and use of different lines due to drug compatibility. We used to place three lines (two venous, one arterial) 1 week prior to transplantation. The arterial line is used primarily for blood draws, and the two venous lines are used for drug delivery. However, we have recently started placing only two venous lines in order to lessen the risk of infection by minimizing indwelling lines with success. We have found that continuous blood pressure monitoring is generally not needed once the animal is recovered from anesthesia, and blood draws are able to be performed from one of the two venous lines. Importantly, soluble MMF must be placed on its own dedicated D5%W carrier line and not mixed with normal saline (*see* Fig. 1). The following is our procedure for placing two lines into the jugular veins (the procedure for placing three lines is described in the Chap. 11.

1. Generally, central lines are placed in the great vessels of the neck, on the left side.

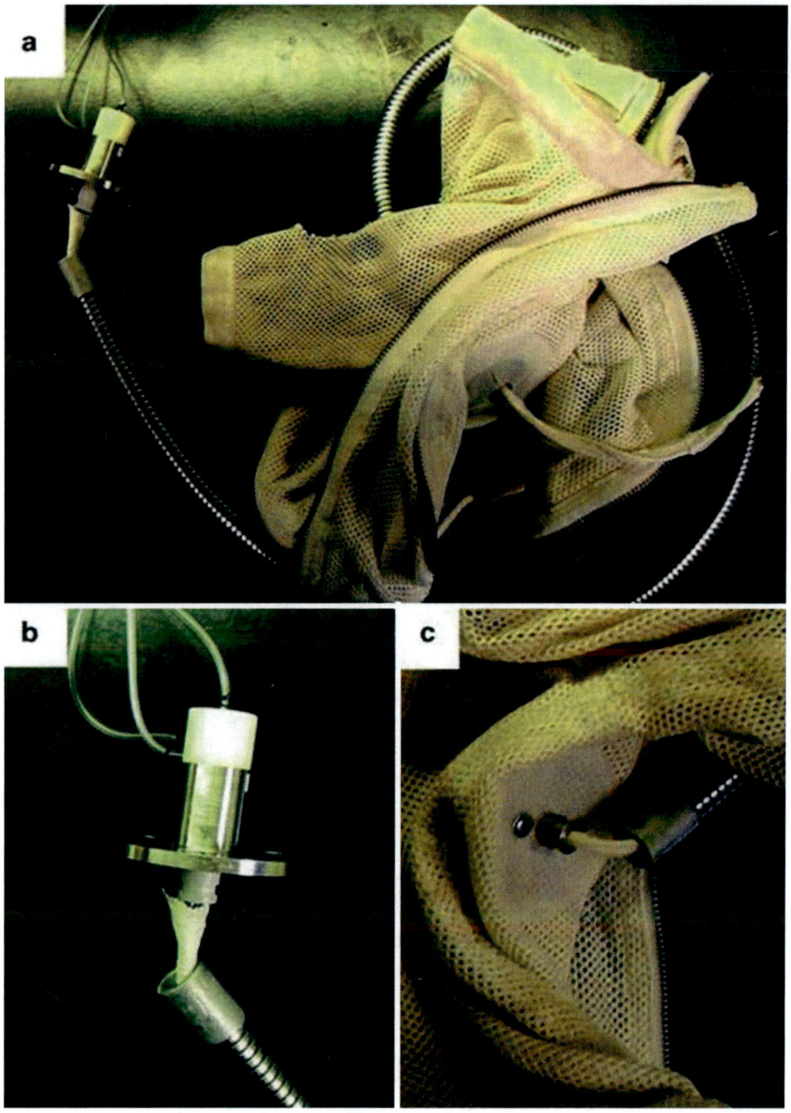

Fig. 1 (**a**) A picture of the line system used for baboon recipients of xenografting procedures. The mesh jacket is rigidly attached to an aluminum tether through which 1–3 catheters are placed. The tethers are 4–6 ft. in length. (**b**) A close-up view of the swivel shown in (**a**), which has been unattached from the aluminum tether. As many as three catheters are securely affixed to each of three ports on the swivel. This allows the animal to move within the cage without twisting the lines. (**c**) A close-up view of the connection between the jacket and the tether. In this image, the jacket has been detached from the tether, showing the tape-wrapped catheters traveling through the jacket and into the tethering system

2. Make a 3-cm transverse incision sharply, one fingerbreadth above the clavicle, starting from the midline and extending across the anterior border of the sternocleidomastoid (SCM).

3. Electrocautery and blunt dissection should be used to transect the platysma and isolate the external jugular (EJ) vein and the internal jugular (IJ) vein.
4. Make a small incision in the animal's back (between shoulder blades), and tunnel the lines through the subcutaneous tissue and into the open neck wound.
5. Dissect the sternocleidomastoid and retract medially or laterally for access to the IJ followed by the EJ.
6. Place a small bulldog clamp on the vessels or tie the vessels, 1 cm proximally.
7. Create a venotomy using tenotomy scissors and then cannulate each vessel.
8. Release clamps and advance the venous line (if clamped).
9. Close the skin with buried subcuticular sutures.
10. Place a protective line jacket on the animal in standard fashion.

3.2 Splenectomy

1. A left abdominal incision is made from the left subcostal margin to the middle of the abdomen and is carried down through the peritoneum.
2. Gently retract the spleen and carefully dissect the pancreatic tail from the spleen. Rotate the spleen to release the splenocolonic and retroperitoneum attachments.
3. Expose the spleen and isolate the splenic vein and artery.
4. Transect splenic vein and artery between ligature ties.
5. Expose short gastric arteries and transect between ligature ties.
6. After removing the spleen and ensuring hemostasis, close the abdomen in 3–4 layers. Our practice is to close the peritoneum with absorbable suture, followed by closure of the fascia with a running permanent suture, and then closing the superficial fascia and skin with a running absorbable suture.
7. Wean off the ventilator and replace the jacket. Once recovered from anesthesia, the animal should be perched within 2 h of extubation. *See* **Note 13** for animal care recommendations.

3.3 Irradiation for NHP

1. Animal is first sedated and placed in a transport cage for traveling to the irradiation center.
2. All animals receive whole body irradiation and thymic irradiation (thymic area is marked and other areas are shielded).
3. After the irradiation, animals are immediately brought back to the cage. Animals should be perched within 2 h after recovery from anesthesia.

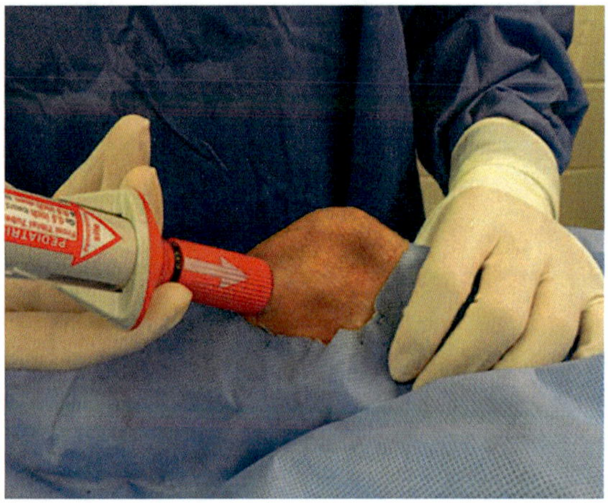

Fig. 2 A picture of the proper positioning for IBBM injection. Rt. tibial tuberosity was used for IBBM injection

3.4 Bone Marrow Transplantation

3.4.1 Intra-bone Bone Marrow Transplantation (IBBMTx. Yamada Method [9])

1. The animal is first brought to the OR and intubated. The animal is placed in the supine position, and the legs are prepped and draped in the usual sterile fashion. It is our practice to clip the hair from the mid thigh distal to the mid tibia and prep the entire leg. Once draped, the tibial tuberosity is palpated (Fig. 2).
2. It is important to ensure the legs are stable for the procedure. This can be aided by first tying the legs to the OR table in slight flexion using soft restraints. Grasp the leg firmly, and discharge the biopsy needle gun flush against the medial tibia, approximately one fingerbreadth beneath the tibial tuberosity. Remove the cannula and draw back 5–10 mL of BM.
3. Inject donor BM through the injection spike, which has been pre-mixed with collagen solution. At this time, it is extremely important to inject the BM slowly as well as communicate with the anesthesiologist. It is very common for the animal to display an anaphylactic reaction to the IBBMTx, and it is good practice to premedicate with steroids and diphenhydramine. Anticipate that the animal may require vasopressor support during the procedure, which is not uncommon.
4. Remove injection spike and hold firm compression over the wound for approximately 30 min to prevent hematoma formation.

3.4.2 IBBMTx

1. Peripheral line insertion is performed with an 18–22G angiocatheter before the procedure begins.
2. Flush the line with normal saline.
3. Slowly inject diluted BM through the line.
4. After BM injection, flush the line again with normal saline.

5. If the animal displays anaphylaxis to the IBBMTx, the peripheral line can be used for emergency drug injection.

3.5 Flow Cytometry to Assess T-Cell and B-Cell Depletion and Macro-chimerism

1. Stain 75 μL of whole blood with anti-Pan pig and anti-pig CD45 antibodies. Incubate for 30 min at room temperature in the dark. Use conjugated anti-mouse IgG1 and anti-mouse IgG2a as controls for nonspecific binding.
2. Add 2 mL of FACS Lysing Buffer to each tube and vortex well.
3. Incubate for 5 min at room temperature in the dark.
4. Vortex each tube once more.
5. Centrifuge 5 min at 260 × *g* at room temperature.
6. Add 10 μL of secondary reagent (i.e., PE-avidin), if necessary.
7. Vortex gently.
8. Incubate for 15–30 min at room temperature in the dark.
9. Wash twice with FACS buffer.
10. Acquire cells using flow cytometer.

3.6 CFU Assay

1. Obtain a BM sample and drain away most of the liquid, and then place into a medium Petri dish.
2. Filter through a 40-micron filter into a 50-mL conical tube.
3. Centrifuge 10 min at 350 × *g* at room temperature.
4. Discard supernatant and resuspend pellet in 5 mL of PBS.
5. Count cells using standard cell counting technique.
6. Dilute with PBS to make a suspension of 2.5×10^5 cells/mL.
7. Take 200 μL (5×10^4 cells/mL) and add to each 3-mL aliquot of methylcellulose media.
8. Incubate the plate in the CO_2 incubator for 2 weeks at 37 °C.
9. Pick colonies under an inverted microscope using a pipette tip.
10. Spin down at highest speed (at least 16k) on a standard tabletop microcentrifuge for 10 min.
11. Extract DNA using the blood DNA extraction kit.
12. PCR assay to amplify the porcine cytochrome *b* gene.

4 Notes

1. For animal use in the OR, ensure all equipment is sterilized before animal sedation. Surgeons should scrub and maintain sterile technique throughout the procedure.
2. The EJ should be wrapped with a vessel loop. This is done because the vein is very small and frequently experiences

vasospasm. To prevent visual obstruction, the EJ catheter is placed after the IJ vein and arterial catheters.

3. The arterial catheter should not be advanced more than 6 cm in a 6-kg baboon.

4. The IJ and EJ are low pressure vessels. After distal ligation, a proximal clamp is typically unnecessary prior to venotomy. Because, anatomically, the EJ communicates with the subclavian vein at an acute angle, the EJ line is typically only advanced 1–2 cm. The catheters are tacked to the SCM, deep to the reapproximated platysma. This may help to avoid contamination if a wound infection (rare) occurs.

   Catheter-related issues: if line infection occurs or the lines thrombose, they must be removed. In order to do so, the animal is taken to the OR for line removal through the same incision used for placement. Expect dense scar tissue around the lines, and extreme care must be taken to avoid puncture or damaging the lines during dissection, as this would introduce a potentially fatal air embolus or hemorrhage. If after removal additional lines are required, the femoral vessels can be used. In doing so, the femoral artery will need to be tied distally, which is tolerated remarkably well in baboons [37]. Both ipsilateral femoral vessels can be used during a single line placement. In some cases, it may be necessary to subcutaneously tunnel a second venous line across the animal's anterior pelvis. Two and sometimes three lines can be tunneled safely through the subcutaneous tissues over the animal's back (at the level of L1). The longitudinal groin incisions, despite not being covered by the jacket, are rarely a problem. If needed, all four vessels (two femoral arteries and two femoral veins) can be ligated without causing clinically significant morbidity.

5. The tail of the pancreas usually overlies the splenic hilum, and the surgeon should take great care in avoiding injury to the pancreas during splenectomy.

6. All splenic vessels including the short gastrics should be identified, clipped, or ligated between silk sutures.

7. If BM biopsy is performed in the correct position on the tibia, BM is easily aspirated back through the injection spike. If BM does not aspirate easily, researchers should remove the spike and hold pressure for 30 min. Attempt biopsy from a different site.

8. It is of tantamount importance to communicate with the anesthesiologist and inject BM slowly. Note the blood pressure, heart rate, and respiratory rate before and after BM injection. Acute anaphylaxis commonly occurs during injection.

9. To avoid or minimize anaphylaxis, BM should be injected slowly. Ideally, the BM should be injected 1 mL at a time,

with pauses between each mL to monitor the animal's condition.

10. Compression of needle incision site should be held for 30 min. If bleeding is still noted, additional compression is recommended to avoid hematoma.
11. For intravenous BM injection, BM should be diluted in NS and slowly injected through the peripheral line.
12. Animal care: following line placement and prior to transplantation, the animals undergo a brief but stringent induction using T-cell and B-cell depleting agents. These are generally monoclonal or polyclonal antibodies that are commercially available. Administration of any cellular depleting antibody should be preempted by flow-cytometric T- and B-cell subset phenotyping. T-cell depleting agents have been tested at multiple doses and at multiple time points; however, the general goal is to achieve near-complete T-cell depletion (50–150 cells/μL) for the first 2 weeks which protects the transplanted BM cells in the induction period.

Acknowledgments

We thank Ms. Haruna Shimizu for her editorial assistance. This research was supported by NIH grant (NIAID 5P01AI045897). All procedures and animal care were performed in accordance with the Principles of Laboratory Animal Care formulated by the National Society for Medical Research and the Guide for the Care and Use of Laboratory Animals prepared by Columbia University Medical Center.

References

1. Shin JS, Kim JM, Kim JS et al (2015) Long-term control of diabetes in immunosuppressed nonhuman primates (NHP) by the transplantation of adult porcine islets. Am J Transplant 15(11):2837–2850. https://doi.org/10.1111/ajt.13345
2. Shin JS, Min BH, Kim JM et al (2016) Failure of transplantation tolerance induction by autologous regulatory T cells in the pig-to-non-human primate islet xenotransplantation model. Xenotransplantation 23(4):300–309. https://doi.org/10.1111/xen.12246
3. Higginbotham L, Mathews D, Breeden CA et al (2015) Pre-transplant antibody screening and anti-CD154 costimulation blockade promote long-term xenograft survival in a pig-to-primate kidney transplant model. Xenotransplantation 22(3):221–230. https://doi.org/10.1111/xen.12166
4. Iwase H, Hara H, Ezzelarab M et al (2017) Immunological and physiological observations in baboons with life-supporting genetically engineered pig kidney grafts. Xenotransplantation 24(2):e12293. https://doi.org/10.1111/xen.12293
5. Mohiuddin MM, Singh AK, Corcoran PC et al (2016) Chimeric 2C10R4 anti-CD40 antibody therapy is critical for long-term survival of GTKO.hCD46.hTBM pig-to-primate cardiac xenograft. Nat Commun 7:11138. https://doi.org/10.1038/ncomms11138
6. Iwase H, Liu H, Wijkstrom M et al (2015) Pig kidney graft survival in a baboon for 136 days: longest life-supporting organ graft survival to

date. Xenotransplantation 22(4):302–309. https://doi.org/10.1111/xen.12174

7. Hering BJ, Wijkstrom M, Graham ML et al (2006) Prolonged diabetes reversal after intraportal xenotransplantation of wild-type porcine islets in immunosuppressed nonhuman primates. Nat Med 12(3):301–303. https://doi.org/10.1038/nm1369
8. Cardona K, Korbutt GS, Milas Z et al (2006) Long-term survival of neonatal porcine islets in nonhuman primates by targeting costimulation pathways. Nat Med 12(3):304–306. https://doi.org/10.1038/nm1375
9. Yamada K, Sachs DH, DerSimonian H (1995) Human anti-porcine xenogeneic T cell response. Evidence for allelic specificity of mixed leukocyte reaction and for both direct and indirect pathways of recognition. J Immunol 155(11):5249–5256
10. Murray AG, Khodadoust MM, Pober JS, Bothwell AL (1994) Porcine aortic endothelial cells activate human T cells: direct presentation of MHC antigens and costimulation by ligands for human CD2 and CD28. Immunity 1 (1):57–63
11. Yamada K, Yazawa K, Shimizu A et al (2005) Marked prolongation of porcine renal xenograft survival in baboons through the use of alpha1,3-galactosyltransferase gene-knockout donors and the cotransplantation of vascularized thymic tissue. Nat Med 11(1):32–34. https://doi.org/10.1038/nm1172
12. Nikolic B, Gardner JP, Scadden DT, Arn JS, Sachs DH, Sykes M (1999) Normal development in porcine thymus grafts and specific tolerance of human T cells to porcine donor MHC. J Immunol 162(6):3402–3407
13. Kalscheuer H, Onoe T, Dahmani A et al (2014) Xenograft tolerance and immune function of human T cells developing in pig thymus xenografts. J Immunol 192(7):3442–3450. https://doi.org/10.4049/jimmunol.1302886
14. Shimizu I, Fudaba Y, Shimizu A, Yang YG, Sykes M (2008) Comparison of human T cell repertoire generated in xenogeneic porcine and human thymus grafts. Transplantation 86 (4):601–610. https://doi.org/10.1097/TP.0b013e318182d47a
15. Zhao Y, Swenson K, Sergio JJ, Arn JS, Sachs DH, Sykes M (1996) Skin graft tolerance across a discordant xenogeneic barrier. Nat Med 2(11):1211–1216
16. Lee LA, Gritsch HA, Sergio JJ et al (1994) Specific tolerance across a discordant xenogeneic transplantation barrier. Proc Natl Acad Sci U S A 91(23):10864–10867
17. Yamada K, Shimizu A, Utsugi R et al (2000) Thymic transplantation in miniature swine. II. Induction of tolerance by transplantation of composite thymokidneys to thymectomized recipients. J Immunol 164(6):3079–3086
18. Kamano C, Vagefi PA, Kumagai N et al (2004) Vascularized thymic lobe transplantation in miniature swine: thymopoiesis and tolerance induction across fully MHC-mismatched barriers. Proc Natl Acad Sci U S A 101 (11):3827–3832. https://doi.org/10.1073/pnas.0306666101
19. Rivard CJ, Tanabe T, Lanaspa MA et al (2018) Upregulation of CD80 on glomerular podocytes plays an important role in development of proteinuria following pig-to-baboon xenorenal transplantation - an experimental study. Transplant international : official journal of the European Society for Organ Transplantation 31(10):1164–1177. https://doi.org/10.1111/tri.13273
20. Huang CA, Fuchimoto Y, Scheier-Dolberg R, Murphy MC, Neville DM Jr, Sachs DH (2000) Stable mixed chimerism and tolerance using a nonmyeloablative preparative regimen in a large-animal model. J Clin Invest 105 (2):173–181. https://doi.org/10.1172/JCI7913
21. Kawai T, Cosimi AB, Colvin RB et al (1995) Mixed allogeneic chimerism and renal allograft tolerance in cynomolgus monkeys. Transplantation 59(2):256–262
22. Cretin N, Bracy J, Hanson K, Iacomini J (2002) The role of T cell help in the production of antibodies specific for gal alpha 1-3Gal. J Immunol 168(3):1479–1483
23. Tonomura N, Shimizu A, Wang S et al (2008) Pig islet xenograft rejection in a mouse model with an established human immune system. Xenotransplantation 15(2):129–135. https://doi.org/10.1111/j.1399-3089.2008.00450.x
24. Lan P, Tonomura N, Shimizu A, Wang S, Yang YG (2006) Reconstitution of a functional human immune system in immunodeficient mice through combined human fetal thymus/liver and CD34+ cell transplantation. Blood 108(2):487–492. https://doi.org/10.1182/blood-2005-11-4388
25. Aksentijevich I, Sachs DH, Sykes M (1992) Humoral tolerance in xenogeneic BMT recipients conditioned by a nonmyeloablative regimen. Transplantation 53(5):1108–1114
26. Sharabi Y, Aksentijevich I, Sundt TM 3rd, Sachs DH, Sykes M (1990) Specific tolerance induction across a xenogeneic barrier: production of mixed rat/mouse

lymphohematopoietic chimeras using a nonlethal preparative regimen. J Exp Med 172 (1):195–202

27. Griesemer A, Liang F, Hirakata A et al (2010) Occurrence of specific humoral non-responsiveness to swine antigens following administration of GalT-KO bone marrow to baboons. Xenotransplantation 17 (4):300–312. https://doi.org/10.1111/j.1399-3089.2010.00600.x
28. Tseng YL, Dor FJ, Kuwaki K et al (2004) Bone marrow transplantation from alpha1,3-galactosyltransferase gene-knockout pigs in baboons. Xenotransplantation 11 (4):361–370. https://doi.org/10.1111/j.1399-3089.2004.00151.x
29. Liang F, Wamala I, Scalea J et al (2013) Increased levels of anti-non-gal IgG following pig-to-baboon bone marrow transplantation correlate with failure of engraftment. Xenotransplantation 20(6):458–468. https://doi.org/10.1111/xen.12065
30. Cui J, Wahl RL, Shen T et al (1999) Bone marrow cell trafficking following intravenous administration. Br J Haematol 107 (4):895–902
31. Tasaki M, Wamala I, Tena A et al (2015) High incidence of xenogenic bone marrow engraftment in pig-to-baboon intra-bone bone marrow transplantation. Am J Transplant 15 (4):974–983. https://doi.org/10.1111/ajt.13070
32. Ide K, Wang H, Tahara H et al (2007) Role for CD47-SIRPalpha signaling in xenograft rejection by macrophages. Proc Natl Acad Sci U S A 104(12):5062–5066. https://doi.org/10.1073/pnas.0609661104
33. Wang H, VerHalen J, Madariaga ML et al (2007) Attenuation of phagocytosis of xenogeneic cells by manipulating CD47. Blood 109 (2):836–842. https://doi.org/10.1182/blood-2006-04-019794
34. Wang C, Wang H, Ide K et al (2011) Human CD47 expression permits survival of porcine cells in immunodeficient mice that express SIRPalpha capable of binding to human CD47. Cell Transplant 20(11–12):1915–1920. https://doi.org/10.3727/096368911X566253
35. Yang YG, deGoma E, Ohdan H et al (1998) Tolerization of anti-Galalpha1-3Gal natural antibody-forming B cells by induction of mixed chimerism. J Exp Med 187 (8):1335–1342
36. Tena AA, Sachs DH, Mallard C et al (2017) Prolonged survival of pig skin on baboons after Administration of pig Cells Expressing Human CD47. Transplantation 101(2):316–321. https://doi.org/10.1097/TP.0000000000001267
37. Sengupta D, Harper M, Jennett B (1974) Effect of carotid ligation on cerebral blood flow in baboons. 2. Response to hypoxia and haemorrhagic hypertension. J Neurol Neurosurg Psychiatry 37(5):578–584

Chapter 15

Studying Xenograft Rejection of Bioprosthetic Heart Valves

Rizwan A. Manji and Jacqueline S. Manji

Abstract

Millions of patients with valvular heart disease have benefitted from heart valve replacement since the procedure was first introduced in the 1960s; however, there are still many patients who get early structural valve deterioration (SVD) of their bioprosthetic heart valves (BHV). BHV are porcine, bovine, or equine tissues that have been glutaraldehyde fixed to preserve the tissue and presumably make the tissue immunologically inert. These glutaraldehyde-fixed BHV with anti-calcification treatments last long periods of time in older adults but develop early SVD in younger patients. The consensus at present is that the early SVD in younger patients is due to more "wear and tear" of the valves and higher calcium turnover in younger patients. However, as younger patients likely have a more robust immune system than older adults, there is a new hypothesis that BHV xenografts may undergo xenograft rejection, and this may contribute to the early SVD seen in younger patients.

At present, the technology to noninvasively study in vivo whether an implanted BHV in a human patient is undergoing rejection is not available. Thus, a small animal discordant xenotransplant model in young rodents (to match the young patient getting a pig/bovine/equine BHV) was developed to study whether the hypothesis that glutaraldehyde-fixed BHV undergo xenograft rejection had any merit. In this chapter, we describe our model and its merits and the results of our investigations. Our work provides clear evidence of xenograft rejection in glutaraldehyde-fixed tissue, and our small animal model offers an opportunity to study this process in detail.

Key words Bioprosthetic heart valve, Glutaraldehyde, Cardiac surgery, Calcification, Xenograft rejection, Young patient

1 Introduction

Heart valve surgery began in the 1960s with many challenges [1]. Surgeons initially tried to repair heart valves or make new heart valves in the operating room with the patient's own pericardial tissue, but these procedures had high failure rates [1]. Surgeons also tried to use cadaveric tissue with better success, but availability and storage of these cadaveric valves were challenging in the 1960s [2, 3]. Eventually, the decision was made to make and use artificial heart valves, and this allowed the surgery for heart valve diseases in

Cristina Costa (ed.), *Xenotransplantation: Methods and Protocols*, Methods in Molecular Biology, vol. 2110,
https://doi.org/10.1007/978-1-0716-0255-3_15, © Springer Science+Business Media, LLC, part of Springer Nature 2020

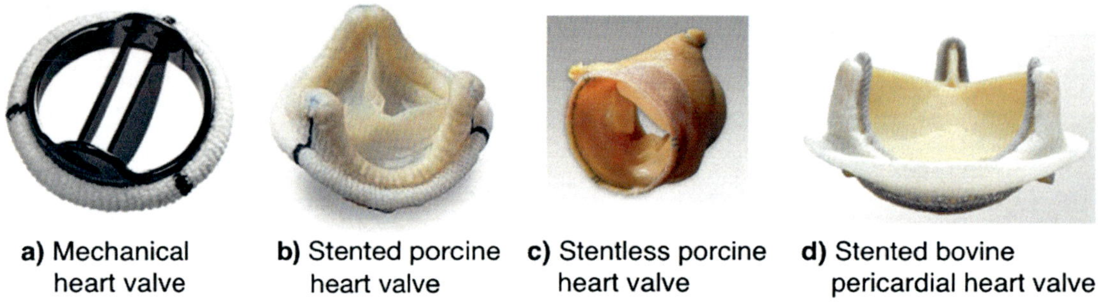

Fig. 1 Examples of heart valves implanted into patients. Heart valves can be mechanical (**a**), porcine valve that is mounted on a stent (**b**), porcine aortic valve and ascending aorta without any stent (**c**), and bovine pericardium that has been formed into a valve and mounted on a stent (**d**)

humans possible and sustainable. At present, over 250,000 heart valves are replaced worldwide each year [4].

There are generally two categories of artificial heart valves—a mechanical heart valve (MHV) and a bioprosthetic heart valve (BHV) (Fig. 1). MHV have long-term durability of at least 18–20 years [5] but need anticoagulation (blood thinners) which poses a significant risk to patients both in terms of potential fatal bleeding if blood thinner level is too high and potential fatal clotting (e.g., stroke) if blood thinner level is too low [4, 6]. There are different BHV on the market (Fig. 1) with their specific surgical advantages and disadvantages. BHV do not need anticoagulation and are thus safer from a bleeding/clotting perspective, but BHV break down much quicker—a process called structural valve deterioration (SVD)—that can be seen as early as 6–8 years after implant [7]. SVD involves thickening and calcification of the valve which narrows the valve's opening leading to obstruction of blood flow (stenosis). The thickening and calcification can also lead to tears in the valve tissue making the valve leak (regurgitation). SVD can thus lead to valve failure and often necessitates repeat cardiac surgery to replace the valve, and this has its own risks [8].

BHV are xenografts as they are made from porcine, bovine, or equine tissues, but they are fixed with glutaraldehyde. Glutaraldehyde fixation gives the tissue some stability, but it is also supposed to make the xenograft tissue essentially immunologically inert [9]. The thickening and calcification are thought to occur due to chemical processes related to the glutaraldehyde fixation [5, 10]. Indeed, valve companies have developed multiple proprietary methods to decrease the calcification of BHV with great success but mostly in older adults—young patients still get early SVD and failure of the valves despite the anti-calcification procedures used [7, 10–13]. The general understanding at present is that young patients have increased "wear and tear" of the BHV as well as increased calcium turnover, and that is the reason that younger patients get early SVD of BHV [4, 14]. However, patients after their teenage years are not growing fast enough to have high calcium turnover, their heart rates are probably similar or maybe even slower (if athletic) than older adults, and the pressures generated inside their hearts (opening and closing their heart valves) are likely similar or lower than older adults with high blood pressure and heart muscle hypertrophy [15]. Thus, one wonders if there is another process besides "wear and tear" and high calcium turnover to explain the higher rate of SVD in younger patients. Younger patients likely have a more robust immune system than older adults [16–18], and there are many chronic inflammatory/immune diseases such as atherosclerosis, sarcoidosis, and tuberculosis where one gets calcification [19–21]. Perhaps xenograft rejection of the glutaraldehyde-fixed tissue leads to the thickening/calcification seen in the BHV that have SVD.

There is some evidence to suggest that there may be a rejection process (cellular infiltrate and antibody binding) to BHV implanted into humans, but much of this evidence comes from patients after their valve has failed and been removed [22–26]. This does not allow for in-depth study of the process. In addition, cardiac surgeons and cardiologists understand the failure of BHV to be dependent on calcification of the valve (they do not necessarily think xenograft rejection). At present we do not have any good technology to be able to detect rejection occurring on the valve in the patient in real time and to be able to correlate the immune response to the valve with calcification of the valve. Demonstrating that there is a rejection process, that there is a correlation between the immune response and calcification, and that there may be ways to mitigate the response would be important for clinicians. Showing these factors and relationships may help to change the way clinicians think about the disease process and potentially lead to new therapies for patients (especially young patients) with heart valve disease (e.g., using genetically modified xenograft valves).

As BHV (porcine/bovine/equine tissue) implantation into humans is a discordant xenotransplant and as SVD is mostly a problem in young patients, we have developed a small animal discordant xenotransplantation model (guinea pig to rat) using young rats as recipients of the xenograft. This model has the advantages of using syngeneic (genetically identical) Lewis rats to minimize variability in the experiments, being relatively inexpensive allowing one to test a number of variables if needed, giving one the ability to do a detailed study of the time course of the immune response if needed, and giving one the ability to introduce different therapies to see the different responses. Disadvantages include the fact that the implanted valve does not open and close like it would in a normal heart (since the implant is put into the abdominal aorta where the valve stays open the whole time) and important antigens in the porcine/bovine/equine to human combination such as galactose α1,3 galactose (αGal) and N-glycolylneuraminic acid (Neu5Gc) cannot be examined in this model as those xenoantigens would not be relevant in the guinea pig-to-Lewis rat model. However, the model we describe here still has a lot of important benefits as outlined above.

In a study we conducted using the model [27], aortic valve aorta grafts from rats (syngeneic) or guinea pigs (xenogeneic) were implanted into the infrarenal aortas of young rat recipients either fresh or after 48 h of glutaraldehyde fixation. Recipient rat serum was collected at the time of the graft implant for measurement of rat IgG levels. The grafts were harvested at 20 days postoperatively, at which time rat serum was again collected for rat IgG measurements, and tissue was collected for histology and immunohistochemistry (IHC) studies and calcium determinations. The experimental groups were fresh rat aortic-valve and aorta graft into infrarenal aorta of rat (control for surgery), glutaraldehyde-fixed rat to rat (syngeneic control for glutaraldehyde), glutaraldehyde-fixed guinea-pig aortic-valve and aorta graft into infrarenal aorta of rat (xenogeneic experimental group), and glutaraldehyde-fixed guinea pig to rat that was treated with steroids from time of implant to time of harvest at 20 days (xenogeneic intervention group).

Our results showed that there were over two times more adventitial T-cell/macrophage infiltrate, over three times more valve inflammation, over 10 times more valve T-cell/macrophage infiltrate, and over three times antibody rise in the guinea-pig-to-rat xenogeneic experimental group compared to the syngeneic control groups (fresh or glutaraldehyde fixed) with the results being statistically significant ($p < 0.05$). Steroid treatment decreased antibody rise and inflammation in the guinea-pig-to-rat xenogeneic groups ($p < 0.05$). Media/adventitia inflammation ($p = 0.02$) and percent macrophage infiltration ($p = 0.01$) correlated to the amount of calcification establishing a crucial link between the immune response and calcification [27].

There have been several publications since our paper was published confirming that indeed there is evidence of xenograft rejection in glutaraldehyde-fixed tissue and that there is a relationship between the immune response and calcification. There are publications looking at current day BHV and the role that xenograft rejection may play in the SVD of the valves. There has been demonstration of αGal and Neu5Gc antigen on the valves and that these antigens are recognized by human antibody with diminution of the antibody recognition in pig valves that have had αGal and Neu5Gc knocked out [28–33]. Studies have also been reported demonstrating an increase in serum anti-αGal antibody in patients implanted with BHV [30, 34, 35]. There have been studies demonstrating a connection between anti-αGal antibody and calcification [36, 37] and multiple studies trying to decrease the antigenicity and calcification of xenograft valves including the creation of genetically engineered animals for heart valve replacement [38–42]. Limited testing of genetically engineered BHV has also been done in non-human primates (NHP) implanting the valve into the heart where it would open and close and need to work properly for the NHP's health and survival [43]. It was found in these NHP that αGal antibody levels were elevated postoperatively [43]. As NHP do not make anti-Neu5Gc antibody, the Neu5Gc antigen-antibody interaction (which is presumed to be of importance in humans) could not be tested. Even though the NHP model would be closer to the human situation, that model is much more expensive and much more complex (need for specialized equipment to do open heart surgery) and requires much higher specialized skill sets and human resources to be practical to use. Our model is much more practical and affordable allowing one to still study important concepts in the xenograft rejection of glutaraldehyde-fixed tissue. We provide here details of this model.

2 Materials

2.1 Anesthetic and Post-anesthetic Care

1. Isoflurane and anesthetic vaporizer and delivery system.
2. Oxygen tank.
3. Warming water blanket and warm blanket.
4. Heating lamp.
5. Buprenorphine analgesic.
6. 0.9% saline or Ringer's lactate (warm).
7. 25–27-gauge needles and 1-mL syringes.
8. Methylprednisolone (0.5 mg/kg).

2.2 Surgical Procedure

1. 6- to 8-week-old inbred Lewis rats weighing 150–200 g (*see* **Note 1**).
2. Guinea pigs weighing 160–180 g (*see* **Note 2**).
3. Zeiss operating microscope.
4. Standard microsurgical instruments (11 blade and 15 blade scalpels, ophthalmic iris scissors, straight and curved microvascular scissors, curved microvascular needle driver, straight and curved microvascular forceps, Halsted-mosquito hemostats, microvascular clips).
5. 10-0 nylon sutures (Sharpoint, Reading, PA); 2-0 silk sutures.
6. Sterile gauze; sterile cotton tip sticks.
7. Sterile basins.
8. Heparin sodium injection.
9. Surgicel absorbable hemostatic agent (Ethicon).

2.3 Aortic Valve/Aorta Treatment

1. Glutaraldehyde (Sigma-Aldrich) (*see* **Note 3**).
2. 10% buffered formalin (Sigma-Aldrich) (*see* **Note 4**).

2.4 Histological Analyses

1. Hematoxylin-eosin (H&E).
2. Primary antibody anti-rat CD68 (MCA341R; Serotec, Raleigh, NC) (*see* **Note 5**).
3. Primary antibody anti-rat CD3 (MCA7722; Serotec, Raleigh, NC) (*see* **Note 6**).
4. Secondary antibody biotinylated goat anti-mouse antibody (Jackson ImmunoResearch, West Grove, PA) (*see* **Note 7**).
5. ABC reagent (Vector Laboratories, Burlingame, CA); DAB (BioGenex, San Ramon, CA).
6. AxioVision software program (Carl Zeiss Canada Ltd., Toronto, Ontario) or equivalent.
7. MetaMorph software program (Universal Imaging Corporation, Downingtown, PA) or equivalent.

2.5 Calcium Assay

1. Polypropylene round bottom tubes (5 mL); microfuge tubes (1.5 mL).
2. Centrifuge (10,000 × *g*); 530 nm on a Dynex MRII plate reader.
3. Calcium chloride (to make standard); 0.1 M HCl; 0.4 M HCl; 0.15 M NaOH.
4. o-Cresolphthalein complexone; 8-hydroxyquinoline; 2-aminoethanol (ethanolamine); KCN.
5. Copper sulfate pentahydrate (molecular weight 249.7 g/mol); sodium citrate (molecular weight 294.1 g/mol); sodium carbonate, sodium hydroxide.

2.6 Rat IgG ELISA

1. 96-well ELISA plates.
2. Rat IgG ELISA kit from Bethyl Laboratories Inc. (Montgomery, TX).
3. Coating antibody—goat anti-rat IgG-Fc (A110-136A) in carbonate-bicarbonate buffer (*see* **Note 8**).
4. Control rat sera (RS10-100) (*see* **Note 9**).
5. Secondary antibody—goat anti-rat IgG-Fc HRP conjugate (A110-136P) at 1:10,000 dilution (*see* **Note 10**).
6. TMB substrate and peroxidase, as well as plate reader to read at 450 nm.

3 Methods

3.1 Surgical Procedure, Postoperative Care, and Valve/Graft Harvest

The procedure followed was a modification of the technique by Yankah et al. [44].

3.2 Donor Preparation

1. Administer subcutaneously 5 mL of warm Ringer's lactate solution to the donor rat or guinea pig (which was ~50 g smaller than the recipient rat) immediately before the surgery. Use a warm heating blanket under the animal during the surgery.
2. Induce anesthesia of the donor animal using a mixture of oxygen and 4–5% isoflurane in the anesthesia system. Once induced (as assessed by loss of digital sensitivity), decrease the isoflurane to 1.5% for maintenance of anesthesia.
3. Perform a median sternotomy exposing the heart, aorta, and lungs (*see* **Note 11**).
4. Administer heparin (500 units) distal to the aortic arch to anticoagulate the donor and avoid any clotting around the aortic valve or the aortic graft.
5. Excise the heart and the ascending aorta to the aortic arch (*see* Fig. 2a) using scissors.
6. Cut the aorta proximally just beneath the aortic valve taking a small amount of muscle from the left ventricular outflow tract and cut also the aorta distally just proximal to the take-off of the head vessels (*see* Fig. 2a).
7. Ligate to close the two coronary arteries with 10-0 nylon suture.
8. Then process the aortic valve aorta graft (*see* Subheading 3.7).

3.3 Recipient Procedure (See Note 12)

1. Administer subcutaneously 5 mL of warm Ringer's lactate solution to the recipient immediately before the surgery. Use a warm heating blanket under the animal during the surgery.
2. Induce anesthesia of the donor animal using a mixture of oxygen and 4–5% isoflurane in the anesthesia system. Once induced (as assessed by loss of digital sensitivity), decrease the isoflurane to 1.5% for maintenance of anesthesia.
3. Perform a midline laparotomy. Retract the bowels to the right and wrap in a warm saline-soaked gauze (*see* **Note 13**).
4. Identify the aorta and inferior vena cava below the renal (kidney) arteries and excise the tissue (retroperitoneal tissue) to allow gentle mobilization of the aorta and inferior vena cava (*see* Fig. 2b).

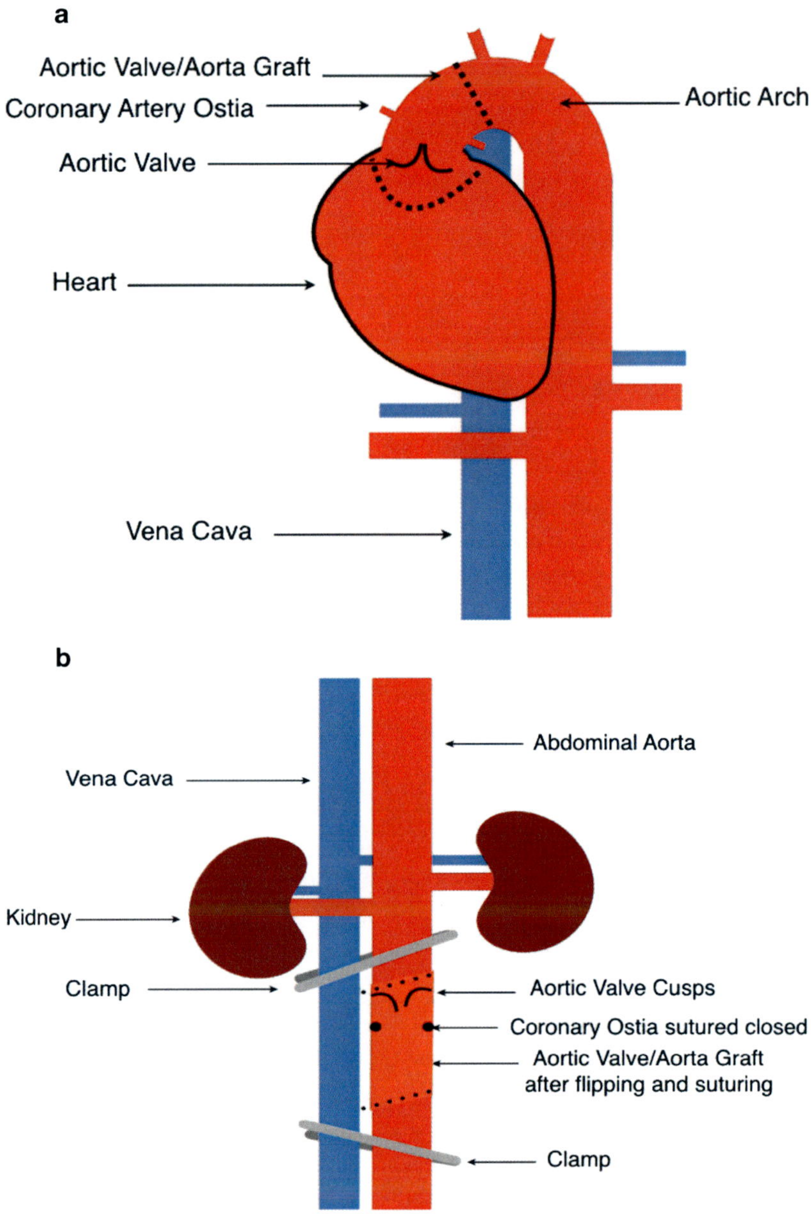

Fig. 2 The donor animal (rat or guinea pig) has the chest cut open through a midline sternotomy. The coronary ostia are sutured closed. The aortic valve is harvested (just below the aortic annulus) as is the ascending aorta up to start of aortic arch (**a**). The aortic valve aorta graft is then flipped and sutured into the infrarenal aorta of the donor rat so that the aortic valve is superior and opens when blood flows by superiorly (**b**). As the ascending aorta is larger than the abdominal aorta, the incisions in the abdominal aorta needed to be slightly diagonal to allow proper matching for anastomosis

5. Administer heparin (500 units) through the infrahepatic vena cava using a 27-gauge needle to anticoagulate the rat and avoid thrombosis of the graft during the time the clamp is applied.
6. After 5–10 min (to ensure the recipient is anticoagulated), use a 25-gauge needle to collect 500–1000 μL of blood from the infra-renal aorta (part of the aorta to be excised).
7. Collect the blood into a microfuge tube and centrifuge at 10,000 × *g*. Obtain the serum from the upper layer and freeze at −80 °C for future rat ELISA studies.
8. Place clamps across the aorta below the renal arteries proximally and distally (*see* Fig. 2b) and excise the segment of the aorta in between. The cut in the recipient aorta should be made slightly diagonally both proximally and distally to allow the smaller abdominal aorta to accommodate the larger ascending aorta/aortic valve (*see* **Note 14**).
9. Then flip the aortic valve aorta graft from the donor so the aortic valve portion is located superior to the rest of the fragment, and the ascending aorta portion remains inferior (*see* **Note 15**).
10. Suture the donor graft into place in an end-to-end fashion using 10-0 nylon suture (*see* **Note 16**).
11. Once the graft was sutured in, gently remove the clamps.
12. Hemostasis is ensured by placing gentle pressure on the suture lines and Surgicel over the graft.
13. Place the bowel back in the abdomen and close the abdomen in one layer using 2-0 silk suture (*see* **Notes 17** and **18**).
14. Discontinue the anesthetic (oxygen can be continued) and allow the recipient to wake up.

3.4 Postoperative Care

1. Monitor closely the recipient rat for the next 24–48 h (depending on how the recipient is doing).
2. Allow the rat to eat and drink ad libitum.
3. Give the rat 5–10 mL of warm saline subcutaneously post-surgery as needed for the first 2–3 days (i.e., if the rat is not drinking well and seems dehydrated).
4. Administer buprenorphine (0.1 mg/kg) to the rat for analgesia twice daily subcutaneously for the first 3 days post-surgery and then as needed if it appears in pain (lethargic, not active, and/or not eating/drinking well).
5. Monitor the rats for wound infection (by examination of the wound for redness and purulent discharge and the overall behavior such as mobility, eating/drinking, and interactions) (*see* **Note 19**).

3.5 Steroid Treatment

For the recipient getting steroid, methylprednisolone (0.5 mg/kg) is dissolved in saline and administered into the peritoneum in the right lower quadrant of the abdomen using a 25-gauge needle daily for the 20 days of treatment (*see* **Note 20**).

3.6 Graft Harvest

1. After 20 days from graft implant, harvest the aortic valve aorta graft.
2. Induce anesthesia of the recipient animal by using a mixture of oxygen and 4–5% isoflurane. Once induced, decrease the isoflurane to 1.5% for maintenance of anesthesia.
3. Perform a midline laparotomy, retract the bowels to the right, and wrap them in a warm saline-soaked gauze as described above.
4. Administer heparin (500 units) through the abdominal aorta above the graft using a 27-gauge needle.
5. After 5–10 min (to allow the recipient to be properly anticoagulated), insert a 25-gauge needle into the aorta above the graft (to prevent injuring the graft) and collect as much blood as could be collected. This blood is put into a microfuge tube and centrifuged at 10,000 × *g*. Then remove the serum and freeze it at −80 °C for future rat ELISA studies.
6. Identify the previously implanted aortic valve aorta graft and mobilize the tissue.
7. Excise the graft at the previous suture lines and place it in saline to wash off excess blood and debris. Process then the aortic valve aorta graft as outlined below (*see* Subheading 3.7) (*see* **Note 21**).

3.7 Aortic Valve and Aorta Graft Processing

1. After harvest from the donor and prior to implant into the recipient, keep the aortic valve aorta graft either in saline at room temperature until ready for implant into the recipient (fresh group) or placed in 0.6% glutaraldehyde for 48 h at room temperature.
2. Prior to implant of the glutaraldehyde fixed aortic valve aorta graft, thoroughly rinse the graft at least three times in sterile saline to remove any excess glutaraldehyde.
3. At the time of recipient graft harvest, cut transversely the aortic valve segment (superior in the recipient) and place it in 10% formalin for histology and immunohistochemistry studies.
4. Then cut vertically the ascending aorta with half being frozen at −80 °C for calcium studies and the other half placed in 10% formalin for histology and immunohistochemistry studies.

3.8 Histological Analyses

We use standard procedures for processing of the tissue samples, generating the paraffin blocks and basic staining techniques such as hematoxylin-eosin (H&E). Automated histological systems can be used when available. For immunohistochemistry, standard methods are also followed [45].

1. Cut tissue samples (aortic valve and aorta graft) embedded in paraffin blocks into 4-μm sections and stain with H&E.
2. Conduct the histological analysis on two separate occasions in a blinded fashion examining the valve and aortic wall (media and adventitia) for cellularity/inflammation and providing a score for each sample (*see* **Note 22**). An appropriate scoring system is 0, no inflammation; <1, mild inflammation (consistent with postoperative healing); 2, moderate inflammation; 3, severe inflammation; and 4, very severe inflammation (sheets of inflammatory cells). The presence of eosinophils, thrombus, and infection can also be recorded.
3. For immunohistochemistry, use the antibodies outlined in Subheading 2.
4. Quantitation of the immunohistochemistry is conducted by taking four pictures from random parts of each slide using a software program such as AxioVision and then an image analysis software program such as MetaMorph to determine what percentage of positive signal in each picture (brown staining from CD3 and CD68). The various percentages per picture are averaged to determine how much of each section is positive (CD68 or CD3).

3.9 Calcium Content of the Aortic Valve/Aorta Conduit

The calcium determination procedure described is based on the method of Sarkar and Chauhan [46] with protein determination based on the standard Lowry method [47] (*see* **Note 23**). Note that calcium values are expressed as micrograms of calcium per milligram of protein for the sample. Furthermore, the calcium results can be determined from the native valve/aorta tissue from each recipient and this value subtracted from the graft value (as an internal standard control).

1. Calcium assay (**steps 1–13**): Prepare calcium standard by using a 10 mg/mL calcium chloride solution (in distilled water) and dilute 1/1000 to 10 μg/mL.
2. Prepare samples by placing tissues in a 5-mL polypropylene round-bottomed tube.
3. Add 300 μL of 0.1 M HCl.
4. Grind.
5. Put 50 μL of the mixture into a 1.5-mL microfuge tube for protein assay (see later, **step 18**).

6. Place remaining sample into a 1.5-mL microfuge tube and let sit for 15 min at room temperature.
7. Prepare reaction mixture A for calcium assay: Mix 0.0019 g of o-Cresolphthalein complexone (0.3 mM), 0.0260 g of 8-hydroxyquinoline (20 mM), and 10 mL of 0.4 M HCl.
8. Prepare reaction mixture B for calcium assay: Mix 0.96 mL of 2-aminoethanol (ethanolamine), 0.00325 g of KCN, and 9.14 mL of distilled water.
9. Prepare working reagent by mixing mixture A and mixture B 1:1 (*see* **Note 24**).
10. Take the sample that has been sitting at room temp for 15 min and centrifuge it at 10,000 × *g* for 5 min.
11. Pipette 20 μL of each sample into a multi-well plate.
12. Add 200 μL of the reaction mixture (see above) to each well.
13. Read plate at 530 nm.
14. Protein assay (**steps 14–28**): Prepare Lowry A for protein assay: Take 0.5 g of copper sulfate pentahydrate (MW = 249.7 g/mol) and 1 g of sodium citrate (MW = 294.1 g/mol) to a 50-mL graduated cylinder, and dissolve using distilled water. Dilute up to full volume (50 mL) using distilled water.
15. Prepare Lowry B for protein assay: To make 250 mL of Lowry "B," take 10 g of sodium carbonate and 2 g of sodium hydroxide and dissolve in distilled water. Top up to full volume (250 mL) using a graduated cylinder and distilled water.
16. Make a 4% albumin stock solution for the protein assay (weigh out 200 mg of bovine serum albumin (BSA) into a 5-mL volumetric flask, and fill to final volume with sterile saline. Store in −70 °C freezer.
17. Make a 0.04% standard solution of BSA by adding 20 μL of 4% solution to 1980 μL of sterile saline. Store in 5-mL aliquots in −70 °C freezer.
18. Take the 50-μL aliquot collected earlier (in the 1.5-mL microfuge tube) and add 200 μL of 0.15 M NaOH.
19. Incubate at room temp until ready to proceed with assay.
20. Centrifuge at 10,000 × *g* for 5 min.
21. Pipette the BSA standard (10, 20, 30, 40, 50,100 μL in triplicate) into first three columns of a 96-well plate.
22. Pipette 10–30 μL of the diluted homogenate mixture.
23. Prepare reaction mixture: 49 mL of Lowry B + 1 mL of Lowry A (see **note 25**).
24. Add 200 μL of reaction mixture to each well.

25. Leave plate for 10 min.
26. Add 20 μL of Folin reagent to each well and mix well (very important)—use a 200-μL multipipettor.
27. Leave plates for 30 min.
28. Read plate using plate reader with a 750-nm filter on metabolite assay (protein assay).

3.10 Rat IgG ELISA

The rat IgG ELISA is conducted using standard protocols [48] with the antibodies and reagents outlined in Subheading 2. The amount of rat IgG from time of graft implant is then compared to the amount corresponding to the graft harvest for the different groups.

4 Notes

1. All animal procedures need to be approved by the local institution's animal welfare/ethics committees.
2. The sizes of the rats and guinea pigs are important to make sure the vessels are of appropriate size for the vascular anastomoses. The ascending aorta/aortic valve donor graft are generally larger in diameter than the abdominal aorta of the recipient. Thus, the donor rat or guinea pig should be about 50 g smaller than the recipient rat.
3. Glutaraldehyde solution 50% wt% in water needs to be diluted to 0.6% glutaraldehyde for glutaraldehyde fixation of the valves. The solution can be stored at room temperature.
4. 10% buffered formalin can be purchased directly from Sigma-Aldrich.
5. For detection of rat macrophages and monocytes. The 1:100 dilution worked best for our studies.
6. For detection of rat T cells. The 1:200 dilution worked best for our studies.
7. A dilution of 1:200 worked best for our studies.
8. For coating antibody, 1:100 dilution worked best for our studies.
9. Standard rat sera were used to create a standard curve.
10. For secondary antibody, a dilution of 1:10,000 worked best for our studies.
11. A scalpel was used to incise the skin from the neck down to the middle of the abdomen. Scissors were then used to make a hole under the sternum and then the sternum cut open with scissors.

12. In our hands, we could not get the guinea pig to be recipient for any of the procedures as the abdominal aorta of the guinea pig was too small to put in an aortic valve aorta graft.
13. A scalpel was used to make a midline incision from just below the xyphoid process of the sternum down to the pubic area of the rat. Care had to be taken not to puncture the pleural spaces at the top part of the incision and not to injure the bladder at the bottom part of the incision. Scissors were used after the initial skin incision to open the abdomen.
14. It is important to avoid clamping above the renal arteries to avoid causing kidney injury. It is important to not excise aorta above the renal arteries as that procedure increases the risk of compromising blood flow to other abdominal organs (e.g., bowel). Care also needs to be applied in placing the clamps to avoid tearing any vessels.
15. In implanting a glutaraldehyde-fixed valve, it is necessary to thoroughly rinse (at least three times) the aortic valve/aorta conduit in a large volume of saline to remove excess glutaraldehyde.
16. The glutaraldehyde-fixed grafts are harder than non-glutaraldehyde-fixed tissue and could easily tear. Thus, when suturing this tissue, it is important to follow the line of curvature of the needle while putting the suture needle in and taking it out of the tissue; it is important to keep the suture thread parallel to the line of entrance and exit while pulling the thread out of the tissue and be careful when tying the square knots of the sutures not to tear the tissue.
17. It is important to make sure that the bowel is returned to the abdomen in proper orientation (i.e., how it was taken out)—make sure no new twists have been introduced. Also, during the operation, if the gauze covering the bowel was starting to get dry, then a few mL of warm saline or Ringer's lactate solution should be poured onto the gauze and bowel.
18. Care should be taken when closing the abdomen to avoid accidentally catching any bowel or other abdominal contents in the suture.
19. If the rat gets a serious wound infection or becomes ill, the rat should be humanely euthanized following institutional policies.
20. Care should be taken not to have the needle go too deep preventing in this way hitting the bowel. The rat should be observed for evidence of infection (steroids can increase infection risk). If the rat becomes ill, it should be humanely euthanized following institutional policies.

21. Care should be taken not to injure the aortic valve when harvesting the graft.
22. The histological analysis is better conducted by a pathologist or under his/her supervision. Areas with excessive amounts of muscle or suture material should be avoided in the histological scoring.
23. The principle of the calcium determination method is that an acid solution will cause calcium to disassociate from proteins and then the calcium will react with o-cresolphthalein complexone (OCPC). Changing the conditions to alkali will cause the Ca-OCPC complex to form a purple color, which is measured at 530 nm on a plate reader. The color development is proportional to the amount of calcium in the sample. Interference from heavy metals is controlled by KCN, and magnesium interference is blocked by 8-hydroxyquinoline.
24. Must be prepared fresh each time.
25. This mixture is not stable. Prepare just before use.

References

1. Cooley DA, Frazier OH (2000) The past 50 years of cardiac surgery. Circulation 102: IV-87–IV-93
2. Ross D (1962) Homograft replacement of the aortic valve. Lancet 2:487–488
3. Barratt-Boyes BG (1965) A method for preparing and inserting a homograft aortic valve. Br J Surg 152:847–856
4. Siddiqui RF, Abraham JR, Butany J (2009) Bioprosthetic heart valves: modes of failure. Histopathology 55:135–144
5. Turina J, Hess OM, Turina M et al (1993) Cardiac bioprostheses in the 1990s. Circulation 88(2):775–781
6. Le Tourneau T, Lim V, Inamo J et al (2009) Achieved anticoagulation vs prosthesis selection for mitral mechanical valve replacement: a population-based outcome study. Chest 136:1503–1513
7. Hammermeister K, Sethi GK, Henderson WG et al (2000) Outcomes 15 years after valve replacement with a mechanical versus a bioprosthetic valve: final report of the veterans affairs randomized trial. J Am Coll Cardiol 36:1152–1158
8. Schoen FJ, Levy RJ (2005) Calcification of tissue heart valve substitutes: progress toward understanding and prevention. Ann Thorac Surg 79:1072–1080
9. Carpentier A, Deloche A, Relland J et al (1974) Six-year follow-up of glutaraldehyde-preserved heterografts. With particular reference to the treatment of congenital valve malformations. J Thorac Cardiovasc Surg 68:771–782
10. Piazza N, Bleiziffer S, Brockmann G et al (2011) Transcatheter aortic valve implantation for failing surgical aortic bioprosthetic valve from concept to clinical application and evaluation (Part 1). J Am Coll Cardiol Intv 4:733–744
11. Yun KL, Miller DC, Moore KA et al (1995) Durability of the Hancock MO bioprosthesis compared with the standard aortic valve bioprosthesis. Ann Thorac Surg 60:5221–5228
12. Huang G, Rahimtoola SH (2011) Prosthetic heart valve. Circulation 123:2602–2605
13. Bottio T, Rizzoli G, Caprili L et al (2005) Biological versus mechanical aortic prosthesis? A nineteen-year comparison in a propensity-matched population. J Heart Valve Dis 14:493–500
14. Simionescu DT (2004) Prevention of calcification in bioprosthetic heart valves: challenges and perspectives. Expert Opin Biol Ther 4:1971–1985
15. Manji RA, Menkis AH, Ekser B et al (2014) Bioprosthetic heart valves of the future. Xenotransplantation 21:1–10
16. Nikolich-Žugich J (2014) Aging of the T cell compartment in mice and humans: from no naive expectations to foggy memories. J Immunol 193:2622–1629

17. Manji RA, Lee W, Cooper DKC (2015) Xenograft bioprosthetic heart valves: past, present and future. Int J Surg 23:280–284
18. Manji RA, Menkis AH, Ekser B et al (2012) Porcine bioprosthetic heart valves: the next generation. Am Heart J 164:177–185
19. Doherty TM, Asotra K, Fitzpatrick LA et al (2003) Calcification in atherosclerosis: bone biology and chronic inflammation at the arterial crossroads. Proc Natl Acad Sci U S A 100:11201–11206
20. Basaraba RJ, Bielefeldt-Ohmann H, Eschelbach EK et al (2008) Increased expression of host iron-binding proteins precedes iron accumulation and calcification of primary lung lesions in experimental tuberculosis in the Guinea pig. Tuberculosis 88:69–79
21. Das BB, Shoemaker L, Kim E et al (2012) Severe calcification of the aorta (porcelain aorta) associated with sarcoidosis in a pediatric heart transplant recipient. Pediatr Transplant 16:E162–E166
22. Talbert WM Jr, Wright P (1982) Acute aortic stenosis of a porcine valve heterograft apparently caused by graft rejection: case report with discussion of immune mediated host response. Tex Heart Inst J 9:225–229
23. Stein PD, Wang CH, Riddle JM, Magilligan DJ Jr (1988) Leukocytes, platelets, and surface microstructure of spontaneously degenerated porcine bioprosthetic valves. J Card Surg 3:253–261
24. Riddle JM, Magilligan DJ Jr, Stein PD (1981) Surface morphology of degenerated porcine bioprosthetic valves four to seven years following implantation. J Thorac Cardiovasc Surg 81:279–287
25. Rocchini AP, Weesner KM, Heidelberger K et al (1981) Porcine xenograft valve failure in children: an immunologic response. Circulation 64:II162–II171
26. Manji RA, Hara H, Cooper DK (2015) Characterization of the cellular infiltrate in bioprosthetic heart valves explanted from patients with structural valve deterioration. Xenotransplantation 22:406–407
27. Manji RA, Zhu LF, Nissar NK et al (2006) Glutaraldehyde-fixed bioprosthetic heart valve conduits calcify and fail from xenograft rejection. Circulation 114:318–327
28. Naso F, Gandaglia A, Bottio T et al (2013) First quantification of alpha-Gal epitope in current glutaraldehyde-fixed heart valve bioprostheses. Xenotransplantation 20:252–261
29. Reuven EM, Leviatan Ben-Arye S et al (2016) Characterization of immunogenic Neu5Gc in bioprosthetic heart valves. Xenotransplantation 23:381–392
30. Park CS, Park SS, Choi SY et al (2010) Anti alpha-gal immune response following porcine bioprosthesis implantation in children. J Heart Valve Dis 19:124–130
31. Lee W, Hara H, Cooper DK et al (2015) Expression of NeuGc on pig heart valves. Xenotransplantation 22:153–154
32. Lee W, Long C, Ramsoondar J et al (2016) Human antibody recognition of xenogeneic antigens (NeuGc and Gal) on porcine heart valves: could genetically modified pig heart valves reduce structural valve deterioration? Xenotransplantation 23:370–380
33. Naso F, Gandaglia A, Iop L et al (2011) First quantitative assay of alpha-Gal in soft tissues: presence and distribution of the epitope before and after cell removal from xenogeneic heart valves. Acta Biomater 7:1728–1734
34. Park CS, Oh SS, Kim YE et al (2013) Anti-alpha-Gal antibody response following xenogeneic heart valve implantation in adults. J Heart Valve Dis 22:222–229
35. Konacki KZ, Bohle B, Blumer R et al (2005) Alpha-Gal on bioprostheses: xenograft immune response in cardiac surgery. Eur J Clin Investig 35:17–23
36. Mcgregor CG, Carpentier A, Lila N et al (2011) Cardiac xenotransplantation technology provides materials for improved bioprosthetic heart valves. J Thorac Cardiovasc Surg 141:269–275
37. Lila N, Mcgregor CG, Carpentier S et al (2010) Gal knockout pig pericardium: new source of material for heart valve bioprostheses. J Heart Lung Transplant 29:538–543
38. Lim HG, Choi SY, Yoon EJ et al (2013) In vivo efficacy of alpha-galactosidase as possible promise for prolonged durability of bioprosthetic heart valve using alpha1,3-galactosyltransferase knockout mouse. Tissue Eng A 19:2339–2348
39. Lim HG, Kim GB, Jeong S et al (2015) Development of a next-generation tissue valve using glutaraldehyde-fixed porcine aortic valve treated with decellularization, α-galactosidase, space filler, organic solvent and detoxification. Eur J Cardiovasc Surg 48:104–113
40. Kim MS, Jeong S, Lim HG et al (2015) Differences in xenoreactive immune response and patterns of calcification of porcine and bovine tissues in α-Gal knock-out and wild-type mouse implantation models. Eur J Cardiovasc Surg 48:392–399
41. Kim MS, Lim HG, Kim YJ (2016) Calcification of decellularized and alpha-galactosidase-

treated bovine pericardial tissue in an alpha-Gal knock-out mouse implantation model: comparison with primate pericardial tissue. Eur J Cardiovasc Surg 49:894–900

42. Naso F, Stefanelli U, Buratto E et al (2017) Alpha-gal inactivated heart valve bioprostheses exhibit an anti-calcification propensity similar to knockout tissues. Tissue Eng A 23:1181–1195
43. McGregor CG, Kogelberg H, Vlasin M et al (2013) Gal-knockout bioprostheses exhibit less immune stimulation compared to standard biological heart valves. J Heart Valve Dis 22:383–390
44. Yankah AC, Wottge HU, Muller-Ruchholtz W (1988) Antigenicity and fate of cellular components of heart valve allografts. In: Yankah AC, Hetzer R, Yacoub MH (eds) Cardiac valve allografts 1962–1987: current concepts on the use of aortic and pulmonary allografts for heart valve substitutes. Steinkopff Verlag, Darmstadt, pp 77–87
45. Hartmann K-U (1997) Indirect immunostaining, visualization by the horse- radish peroxidase (HP) reaction. In: Lefkovits I (ed) Immunology methods manual, vol 3. Academic Press pp, San Diego, pp 1229–1230
46. Sarkar BCR, Chauhan UPS (1967) A new method for determining micro quantities of calcium in biological materials. Anal Biochem 20:155–166
47. Lowry OH, Rosebrough NJ, Farr AL et al (1951) Protein measurement with the Folin phenol reagent. J Biol Chem 193:265–275
48. Mellado M, Kremer L, Manes S et al (1997) Screening in ELISA. In: Lefkovits I (ed) Immunology methods manual, vol 2. Academic Press, San Diego, pp 1146–1148

Chapter 16

Corneal Xenotransplantation: Anterior Lamellar Keratoplasty

Bertrand Vabres, Bernard Vanhove, and Gilles Blancho

Abstract

Corneal transplantation for the treatment of corneal blindness is challenging in many countries due to the shortage of graft procurement. Xenocorneal transplantation is an interesting alternative to explore despite immunologic rejection, which mainly involves endothelial cells. As anterior lamellar keratoplasty, when indicated, shows less immunologic reaction, we developed and describe below a pig-to-non-human-primate model of anterior lamellar corneal xenograft. This model can be used to assess the efficacy of corneas from genetically modified pigs.

Key words Xenotransplantation, Corneal transplantation, Pig, Primate, Anterior lamellar keratoplasty, CTLA4Ig

1 Introduction

Corneal diseases are the second leading cause of blindness and visual impairment. Although corneal grafting provides good results, the global shortage of donor tissue leaves untreated most of cases worldwide [1]. Furthermore, although corneal tissue procurement is effective in most developed countries, procurement remains challenging due to psychological or cultural conditions. Thus, active efforts are performed to develop alternatives to allogenic tissue donor, such as synthetic prosthesis (keratoprothesis) and more recently developments in bioengineered corneal substitutes [2]. In spite of these strategies being promising to correct corneal stromal dysfunction, they cannot nowadays pretend to correct corneal cellular (endothelial) dysfunction. Another option is exploring non-allogeneic corneal tissue procurement and xenografting, the historical first step to modern keratoplasties. With recent developments in immunology, a better understanding of the rejection process and gene therapy, as well as the availability of many genetically modified source pigs, xenotransplantation has gained acceptance. In order to provide sufficient amount of tissue

Cristina Costa (ed.), *Xenotransplantation: Methods and Protocols*, Methods in Molecular Biology, vol. 2110,
https://doi.org/10.1007/978-1-0716-0255-3_16,

at an acceptable cost and management burden, the pig appears as the most effective source after considering previous developments. Although graft failure by non-acute rejection occurs in full-thickness xenografts [3], lamellar graft shows better prognosis [4]. Preclinical studies have shown encouraging cornea xenograft survival [5, 6], including some studies performed in vivo in pig-to-nonhuman-primate (NHP) models. Penetrating and lamellar conventional pig cornea in NHP survive up to 3 months without any treatment and up to 4 months when local subconjunctival corticosteroid injections are administered at the time of transplantation [7]. Anterior lamellar keratoplasty is a partial corneal graft that spares the recipient endothelium and is indicated when the stroma is damaged (without endothelial alteration). It represents about 40% of all cornea allografts. Whenever possible, this partial cornea graft is of benefit compared with penetrating keratoplasty (or full-thickness corneal transplantation) as it avoids rejection of corneal endothelium and leads to better outcome [4]. The model described here is a model of pig-to-NHP anterior lamellar corneal xenograft that provides the opportunity to explore the use and outcome of different types of donors such as wild-type (WT) pigs as controls or genetically modified pigs such as the Gal KO multiple-transgenic and hCTLA4-Ig transgenic pigs our team has studied [8].

2 Materials

2.1 Animals

All procedures in this study were in strict compliance with the framework of European and French regulation for animal experimentation and animal welfare. All the procedures met the requirements of the ARVO Statement for the Use of Animals in Ophthalmic and Vision Research (http://www.Arvo.Org/About.arvo/policies/statement-for_the_use_of_animals_in_ophthalmic_and_visual_research/).

2.1.1 Recipients

On-purpose-bred adult male Macaca fascicularis, 21–33 months old, weighing from 2.4 to 5.5 kg, originating from Mauritius and obtained from Bioprim (Baziege, France) after quarantine. We use healthy animals negative for two cutaneous tuberculin challenges, for rabies and pathogenic enteric bacteria. The model implies selection of monkeys whose corneal sizing comes closest to humans for ocular features (no expression of alpha-galactosyl epitope, non-regenerative endothelium). The species is widespread and is commonly used in acceptable ethical conditions.

2.1.2 Donors

Here we provide details of donor pigs we have used to obtain corneas for this type of studies as an example of what can be done with this model.

1. Wild-type pig such as WT Large White pigs (WT, EARL du Pont Romain, Surzur, France) or provided by INRA Centre de Tours pig herd (6 months old).
2. GTKO/ hCD39/hCD55/hCD59/fucosyl transferase transgenic pigs. These transgenic pigs have knocked out the galactosyl epitope (Gal KO), the main xenogeneic epitope unexpressed in old-world monkeys and humans and also the main target of preformed xeno-antibodies. The GTKO/ hCD39/hCD55/hCD59/fucosyl transferase transgenic pigs we use were generated by somatic cell nuclear transfer [9] at Avantea (Cremona, Italy) using fibroblasts supplied by Peter Cowan and Tony d'Apice (St Vincent's Hospital, Melbourne, Australia) and have been previously described [10] (*see* **Note 1**).
3. hCTLA4Ig transgenic pig. The hCTLA4-Ig transgene construction and hCTLA4-Ig genetically modified pig production have been previously described [11]. Transgenic pigs used by our team are from the # NSE-18108 line. Breeding, euthanasia, and tissue sampling of hCTLA4-Ig transgenic pigs are performed at the INRA Centre de Tours animal facility (Nouzilly, France, $n = 2$; age $= 6$ and 12 months).

2.2 Harvesting Cornea

1. Single-use microsurgical instruments: toothed forceps, 15-mm shaft, single-use Castroviejo scissors (BVI, Waltham, MA 02451 USA).
2. Cornea organ-cultured media Cornea Prep II® (Eurobio, Courtaboeuf, France).
3. Sterile drape.
4. Povidone iodine 5% ophthalmic solution (Bétadine® 5%).
5. Balanced salt solution.
6. Sterile gloves.

2.3 Grafting

1. Anesthetics: Zoletil (Tiletamine + Zolazepam, Virbac, Carros, France) and isoflurane.
2. Operating microscope (such as LEICA M820 Statif F19, Leica Microsystems, Wetzlar, Germany).
3. Sterile gloves.
4. Sterile drape.
5. Eye speculum.
6. Microsurgical instruments: Bonn forceps, Troutman forceps, Vannas scissors, Barraquer needle holder (MORIA S.A. 92160 Antony—France).
7. Sutures: nylon 10.0 sutures (Ethilon®10.0, Ethicon, Blue Ash, OH, USA).

8. Crescent knives (Crescent® double beavel, Alcon, Hünenberg, Swizerland).
9. 6.5-mm-diameter single-use Hessburg-Barron® donor/recipient vacuum trephine (Moria, Antony, France).

2.4 Postoperative Care

1. Ophthalmic ointment containing oxytetracycline and dexamethasone (STERDEX®; Laboratoires Thea, Clermont-Ferrand, France).

3 Methods

3.1 Harvesting Cornea

For this procedure, pigs are euthanized by electrocution and operated for corneoscleral button excision of both eyes under sterile operative conditions. To this end, the following steps are carried out:

1. After conjunctiva resection, manually trephinate the sclera at about 1 mm from the limbus.
2. Then gently remove the iris from the corneoscleral button.
3. After removal, store the porcine corneoscleral buttons at room temperature in cornea organ-cultured media Cornea Prep II_® for up to 24 h.
4. In operating theater, remove the corneal grafts from culture media and prepare the grafts in sterile conditions by trephination (endothelial route) of the sclerocorneal button using a single-use punch trephine 6.5 mm (*see* **Note 2**) for the procurement of a graft presenting a thickness of 350–400 μm.

3.2 Grafting

Right eye surgeries are performed under general anesthesia induced by Zoletil followed by isoflurane inhalation maintenance using similar techniques to those used for humans. The following steps are performed for grafting:

1. Under operating microscope, trephinate the recipient monkey cornea circumferentially at 6.5 mm diameter using a single-use Hessburg-Barron recipient vacuum trephine (*see* **Note 3**).
2. Perform lamellar dissection with single-use Crescent knives to preserve a 50 μm bed including the thin posterior stroma and epithelium (*see* **Note 4**).
3. Then strip the donor pig corneal button of its Descemet's membrane (*see* **Note 5**) and suture the corneal button into the recipient bed with 16 nylon 10.0 single sutures (*see* **Notes 6** and **7**). Figures 1 and 2 illustrate the aspect of the graft at the end of surgery and 1 week after implantation.

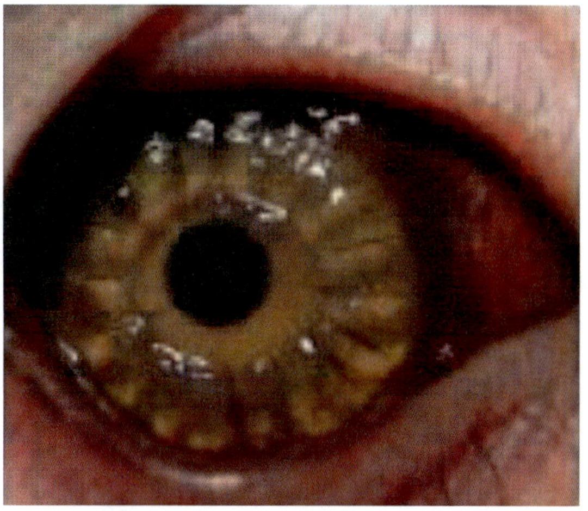

Fig. 1 Aspect of the graft at the end of surgery

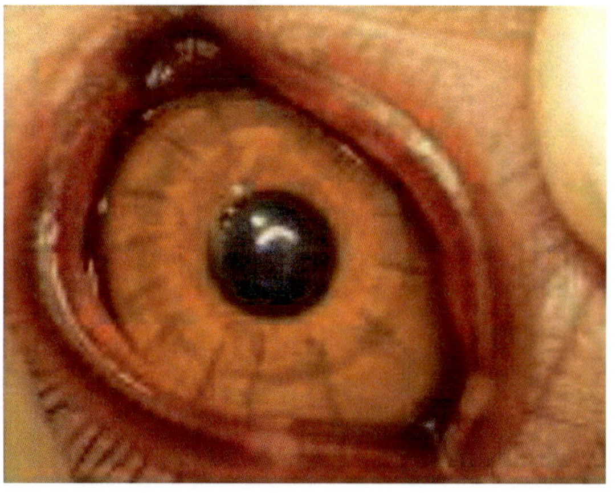

Fig. 2 Graft at 1 week after implantation

3.3 Postoperative Care

1. Apply ophthalmic ointment containing oxytetracycline and dexamethasone (STERDEX®) daily until rejection occurs (*see* **Note 8**).

4 Notes

1. The absence of expression of Gal epitopes in the corneas we have used was verified by immunohistology using Bandeiraea simplicifolia-Isolectin B4 (IB4)-FITC conjugate (Sigma, Saint-Louis, MI, USA). Expression of hCD55, hCD59, and hCD39 was verified with mouse anti-hCD55 (67), mouse anti-hCD59 (MEM-43), and mouse anti-hCD39 (A1) all from AbD Serotec (Oxford, UK). Whereas expression of hCD55 and hCD59 could be evidenced by immunohistology, the signal obtained with the hCD39 antibody was not different from background.

2. As the Macacca corneal diameter is around 10 mm, we have chosen a diameter of graft at 6.5 mm to leave at least 1.5 mm from edge graft to limbus to avoid reaction from limbal vessels.
3. Deep nonpenetrating trephination: the Hessburg-Barron recipient vacuum trephine allows depth management trephination by adjustable preset depth setting.
4. Deep anterior lamellar dissection of the recipient cornea may be facilitated by blunt dissection.
5. Due to strong adherence of the pig Descemet's membrane to the corneal stroma, stripping off the membrane may need further dissection with Vannas scissor or Crescent Blade.
6. Single sutures are preferred to running sutures. Loosing sutures are managed easier by simple removal if needed. All sutures should be removed in the next 3 months.
7. Avoid penetrating suture which results in an inflammatory response in anterior chamber.
8. Closing lids at the end of the surgery is not necessary to promote epithelialization that happens quickly in the next 24 h.

Acknowledgments

The author would like to thank Stephanie Le Bas-Bernardet, David Minault, Jeremy Hervouet, Veronique Daguin, and Annaıck Pallier, for their collaboration; Simon C. Robson, Mark B. Nottle, Peter J. Cowan, Eric Venturi, Pascal Mermillod, Philippe Brachet, Cesare Galli, Irina Lagutina, and Roberto Duchi, for allowing us to work on transgenic pig lines; and Gilles Blancho, Jean-Paul Soulillou, and Bernard Vanhove for their active support and helpful advices.

References

1. Whitcher JP, Srinivasan M, Upadhyay MP (2001) Corneal blindness: a global perspective. Bull World Health Organ 79:214–221
2. Fagerholm P, Lagali NS, Merrett K et al (2010) A biosynthetic alternative to human donor tissue for inducing corneal regeneration: 24-month follow-up of a phase 1 clinical study. Sci Transl Med 2(46):46–61
3. Oh JY, Kim MK, Ko JH et al (2009) Acute cell-mediated rejection in orthotopic pig-to-mouse corneal xenotransplantation. Xenotransplantation 16(2):74–82
4. Oh JY, Kim MK, Ko JH et al (2009) Histological differences in full-thickness vs. lamellar corneal pig-to-rabbit xenotransplantation. Vet Ophthalmol 12(2):78–82
5. Hara H, Cooper DK (2011) Xenotransplantation—the future of corneal transplantation? Cornea 30:371–378
6. Kim MK, Wee WR, Park CG, KIm SJ (2011) Xenocorneal transplantation. Curr Opin Organ Transplant 16:231–236
7. Zhiqiang P, Cun S, Ying J et al (2007) WZS-pig is a potential donor alternative in corneal xenotransplantation. Xenotransplantation 14:603–611
8. Vabres B, Le Bas-Bernardet S, Riochet D et al (2014) hCTLA4-Ig transgene expression in keratocytes modulates rejection of corneal xenografts in a pig to non-human primate anterior lamellar keratoplasty model. Xenotransplantation 21(5):431–443

9. Lagutina I, Lazzari G, Galli C (2006) Birth of cloned pigs from zona-free nuclear transfer blastocysts developed in vitro before transfer. Cloning Stem Cells 8:283–293
10. Le Bas-Bernardet S, Tillou X, Poirier N et al (2011) Xenotransplantation of galactosyl-transferase knockout, CD55, CD59, CD39, and fucosyl-transferase transgenic pig kidneys into baboons. Transplant Proc 43:3426–3430
11. Martin C, Plat M, Nerriere-Daguin V et al (2005) Transgenic expression of CTLA4-Ig by fetal pig neurons for xenotransplantation. Transgenic Res 14:373–384

Chapter 17

Rat Model of Intra-articular Chondrocyte Xenotransplantation

Maribel Marquina, Magdiel Pérez-Cruz, and Cristina Costa

Abstract

Current treatment options are not providing an adequate solution for cartilage defects. Articular cartilage lesions in particular are not able to repair spontaneously and progressively degenerate with an arthrosic pattern. Aiming to solve this pressing medical need, xenotransplantation of porcine chondrocytes could be developed as a new therapeutic approach. Xenotransplantation is gaining much attention, thanks to the advances in animal genetic engineering and progress in the characterization of the rejection mechanisms that prevent long-term graft survival. In this regard, our team has identified various targets for intervention that should be tested in a meaningful animal model to prove their relevance in rejection of xenogeneic cartilage. To this end, we have recently established a discordant xenotransplantation model by injecting three million porcine articular chondrocytes (PAC) into the femorotibial joint of Lewis rats. This chapter describes this new model, which can be used to assess the immunoregulatory effect of a variety of strategies designed to inhibit rejection of xenogeneic PAC both at the humoral and cellular levels.

Key words Articular cartilage, Chondrocytes, Xenotransplantation, Antibody response, Cellular immune response

1 Introduction

Xenotransplantation of porcine chondrocytes could be a novel therapeutic approach for multiple cartilage defects [1]. Articular cartilage lesions in particular, produced by either traumatic injury or disease (such as osteoarthritis (OA)), are highly prevalent in the general population. Notably, current treatment options are far from being adequate. The use of human chondrocytes for autologous chondrocyte implantation (ACI and related methods) is restricted to very few patients due to limited tissue availability and high technical complexity and cost. Ordinarily, patients are subjected to various treatments and interventions which do not prevent disease progression, and the only definitive solution is applying the highly traumatic and painful joint replacement using a prosthesis.

Cristina Costa (ed.), *Xenotransplantation: Methods and Protocols*, Methods in Molecular Biology, vol. 2110,
https://doi.org/10.1007/978-1-0716-0255-3_17,

Cell therapies based on xenogeneic cells could, on the contrary, be developed to provide an early curative treatment [1, 2]. Porcine articular chondrocytes (PAC), or chondrocytes from other equivalent sources, could be genetically engineered to allow long-term survival and obtained in sufficient quantity with protocols of high quality and control [1–3]. Furthermore, previous reports have demonstrated the capacity of PAC to form hyaline cartilage in in vitro and in vivo models [4, 5]. However, cellular and humoral mechanisms lead to rejection of xenogeneic chondrocytes in the current conditions [1, 6]. Therefore, there is a need to establish relevant preclinical models that allow assessment of various genetic modifications and/or strategies needed to prevent PAC rejection. In this regard, the initial in vivo xenotransplantation studies showed rejection of porcine cartilage in non-articular sites [2, 7]. Recently, we developed a simple rat model that would take into account the particularities of the joint in terms of structure and immunobiology in consideration of the great medical needs specifically affecting articular cartilage. A xenogeneic immune response against PAC was confirmed in this setting.

Here, we describe this small animal model of discordant xenotransplantation for the study of the immune response triggered against pig chondrocytes after intra-articular injection in the femorotibial (knee) joint. This xenotransplantation model is designed to provide preclinical information for future xenogeneic therapies of articular cartilage repair.

2 Materials

2.1 Culture of Pig Chondrocytes

1. PAC (or pig chondrocytes from other sources) freshly isolated [2, 4] or kept frozen at low passage in liquid nitrogen for long-term storage.
2. Dulbecco's Modified Eagle's Medium (DMEM) supplemented with 10% fetal bovine serum (FBS), 200 U/mL penicillin, and 200 μg/mL streptomycin.
3. Complete medium: DMEM supplemented with 10% FBS, 200 U/mL penicillin and 200 μg/mL streptomycin, and 25 mg/L of Endothelial Mitogen Growth Supplement (Merck Millipore) (*see* **Note 1**).
4. Phosphate-buffered saline (PBS) buffer of tissue-culture grade.
5. Trypsin-EDTA solution (1×) or TrypLE™ Express (Invitrogen™), which works especially well.
6. Sterile tissue-culture supplies: 75-cm^2 culture flasks, 15-mL conical-bottom tubes and graduated pipettes.
7. Tissue-culture equipment: Pipettors, laminar-flow cabin, aspiration pump or source of vacuum, incubator at 37 °C and humidified 5% CO_2 atmosphere, centrifuge, inverted microscope.

2.2 Intra-articular Injection of Porcine Chondrocytes

2.2.1 Cell Preparation for Injection

1. PAC expanded in sufficient numbers.
2. PBS of tissue-culture grade.
3. Trypsin-EDTA solution (1×) or TrypLE™ Express (Invitrogen™).
4. 0.4% trypan blue solution in PBS.
5. DMEM with high glucose supplemented with 200 U/mL penicillin and 200 μg/mL streptomycin.
6. Sterile tissue-culture supplies: 15-mL conical-bottom tubes, graduated pipettes, 1.5-mL microfuge tubes, micropipettes, tips.
7. Tissue-culture equipment: pipettors, laminar-flow cabin, aspiration pump or source of vacuum, incubator at 37 °C and humidified 5% CO_2 atmosphere, centrifuge, inverted microscope and Neubauer chamber.

2.2.2 Intra-articular Injection of Cells

1. PAC counted and resuspended in DMEM as described in Subheading 3.2.1.
2. Lewis rats weighing 200–250 g (*see* **Note 2**). Rats should be kept under controlled conditions of light and temperature with food and water ad libitum at the animal facilities.
3. Small animal razor/trimmer.
4. 70% ethanol.
5. 0.3-mL insulin syringes.
6. Inhalatory anesthetic equipment: oxygen bottle, anesthetic machine, and isoflurane.

2.3 Determination of Antibody Responses

2.3.1 Serum Collection and Storage

1. Experimental rats.
2. 1.5-mL or 2-mL blood collection tube (ideally tubes for serum preparation).
3. Dissecting scissors (*see* **Note 3**).
4. Inhalatory anesthetic equipment: oxygen bottle, anesthetic machine, and isoflurane.
5. Sterile 1.5-mL-microtubes, micropipettes, tips.
6. Other equipment: microcentrifuge, −80 °C freezer.

2.3.2 Determination of Anti-PAC Antibody Reactivity by Flow Cytometry

1. PAC expanded in monolayer culture for multiple passages to obtain sufficient cell number and harvested with trypsin as described (*see* **Note 4**).
2. All tissue-culture materials, supplies, and equipment described in Subheading 2.2.1 including tissue-culture tubes (15 or 50-mL tubes depending on amount of cells).
3. PBS/1% bovine serum albumin (BSA). This solution should be filtered sterilized if not used immediately after preparation.

4. Serum collected from experimental rats.
5. Conjugated anti-rat antibodies: anti-IgM, anti-IgG, anti-IgG1, IgG2a, IgG2b, and IgG2c (the antibodies for anti-rat IgG subtypes can be obtained from Southern Biotech Inc., Birmingham, USA).
6. 96-well polystyrene microplates, V-bottom, and plate sealers.
7. FACs tubes.
8. Multicolor flow cytometer and analysis software.

2.4 Determination of Cellular Immune Responses

2.4.1 Blood Collection and Peripheral Blood Lymphocytes (PBL) Isolation

1. Experimental rats.
2. 2-mL or 3-mL blood collection tubes with sodium heparin.
3. Dissecting scissors (*see* **Note 3**).
4. Inhalatory anesthetic equipment: oxygen bottle, anesthetic machine, and isoflurane.
5. PBS.
6. Ficoll-Paque™ Plus (GE Healthcare).
8. PBL media: Roswell Park Memorial Institute (RPMI) 1640 medium with 10% fetal bovine serum (FBS) supplemented with 200 U/mL penicillin and 200 μg/mL streptomycin.
7. 0.4% trypan blue solution in PBS.
8. Sterile tissue-culture supplies: 15-mL conical-bottom tubes, graduated pipettes, T25 flasks, 1.5-mL-microfuge tubes, micropipettes, tips.
9. Tissue-culture equipment: pipettors, laminar-flow cabin, aspiration pump or source of vacuum, incubator at 37 °C and humidified 5% CO_2 atmosphere, centrifuge, inverted microscope, and Neubauer chamber.

2.4.2 Mixed Lymphocyte and PAC Reaction

1. Materials and equipment defined in Subheading 2.1.
2. PBL from experimental rats.
3. Flat-bottom cell-culture 96-well plates.
4. ELISA for rat IL-2 determinations (commercially available, such as Rat IL-2 ELISA kit from RayBiotech, Inc. Norcross, GA, USA).

2.4.3 Harvest of Popliteal Lymph Nodes and Lymphocyte Isolation

1. Experimental rats.
2. Euthanasia chamber and compressed CO_2 gas in cylinders.
3. Sterile 1.5- and 15-mL tubes, micropipettes, and tips.
4. Surgical instruments: scissors and forceps including round-tip small tissue forceps.
5. PBL media: RPMI media with 10% fetal bovine serum (FBS).
6. Extraction buffer: PBS / 1% BSA.

7. 60-mm tissue-culture plates.
8. Sterile microscope glass slides with ground edge.
9. Red Blood Cell Lysing Buffer (Sigma-Aldrich).
10. Other equipment: centrifuge, microcentrifuge, Neubauer chamber.

2.4.4 Characterization of Cell Populations in Draining Lymph Nodes

1. PBL from experimental rats.
2. PBS/1% BSA.
3. Fixation/permeabilization solution (Milteny Biotec GmbH, Germany).
4. Permeabilization buffer (Milteny Biotec GmbH, Germany).
5. Conjugated anti-rat antibodies (all from Myltenyi Biotec): CD4 (Domain 1)-FITC (clone: REA482), CD8a-FITC (clone: REA437), IgG1-PE (isotype control FoxP3), Ki67-APC (clone: REA183), FoxP3-PE (clone 3G3).
6. Equipment: multicolor flow cytometer and analysis software.
7. Other equipment: Neubauer chamber, microcentrifuge.

3 Methods

3.1 Culture and Harvest of Pig Chondrocytes

Carry out the cell-culture procedures in a laminar-flow cabinet.

1. Seed freshly isolated PAC in a 75-cm^2 culture flask in 10–12 mL of complete media. We identify these cells as passage 0 or p0.
2. Culture the PAC in an incubator at 37 °C in a humidified atmosphere of 5% CO_2 changing the medium every 3 days until cells are almost confluent.
3. Then, remove the culture medium, wash PAC monolayers once with PBS (by adding a few mL to cover and remove completely), and treat them with 1.5 mL of trypsin-EDTA solution at 37 °C for a few min until complete dissociation. This can be best assessed by observation under an inverted microscope.
4. Resuspend cells in complete medium to stop trypsin activity (*see* **Note 5**).
5. Split the chondrocyte culture 1–2. To this end, seed on two culture flasks per each flask harvested (*see* **Note 6**). PAC at low passage can also be frozen after harvest following standard freezing procedures for mammalian cells.
6. Culture the chondrocytes until almost confluence and repeat the disassociation procedures to obtain the next passage.

7. Keep expanding the cultured PAC for a few passages (usually up to 3 or 4) until sufficient cells are obtained for the desired experiment. DMEM/10%FBS (without growth factors) can be used at this stage.
8. For harvest, repeat **steps 3** and **4** and transfer resuspended cells to a 15-mL tube for centrifugation at 260 × *g* for 5 min.
9. Resuspend the cells in media and determine the cell number using a Neubauer chamber. PAC can be counted using a ½ dilution with trypan blue (very high viability expected).

3.2 Intra-articular Injection of Porcine Chondrocytes

3.2.1 Cell Preparation for Injection

1. After expansion, wash the PAC in PBS and harvest by trypsinization, incubating the cultures for a few min with trypsin-based solution.
2. Resuspend the PAC by pipetting up and down in high-glucose DMEM supplemented with antibiotics, and transfer to a 15-mL conical tube. Spin the cells at 260 × *g* for 5 min.
3. Discard the supernatant, resuspend the cells in 5 mL of DMEM, and determine the cell number using a Neubauer chamber.
4. For each rat, place the volume containing 3×10^6 PAC in a 1.5-mL microfuge tube and centrifuge and discard the supernatant.
5. Each cell pellet is then mixed with 15 μL of high-glucose DMEM supplemented with antibiotics for i.a. injection in femorotibial joint of Lewis rats. Keep the vial at room temperature until injection.

3.2.2 Intra-articular Injection of Cells

Carry out all animal procedures in a registered animal facility (*see* **Note 2**).

1. A few days prior to surgery, distribute evenly the Lewis rats between experimental groups considering the body weight. The body weight and animal well-being should be monitored throughout the study.
2. In preparation for surgery, place each animal individually in an induction chamber, and induce anesthesia with 4–5% isoflurane in 100% oxygen until loss of podal reflex.
3. After induction, move the animal to a homeothermic blanket and place it in dorsal recumbence. Then maintain anesthesia with 2.5% isoflurane in a flow of 1.5 L/min oxygen administered by means of a facemask connected to a coaxial circuit.
4. Shave the animal leg to be injected at the level of the knee with an electric clipper, and prepare with alcohol at 70%.
5. Use an empty insulin syringe to puncture the knee joint accessing from one side below the patellar ligament, reaching the synovial cavity and extracting by absorption a few μL of synovial fluid.

6. Resuspend well the prepared PAC and load them in a sterile insulin syringe, removing air bubbles.
7. Inject slowly the cells using the same route described for emptying the synovial cavity (*see* **Note 7**).
8. Remove gently the needle and keep under observation for a couple of min for possible leaks. This is very infrequent in our hands.
9. Stop anesthesia and allow the rat to recover under direct supervision.
10. Conduct the follow-up based on the experimental design and use selected time-points for the sample collection as described below (*see* **Note 8**).

3.3 Determination of Antibody Responses

3.3.1 Serum Collection and Storage

Carry out blood collection in an animal facility via one of the recommended routes for rats. We describe a simple procedure for collection from the tail veins, but other routes are accepted by ethical committees (*see* **Note 3**). Select the type of collection tube depending on the purpose of the sample, a sterile 1.5-mL microfuge tube suffices for serum preparation.

1. Collect blood (0.5 mL is appropriate) from experimental rats from tail vein after anesthesia induction by clipping the tail tip with scissors and letting the blood flow into the appropriate tube.
2. To obtain serum, incubate blood (without anticoagulant) at room temperature (22–25 °C) for up to 45 min to enable clotting.
3. Collect serum by mild centrifugation (10 min, 2000 × *g* at 4 °C) and transfer the supernatant (serum) into a new tube. This procedure can be repeated if serum is reddish (*see* **Note 9**).
4. Use samples immediately or store frozen (at −80 °C for long-term storage) for further analysis.

3.3.2 Determination of Anti-PAC Antibody Reactivity by Flow Cytometry

Serum dilution and incubation, and subsequent staining, is better conducted in 96-well plates to facilitate working with a high number of samples for each assay. A layout of each assay is prepared in a template and registered. Cells, serum, and antibodies are all resuspended in PBS/1% BSA for the incubations. Make sure that a sufficient amount of cells is grown to conduct the whole assay.

1. Harvest all PAC needed by trypsinization and resuspend first in DMEM/10% FBS and collect in tubes for centrifugation (260 × *g* for 5 min) at room temperature.
2. Discard supernatant and resuspend cell pellet in PBS/1% BSA to wash cells from media. Cells can be counted at this stage. Spin again (260 × *g* for 5 min) and resuspend the cell pellet in PBS/1% BSA for a final solution of 5–10 × 10^6 cells/mL.

3. Add 0.5–1 × 10^6 of PAC in suspension (to be used as target cells) to each empty well from a 96-well plate with V-bottom following the experimental layout. Distribute evenly in the plate by adding the same volume per well (i.e., 100 μL of a 5–10 × 10^6 cells/mL solution).
4. Centrifuge the plate at 260 × *g* for 5 min, 4 °C. Discard the supernatant in the plate without touching the cellular pellet.
5. Add 100 μL of each sample of rat serum properly diluted (i.e., 1%) in PBS/1% BSA.
6. Seal the plate with the cover and incubate at 4 °C for 30 min.
7. Add 100 μL of PBS and repeat **step 2**.
8. Add 100 μL of diluted anti-rat Ig secondary antibodies (with the chosen specificity, i.e., total IgM, total IgG, or IgG subtypes).
9. After incubation for 30 min at 4 °C, wash cells with 100 μL of PBS by centrifugation (**step 2**).
10. Add 100 μL of PBS, transfer to FACs tubes with 200 μL of PBS, and analyze samples by flow cytometry.

3.4 Determination of Cellular Immune Responses

3.4.1 Blood Collection and PBL Isolation

Carry out blood collection in an animal facility and PBL isolation into a laminar-flow cabinet. PBL are isolated from rat blood by Ficoll gradient followed by a 2-h incubation at 37 °C in media to remove plastic-adherent cells.

1. Collect blood from experimental rats (1.5–2 mL) in tubes containing sodium heparin and mix well to prevent clotting.
2. Transfer the blood of each rat into a 15-mL tissue-culture tube.
3. Add 2 mL of PBS to each sample tube and mix gently. Keep at room temperature.
4. Invert the Ficoll-Paque PLUS bottle several times to ensure thorough mixing, and transfer 3 mL of Ficoll into empty 15-mL tissue-culture tubes (under aseptic conditions). Prepare and label one tube for each PBL isolation.
5. Carefully layer the diluted blood samples (3–4 mL) on top of the Ficoll (do not mix blood and Ficoll) (*see* **Note 10**).
6. Centrifuge the gradient tubes at 400 × *g* for 30–40 min at 18–20 °C without brake.
7. Prepare a new set of labeled sterile 15-mL tubes for the next step.
8. After centrifugation, gently place the gradient tubes back into the laminar-flow cabinet, and remove the upper layer using a clean Pasteur pipette, leaving the lymphocyte layer undisturbed at the interface. Care should be taken not to disturb the lymphocyte layer.

9. Using a clean Pasteur pipette, transfer the lymphocyte layer to a clean tube. It is critical to collect all the interface with the minimum possible amount of Ficoll and supernatant (*see* **Note 11**).
10. Add at least three volumes (6 mL) of PBL media to the lymphocytes and resuspend the cells by gentle repeated pipetting.
11. Centrifuge at 100 × *g* for 10 min at 18–20 °C. Remove the supernatant.
12. Repeat the wash (two previous steps) and resuspend in 3 mL of PBL media.
13. Seed each sample into a T25 flask and incubate for 2 h in an incubator at 37 °C in a humidified atmosphere of 5% CO_2.
14. Harvest the supernatants and determine the cell number for each sample. PBL can be counted in a Neubauer chamber using at a 1/10 dilution with trypan blue and media. Automated cell counters can be used if appropriately calibrated.

3.4.2 Mixed Lymphocyte and PAC Reaction

In our laboratory, T-cell activity is determined for individual rats in co-cultures adapted from previously reported techniques [8, 9].

1. Incubate confluent PAC at 37 °C in 96-well plates with PBL at 10:1 effector/target ratio (being the PBL the effector cells and the PAC the target cells). Include controls of each cell type cultured alone.
2. For additional activation, concanavalin A (ConA, Sigma) can be added to selected wells at a final concentration of 10 μg/mL 30 min after initiating the co-culture.
3. Harvest culture supernatants 24 h later and store frozen until the assay.
4. Measure rat IL-2 secretion by ELISA. A commercial kit can be used following the procedure recommended by the manufacturer.

3.4.3 Harvest of Popliteal Lymph Nodes and Lymphocyte Isolation

For isolation of lymphocytes from draining lymph nodes, the popliteal lymph nodes are collected and submitted to mechanical disruption. The isolated cells can be either used for mixed lymphocyte and PAC reaction or characterized by flow cytometry (as described in the corresponding sections).

1. Collect popliteal lymph nodes from rats 1–2 weeks after PAC injection or sham operation. To this end, euthanize the rat following the approved protocol, place on one side, and expose the area by cutting the skin and muscle as illustrated in Fig. 1.
2. Carefully harvest the lymph nodes with fine forceps by separating them from adipose tissue, and place them in microfuge tubes containing PBL media (RPMI/10% FBS).

3. Prepare a 60-mm plate for each rat containing 3 mL of extraction buffer (PBS/1% BSA). Over the plate, place the popliteal lymph node(s) on the ground edge of a microscope glass slide, add a drop of buffer, and crush it with another glass slide. Keep the ground edges always wet and press them against the node causing friction until the tissue is completely broken and cells fall into the media. Discard the node capsule.
4. Transfer the cell suspension into 15-mL tubes with pipette.
5. Wait 3 min to sediment disaggregated cells and transfer the supernatant into another 15-mL tube.
6. Spin at 260 × *g* for 5 min at room temperature and discard supernatant. If the pellet is red, follow **steps 7–9**.
7. Add 3 mL of Red Blood Cell Lysing Buffer and resuspend cells by gently pipetting up and down.
8. Incubate for 1 min, spin at 260 × *g* for 5 min at room temperature, and discard supernatant.
9. Wash the cells with 3 mL of PBS/1% BSA or media followed by centrifugation.
10. Resuspend the cells and determine the cell number using a Neubauer chamber.

3.4.4 Characterization of Cell Populations in Draining Lymph Nodes

To this end, we stain the isolated lymphocytes with anti-rat CD4-FITC and CD8-FITC antibodies alone or in combination with APC-conjugated anti-mouse Ki67, PE-conjugated anti-Foxp3, or isotype control IgG1 for two-color analysis. As some of these markers are intracellular, the FoxP3 Staining Buffer Set is applied to the studied cells for permeabilization and fixation following recommendations of the manufacturer. Cell surface staining of a gated population of lymphocytes is then measured in a flow cytometer.

1. Spin cell suspension at 300 × *g* for 10 min. Use 300,000 cells/mL per well. Aspirate supernatant completely.
2. Resuspend cells in 180 μL of extraction buffer and transfer it into the 96-well conical-bottom plate.
3. Transfer 90 μL of cell suspension to various wells to label with antibodies.
4. Add 10 μL of anti-CD4 and CD8 antibodies (previously diluted 1/10) in each well.
5. Mix well and incubate for 10 min at 4 °C in the dark (*see* **Note 12**).
6. Wash cells with 100 μL of extraction buffer and centrifuge at 300 × *g* for 10 min. Aspirate supernatant completely and place plate on ice. Work on ice from now on unless indicated.

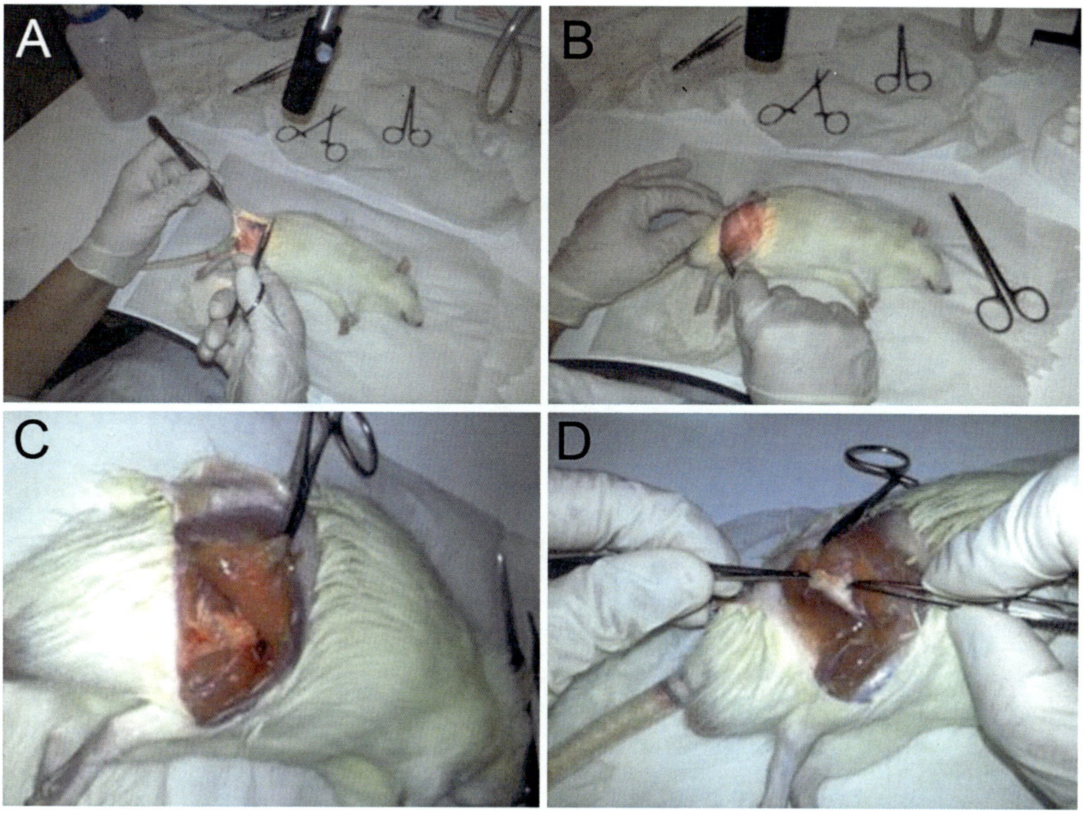

Fig. 1 Harvest of rat popliteal lymph nodes. (**a**) Cutting of the skin. (**b**) Exposure of the muscle. (**c**) Lifting the muscle to expose the lymph nodes. (**d**) Harvest by separation from the fat tissue

7. Resuspend cells in 100 μL of cold/freshly prepared fixation/permeabilization solution.
8. Mix well and incubate for 30 min at 4 °C in the dark. Wash cells with 100 μL of cold extraction buffer (PBS/1% BSA) and centrifuge at 300 × *g* for 5 min at 4 °C. Aspirate supernatant completely.
9. Resuspend cells in 200 μL of cold extraction buffer, and transfer 50 μL of cell suspension into another plate per triplicate.
10. The plate containing 50 μL of stained cells destined to single staining, centrifuge at 300 × *g* for 5 min at 4 °C. Aspirate supernatant completely and resuspend cells in 150 μL of PBS and store at 4 °C until analysis by flow cytometry.
11. The plate destined to double staining for intracellular Ki-67 and FoxP3, centrifuge at 300 × *g* for 5 min at 4 °C. Aspirate supernatant completely and wash with 100 μL of cold 1× permeabilization buffer. Centrifuge at 300 × *g* for 5 min at 4 °C.

12. Aspirate supernatant completely and resuspend cells in 98 μL of cold 1× permeabilization buffer.
13. Add 2 μL of the rat serum. Mix well and incubate for 5 min at 4 °C in the dark.
14. Add 10 μL of the anti-Ki-67, anti-FoxP3, and anti-IgG1 antibodies (previously diluted 1/11) in each corresponding well.
15. Mix well and incubate for 30 min at 4 °C in the dark.
16. Wash cells with 100 μL of cold 1× permeabilization buffer and centrifuge at 300 × *g* for 5 min at 4 °C. Aspirate supernatant completely.
17. Repeat **step 13** once.
18. Resuspend cells in 150 μL of PBS and analyze samples by flow cytometry.

4 Notes

1. We have established in our laboratory the use of a low concentration of Endothelial Mitogen (extract of bovine hypophysis commercialized by a few sources) as a supplement to the media for the culture of pig chondrocytes. In our experience, it helps to grow the pig chondrocytes without causing loss of chondrocyte markers at low passages [8].
2. Animals must be hosted in an experimental facility with surgery unit that functions following the national and international legislation for appropriate use of experimental animals. Both females and males can be used for these experiments. However, they should be both represented in sufficient numbers to have a cohort for each gender as different results should be expected.
3. Blood can be collected from the tail vein by cutting the tip of the tail followed by cauterization after bleeding. This technique works well for repeated bleeding if it is handled gently. Nevertheless, blood can also be collected from other routes such as the femoral vein.
4. The procedure described here can be adapted to other cell types by simply changing the target cell.
5. There is no need for centrifugation for the passage of PAC. We simply use the minimum amount of trypsin solution needed to cover the cells and resuspend in complete medium after finalizing the cell dissociation to block the enzyme activity by dilution and presence of FBS.
6. No major split dilutions are recommended to maintain chondrocyte viability and features, as well as prevent dedifferentiation. Therefore, we recommend PAC are passed at no higher than 1–2 splits.

7. The synovial cavity is emptied to prevent excessive pressure and allow space for the injected cells. However, accessing the joint becomes more difficult once it has been emptied and the needle probably causes some abrasion to the articular cartilage during the injection. We accept this limitation as part of the model as in clinical practice the target joints will be also affected by either traumatic or disease-mediated injury.
8. We have already reported the use of this model [6]. Our results can thus be used as the basis for future studies with information of the expected time course for the immune response.
9. It may be possible to get more serum from each sample by collecting first as much liquid as possible even if some red blood cells are picked up. These are then removed with a second centrifugation. It is not advisable to freeze the sample with residual blood cells as it will cause hemolysis.
10. To facilitate the preparation of the gradient without mixing the layers, bend the tube and put the blood slowly so the blood runs in one side before reaching the Ficoll. In my experience, this works better than placing the Ficoll below the blood.
11. Ficoll is toxic to the cells and should be washed off immediately after the gradient centrifugation.
12. Higher temperatures and/or longer incubation times may lead to nonspecific cell labeling. Working on ice requires longer incubation times.

Acknowledgments

This work was funded by Ministerio de Ciencia e Innovación—Instituto de Salud Carlos III (PI11/02179) and the SLT002/16/00445 grant funded by the Department of Health of the Generalitat de Catalunya, both granted to CC, and cofunded by FEDER funds—a way to build Europe. IDIBELL benefits from the support of CERCA. M.P.-C. was supported by an IDIBELL fellowship.

References

1. Sommaggio R, Uribe-Herranz M, Marquina M, Costa C (2016) Xenotransplantation of pig chondrocytes: therapeutic potential and barriers for cartilage repair. Eur Cell Mater 32:24–39
2. Costa C, Brokaw JL, Wang Y, Fodor WL (2003) Delayed rejection of porcine cartilage is averted by transgenic expression of alpha1,2-fucosyltransferase. FASEB J 17:109–111
3. Sommaggio R, Bello-Gil D, Pérez-Cruz M, Brokaw JL, Máñez R, Costa C (2015) Genetic engineering strategies to prevent the effects of antibody and complement on xenogeneic chondrocytes. Eur Cell Mater 30:258–270
4. Prado D, Fuentes-Boquete IM, Blanco FJ (2012) In vitro repair model of focal articular cartilage defects in humans. Methods Mol Biol 885:251–261. https://doi.org/10.1007/978-1-61779-845-0_16
5. Ramallal M, Maneiro E, López E et al (2004) Xeno-implantation of pig chondrocytes into rabbit to treat localized articular cartilage defects:

an animal model. Wound Repair Regen 12:337–345

6. Marquina M, Collado JA, Pérez-Cruz M et al (2017) Biodistribution and immunogenicity of allogeneic mesenchymal stem cells in a rat model of intraarticular chondrocyte xenotransplantation. Front Immunol 8:1465. https://doi.org/10.3389/fimmu.2017.01465
7. Stone KR, Walgenbach AW, Abrams JT, Nelson J, Gillett N, Galili U (1997) Porcine and bovine cartilage transplants in cynomolgus monkey: I. A model for chronic xenograft rejection. Transplantation 63:640–645
8. Sommaggio R, Máñez R, Costa C (2009) TNF, pig CD86, and VCAM-1 identified as potential targets for intervention in xenotransplantation of pig chondrocytes. Cell Transplant 18:1381–1393. https://doi.org/10.3727/096368909X474249
9. Costa C, Pizzolato MC, Shen Y, Wang Y, Fodor WL (2004) CD86 blockade in genetically modified porcine cells delays xenograft rejection by inhibiting T-cell and NK-cell activation. Cell Transplant 13:75–87

Chapter 18

Porcine Hepatocytes: Isolation and Liver Tissue Engineering for Xenotransplantation

Yi Li, Qiong Wu, Yujia Wang, Hong Bu, and Ji Bao

Abstract

Various liver diseases result in liver failure, and liver transplantation as a definite treatment is limited by the shortage of organs available for transplantation. The use of isolated primary hepatocytes in cell-based therapies including hepatocyte transplantation, tissue engineering liver transplantation, and bioartificial liver support systems has gained increasing interest during the past years. Human hepatocytes are the preferred source of cells. Aside from the organ shortage, the isolation of human liver cells is usually limited by obtaining a sufficient quantity of high-quality, metabolically active cells. Furthermore, livers from which hepatocytes are typically harvested are not suitable for transplantation, with the variability in quantity and quality. Porcine hepatocytes, on the other hand, have the ability to perform complex biological functions and show modifiable behavior. Primary porcine hepatocytes are currently widely used in the investigation of drug metabolism, hepatotoxicity, protein biosynthesis, and gene expression. Primary hepatocytes do not proliferate in vitro and are sensitive to freeze-thaw damage in cryopreservation and thus need to be freshly isolated for each experiment. Consequently, the methods of porcine hepatocyte isolation are being actively sought after. Our laboratories have been involved in various applications of liver cells, and we have long-lasting experiences in liver cell isolation and their application in R&D. We here summarize the present protocol of our laboratories for primary hepatocyte isolation from pig and their liver tissue engineering for xenotransplantation.

Key words Primary porcine hepatocyte, Cell isolation, Cell culture, Liver tissue engineering, Xenotransplantation

1 Introduction

1.1 Hepatocyte Isolation

Primary hepatocyte was first isolated from rats by Berry and Friend in the late 1960s for cell biochemical research and structural studies [1]. With a continuous progress in cell isolation techniques and defined logistics, they were later isolated from many other species, including bovines [2] and humans [3]. The method of pig liver cell isolation is based on the principles of the two-step liver cell isolation using a collagenase perfusion of the rat liver [4], and several modified in situ perfusion techniques have been described by far [5–7]. Isolated primary hepatocytes have been proven to be a valuable

Cristina Costa (ed.), *Xenotransplantation: Methods and Protocols*, Methods in Molecular Biology, vol. 2110,
https://doi.org/10.1007/978-1-0716-0255-3_18,

tool to study drug toxicity, interaction, and metabolism in the liver, but their increased demand is accompanied by their limited availability. Hepatocytes do not proliferate [8] or survive long after isolation [9] and easily lose the expression of liver-specific genes in vitro [10]. The isolation of porcine hepatocytes breaks the barrier of obtaining a sufficient quantity of high-quality, metabolically active cells. The number and metabolic profiles of porcine hepatocytes are basically similar to those of humans, and freshly isolated primary hepatocytes retain most of their liver-specific functions; therefore, porcine hepatocytes are a reasonable alternative to human hepatocytes. The use of genetically engineered fumarylacetoacetate hydrolase (*FAH*)-deficient pigs as incubators for in vivo expansion of human hepatocytes is another promising option. Functional human cell transplantation in utero may lead to large-scale expansion of human hepatocytes in the xenogeneic recipient postnatally [11, 12]. Above all, optimization of primary porcine hepatocyte isolation to increase their number, vitality, and quality has always been in the center of interest and worthy of thorough review.

Whole liver organs or resected liver tissues are both available to isolate hepatocytes, as long as they fulfill the criteria regarding the absence of infectious agents or high level of hepatic lesions such as cyst, steatohepatitis, cirrhosis, fibrosis, or bile stasis. For whole liver isolation, no difference can be observed regarding the gender of the liver donors [13]. The cell yield and possible viability are greatly influenced by the body weight of donor pigs. When the pig weights are less than 5 kg, the livers are too small to produce enough cells. On the contrary, pig weights over 20 kg are difficult to perfuse well in periphery, and stronger flow rate can rupture the tissue, which both can severely reduce yield. High-protein diets (40%) are routinely given to donor pigs for 7 days before the hepatocytes harvests to induce albumin and other liver-specific protein secretion [14]. Various routes can be exploited to acquire hepatocytes in whole liver organs: intrasplenic, directly into the hepatic parenchyma, or in the peritoneal cavity, but more often intraportal [15]. Left lateral sector segments are usually suitable to obtain normal resected tissues, and it is important that the liver lobe should be incised with a single cut and the time between liver resection and organ harvest should be maintained as short as possible. When transport is required, it is mandatory to keep tissue in a transport medium on ice, in a similar way to that of organs transported for liver transplantation, to preserve cells from warm ischemia. With appropriate size and weight of the liver tissue, the lobular blood vessels can be easily exposed for catheterization and perfused. If the liver pieces are too small, it is difficult to perfuse since they have no complete vessel architecture to be entirely perfused. In contrast, big liver pieces are likewise not easy to perfuse since an equal perfusion pressure in the whole tissue requires a pump with a sufficiently strong flow rate. Besides, the ratio of open resection area to liver tissue weight can also affect the perfusion: the smaller the open resection area is, the better the perfusion can be performed. In proper circumstance, size and weight of the liver tissue can give a good approximation of the expected cell yield and possible viability.

Hepatocytes are the most prominent morphological component of the liver, accounting for about 78% of liver volume and more than 65% of cell number [16]. The liver parenchyma is permeated by a collagenous fibrillar network, which is embedded in an extracellular matrix (ECM), mainly composed of a mixture of elastin, heparin sulfate proteoglycan, and surface adhesion molecules, such as laminin and fibronectin. Variations of ECM content between different species have an impact on the degree of difficulty for isolation. For example, the porcine liver contains more collagen and fibrous tissue; as a result, it is more difficult to dissociate than the rodent liver. Mechanical dissociation of the hepatocytes has been replaced by enzymatic digestion, because the former is more aggressive to the liver parenchyma. Enzymatic perfusion method is based on the permeation of the liver with Ca^{2+} chelator solutions to disrupt Ca^{2+}-dependent (cadherin-based) cell junctions and then enzymatic digestion of the liver parenchyma to disrupt cellular-ECM interactions. The enzyme perfusion time required for digestion is not predetermined; rather, perfusion continues until adequate tissue digestion is achieved. The enzyme is temperature sensitive; the activity of the collagenase can be greatly influenced, leading to the decrease of the viability in isolated hepatocytes. The optimum temperature of enzyme solution is 37–38 °C; this temperature is measured from the perfusion site (liver lobes) rather than solution. The temperature of the solution ought to be controlled all the time based on this temperature to obtain the optimal efficiency. Besides, the solutions usually perfuse through hollow-fiber bioreactor to enrich O_2 and reduce the damages of the isolated hepatocytes. Collagenase type IV was a classical and broad-spectrum enzyme used in primary hepatocyte isolation in all species. With upgrade of commercial collagenase products such as Roche Liberase research grades and Serva collagenase NB grades, the isolations are more specific to the tissues and species, while the purity, viability, and yield of isolated primary hepatocytes are highly improved. Except for the primary component, highly purified collagenase I and II mixture, we found that the amount of neutral protease thermolysin also determines the possible viability and cell yield and especially the further success of spheroid formation. N-acetylcysteine (NAC) is an antioxidant that acts through the replenishment of glutathione in the liver. It also has direct antioxidant properties and hepatoprotective effects against liver ischemia/reperfusion injury to improve the viability. It has been proved to improve the viability and conserves the metabolic function of hepatocytes [17]. Some researchers reported that addition of dimethyl sulfoxide (DMSO) to the hepatocyte isolation protocol could also improve the availability and functionality of hepatocytes for transplantation [18]. Iced hepatocyte wash medium containing FBS can terminate enzyme digestion, and mechanical dissociation and filtering are necessary to remove connective tissue. Subsequent centrifugation is required to separate hepatocytes from both dead hepatocytes and non-parenchymal cells. Several washing and centrifugation can (1) enrich hepatocytes, (2) remove part of contamination of other cell types, and (3) remove dead hepatocytes which otherwise can release large amounts of DNA, lactate dehydrogenase, and intracellular antigens to lower cell viability. When the yield is low or selective removal of all non-parenchymal cells, purification is required. Percoll solutions have been developed for purifying suspensions of hepatocytes. The discrepancy of buoyant densities between hepatocytes and non-parenchymal cells facilitates the process of cell separation.

Three quick tests are studied to evaluate the success of hepatocyte isolation, by assessing cell yield, viability, and morphology. The cell yield is usually calculated by the dry weight of hepatocyte mass. The most commonly used method to assess hepatocyte viability is the trypan blue exclusion test. To ensure persistent good viability during culture, at least 90% of the cells should exclude trypan blue on first examination. Light microscopy examination of the hepatocytes provides information on the quality of cells. Blebbing of the plasma membrane is a sign of fluid uptake by damaged or overdigested hepatocytes, which is often followed by cell death. On transmission electron microscopy (TEM), injury of plasma membrane, swelling of mitochondria, degranulation of endoplasmic reticulum and vesiculation of the Golgi apparatus, and vacuolization of the cytoplasm are frequently seen in damaged or overdigested hepatocytes cells, but it is not usually necessary for each isolation.

1.2 Liver Tissue Engineering

Viral hepatitis, fatty liver disease, drug-induced liver injury, liver cirrhosis, hepatic carcinoma, and other liver diseases can cause liver failure. Orthotopic liver transplantation is the only proven therapy for liver failure, but its overall usage is limited by the need for major surgery, lifelong immunosuppression, and the shortage of donor organs. Alternative treatments to liver transplantation have long been sought after, including hepatocyte transplantation and hepatic bioengineered tissue transplantation. When using 2D culture systems, the hepatocytes are not fit for xenotransplantation because (1) they lose their characteristics over culture times and (2) are hard to locate in the transplant site. To overcome this problem, cells can also be cultured in 3D culture systems (bioreactors) or as a sandwich culture surrounded by a gel composed of ECM proteins, e.g., collagen. Unfortunately, sandwich cultures neither provide the complex multicellular environment found in vivo nor mimic the layered liver architecture. The liver-specific functions in sandwich culture decline within the first week, suggesting that significant survival factors are missing. Furthermore, low hepatocyte production in sandwich culture cannot fulfill clinical demands.

Isolated hepatocytes can also be cultured in the medium in the form of suspended spheroids. Spheroids are three-dimensional (3D), multicellular aggregates of hepatocytes that form spontaneously through cell-cell adhesion. Spheroid formation has previously been recognized to provide a favorable microenvironment that protects the hepatocytes from apoptosis, which is associated with high viability and phenotypic stability with high levels of liver-specific functions, such as cytochrome P450 activity, urea synthesis, and albumin production. As spheroids form, hepatocyte polarization develops, leading to the formation of bile canaliculi. Although portal vein thrombosis and increased portal pressures are of great concerns, hepatocyte spheroid transplantation showed increased transplanted cell mass and heterogeneity in distribution, which spread more uniformly throughout the recipient liver, compared to single cell. Most commonly used spheroid culture systems include spinner flasks, rotating cell bioreactors, rocker devices, and plates with nonadherent surfaces. Moreover, coculturing other cell types with primary hepatocytes to form hepatic organoids may be responsible for reestablishing cell junctions such as E-cadherin, which is required for hepatocyte spheroid formation, and protecting hepatocytes from cell death. Thus, it is emerging as a prospective way for long-term culture. Seeding the primary hepatocytes with mesenchymal stem cells, liver sinusoidal endothelial cells, or umbilical vein endothelial cells has been reported to result in enhanced cell-cell interactions and shorten the formation time of spheroids, which leads to improved hepatocyte function and prolongation of the survival time. Cells in these aggregates can maintain their morphology and liver-specific functions for over 1 month.

In liver tissue engineering, it is crucial providing the appropriate three-dimensional ultrastructure and biological activities that include a surface for cell attachment, subsequent tissue development, and cell migration. Biologic scaffolds composed of ECM have been used in reconstructing bioengineered hepatic tissue, combining cells with biomaterial in complex together. Liver-specific ECM, which retains native liver microstructures and related growth factors, provides a more favorable microenvironments for liver cell differentiation [19], hepatocyte growth and proliferation [20, 21], and neovascularization [21, 22]. Moreover, biologic scaffolds have been validated that they could be developed as an alternative carrier for hepatocyte transplantation into heterotopic positions. Decellularized liver ECM was first obtained using ammonium hydroxide and Triton X-100 solution; great progress has been made regarding methods for liver decellularization since then

[23]. Currently, decellularized liver scaffold can be accomplished by vascular perfusion with a cocktail of enzymes, chemical detergents, hypotonic saline rinses, or physical methods that completely remove all cellular elements while largely maintaining the native ultrastructure and active extracellular molecules of the liver matrix [24]. An ideal decellularization protocol includes at least 1 of the 4 approaches listed above or, more commonly, a combination of them in order to maximize their effectiveness. Sodium dodecyl sulfate (SDS), an ionic detergent, has been reported to be very effective in eliminating nuclei from dense organs. Triton X-100 is a milder nonionic detergent than SDS, known to disrupt lipid-lipid and lipid-protein interactions. The immersion and agitation method is quite limited in decellularization on solid organs; perfusion is the preferred method [25, 26]. A combination of SDS, Triton X-100, and PBS perfusion is regarded as an ideal detergent recipe for liver scaffolds harvest [27]. Tissue-engineered ECM scaffold characteristics can be easily post-modified before recellularization for better hemocompatibility, reduced immunogenicity, or immobilization modifications of desired components. Long-term survival of transplanted hepatic bioengineered tissue or whole bioengineered liver is significantly hindered by in vivo thrombogenesis, for which collagen, the major component of ECM, is grossly responsible due to its prothrombotic properties. Modification of heparin mobilization has been proven to prevent thrombus formation [28] and improve the formation of vessels in the hepatic bioengineered tissue after transplantation (unpublished data). Multiple heparin-coating techniques have been developed [29, 30]. Multi-point attachment method can greatly reduce thrombogenicity following blood reperfusion [31]. Endpoint attachment of heparin treated ECM performed higher efficiency of heparin immobilization, ultralow blood component activations, and excellent hemocompatibility [32]. Immune rejection is another limit for the long-term survival of implanted graft. ECM components are highly conserved between different species, so that decellularized liver scaffolds are less associated with acute rejection post-transplantation. Antigenic molecules left behind in the decellularized matrix, such as residual DNA and the Gal epitopes, may be strong immune inducers. Additionally, xenogenic or allogenic ECMs may undergo foreign body degradation and possibly elicit a chronic immune response if unable to be eliminated by the recipient. Crosslinking the liver ECM with fixative agents can contribute to the formation of a covalent bond that generates inter-chain crosslinking, resulting in the alteration of the secretion profile of immune cytokines and the proliferation pattern of human peripheral blood mononuclear cells in vitro. Crosslinking of genipin, a natural crosslinker for proteins, collagen, gelatin, and chitosan derived from gardenia, can lower the immunogenic potential of xenogeneic decellularized liver scaffolds [33]. These features

make the decellularized liver scaffold an ideal carrier for liver tissue engineering.

Insufficient decellularization can trigger undesired host immune responses and lead to difficulty in recellularization and subsequent graft failure. Thus, the decellularized liver scaffold should fulfill the widely accepted criteria to define the satisfactory elimination of cellular components: (1) <50 ng DNA/mg ECM dry weight, (2) <200 bp DNA fragment length, and (3) lack of visible nuclear content with 4,6-diamidino-2-phenylindole (DAPI) or hematoxylin and eosin (HE) staining [34]. Besides, it is important to evaluate the degree of ECM preservation by specific staining (i.e., Masson trichrome, Sirius Red), immunohistochemistry (collagen), and electron microscopy, etc. Critical information such as scaffold porosity, permeability, surface texture, mechanical properties, and compatibility should also be evaluated according to requirements [35]. The level of critical growth factors in the ECM, including hepatocyte growth factor, basic fibroblast growth factors, vascular endothelial growth factor, and insulin-like growth factor ought to be taken into consideration for further supporting of functional hepatocytes [36]. Additional immobilization of desired growth factors or supplement of inadequate growth factors is feasible to enhance implanted cell survival and function.

The ultimate goal in generating a high-quality decellularized liver ECM is successful recellularization. One approach is to drop cell suspension over the matrix until it completely covers the entire sponge-like ECM [37]. Our group observed hepatocyte-like cell differentiation, hepatocyte expansion, and blood flow in the graft and liver injury recovery in the recipient after hepatic bioengineered tissue transplantation (unpublished data). The perfusion of cells through the portal or hepatic veins is another seeding strategy. Compared to direct parenchymal injection and continuous perfusion, multistep infusion showed superior cell engraftment and liver-specific functional capacities [38]. Nyberg et al. successfully decellularized and recellularized a porcine liver and achieved implantation with sufficient blood flow and the absence of thrombosis after 72 h of vascular perfusion [39].

2 Materials

2.1 Hepatocyte Isolation Procedure from Whole Liver

Before starting, all necessary instruments and solution bottles must be prepared sterile and ready to use. Prepare perfusion solution I A (with ethylene glycol tetra acetic acid (EGTA)) and I B (without EGTA) and perfusion solution II before starting the cell isolation. Solutions and medium are sterilely filtered and stored at 4 °C for up to 4 weeks. If the solutions are not prepared freshly, make sure to check pH immediately before the isolation procedure and if necessary adjust the pH.

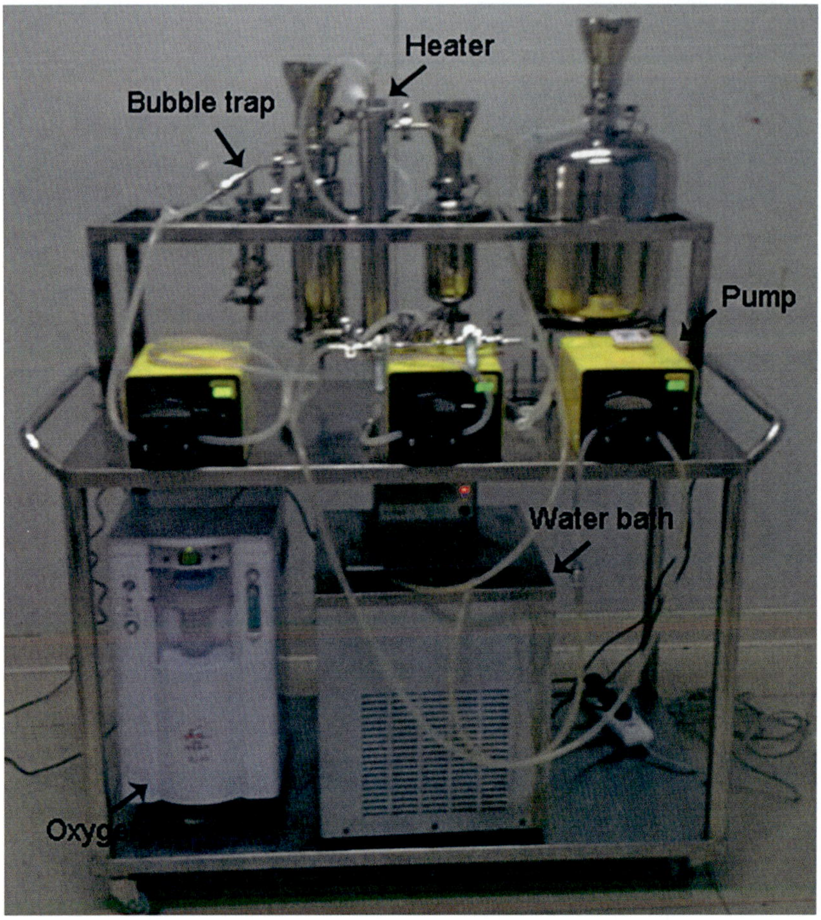

Fig. 1 Perfusion system for hepatocyte isolation

1. Perfusion solution I A (with EGTA) (per 1 L): 8.3 g NaCl, 0.5 g KCl, 2.4 g HEPES, 0.95 g EGTA, pH 7.4.
2. Perfusion solution I B (without EGTA) (per 1 L): 8.3 g NaCl, 0.5 g KCl, 2.4 g HEPES, pH 7.4.
3. Solution 1 of perfusion solution II (per 1 L): 3.9 g NaCl, 0.5 g KCl, 24 g HEPES, 0.8 g NAC, 0.7 g $CaCl_2{\cdot}2H_2O$, pH 7.6. Take out 100 mL solution 1 to prepare solution 2.
4. Solution 2 of perfusion solution II (per 1 L) (prepare on ice): dissolve 4 g bovine serum albumin (BSA) and 130 Wunsch units collagenase in 100 mL of solution 1.
5. Perfusion solution II (per 1 L): Mix solution 1 and solution 2 prior to perfusion start.
6. Hepatocyte wash medium (per 1 L): 10.8 g Williams' E, 2.2 g $NaHCO_3$, 2.6 g HEPES, 1000 IU penicillin, 1000 μg streptomycin, 100 mL fetal bovine serum (FBS), ddH_2O to 1 L, pH 7.3. Adjust the concentration of sodium to 140 mM with distilled water.

7. Hepatocyte culture medium (per 1 L): 10.8 g Williams' E, 2.2 g $NaHCO_3$, 3.6 g HEPES, 1000 IU heparin, 200 IU insulin, 1000 IU penicillin, 1000 μg streptomycin, 50 mL FBS, 3 g BSA, 150 μg linoleic acid, 5 μg epidermal growth factor (EGF), 0.1 μM dexamethasone, 40 μg human glucagon, 6 mg human transferrin, 20 μg Gly-His-Lys, 0.1 μM $CuSO_4 \cdot 5H_2O$, 30 nM Na_2SeO_3, 50 pM $ZnSO_4 \cdot 7H_2O$, 1 g L-carnitine, 0.2 g L-arginine, 0.01 g glycine, ddH_2O to 1 L, pH 7.3. Adjust the concentrate of sodium to 140 mM with distilled water.
8. Equipment: instrument setup as shown in Fig. 1.
9. Water bath set at 39–40 °C for the isolation, depending on the room temperature. Keep the solution temperature at the outflow between 37 and 38 °C to obtain optimal collagenase activity.
10. Connect oxygen to the hollow-fiber bioreactor to enrich O_2. Set the flow rate for the bioreactor for the isolation at 2–5 L/min.
11. Basin with lid for collecting the liver, forceps and scissors for dissociation, a suction bottle for collecting hepatocytes, a four-layer gauze on top of funnel to filter the suspension.
12. Refrigerator at 4 °C (set it up at this exact temperature).

2.2 Resected Liver Tissue Isolation Procedure

All necessary instruments and solution bottles must be prepared sterile and be ready to use. Prepare perfusion solution I A, I B, and II as described above. Prepare a sterile plate, and different sizes of cannulas, wrapping in a surgical drape sheet.

2.3 Hepatocyte Spheroid Culture for Liver Tissue Engineering

All necessary instruments and solution bottles must be prepared sterile and be ready to use.

1. Hepatocyte wash medium and hepatocyte culture medium (as described in Subheading 2.1, filtered sterilely into bottles).
2. Chamber, incubator, and rocker system, setup as shown in Fig. 2.
3. Sterile suction bottle, a two-layer gauze on top of funnel to filter and centrifuge bottles for spheroid harvest.
4. Refrigerator at 4 °C (set it up at this exact temperature).

2.4 Hepatic Bioengineered Tissue Construction

1. Instrument setup as shown in Fig. 3.
2. 1% Triton X-100 solution and 1% SDS solution. Prepared before starting the liver decellularization.
3. Basin for collecting the liver and tubing for perfusion.
4. Heparinization perfusion solution: 3.3 g/L sodium heparin dissolved in distilled water at 4 °C, 33 mg/L $NaNO_2$ then

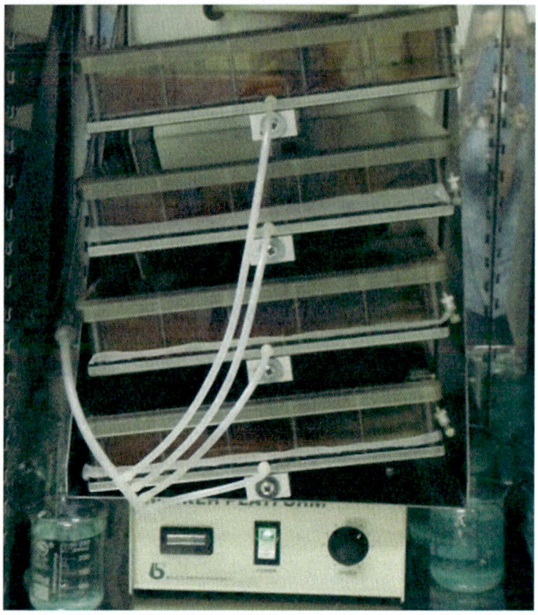

Fig. 2 Rocker system for hepatocyte spheroid formation

Fig. 3 Perfusion system for liver scaffold decellularization

added to the solution, and pH adjusted at 2.7, followed by stirring at 4 °C for 2 h; pH is then adjusted to 7.0 to stabilize the depolymerized heparin in the solution; 0.01 mg/L $NaBH_3CN$ and 0.15 M NaCl are added at pH 3.5 before perfusion.

5. 0.625% genipin solution (for crosslinking process).

3 Methods

3.1 Hepatocyte Isolation Procedure from Whole Liver

1. Administer general anesthesia to pigs (weight 5–8 kg) (*see* **Note 1**).
2. Perform next a midline laparotomy from xiphoid to umbilicus with adequate retraction and exposure. The hepatic artery and the bile duct are then ligated.
3. Inject heparin sodium into the infrahepatic vena cava.
4. Inject nitroglycerin into portal vein 2 min prior to starting the perfusion.
5. Prime the silicon tube with perfusion solution I A. Cannulate the portal vein with silicon tube and fix it (Fig. 4). Flush the liver at the rate of 400 mL/min.
6. Perfuse the liver with at least 5 L of perfusion solution I A to wash out all blood from the tissue; meanwhile dissect the infrahepatic vena cava as outflux (*see* **Note 2**).
7. Perfuse the liver with 1–2 L of perfusion solution I B to wash out EGTA from perfusion solution I A at the rate of 400 mL/ min (*see* **Note 2**).
8. Excise the liver and place it in a basin. Perfuse the liver with 2 L of perfusion solution II to digest the liver at the rate of 500 mL/min (Fig. 5) (*see* **Notes 2–4**).
9. Set up the recirculation tubing. The perfusion solution is then subjected to recycle perfusion at a rate of 500 mL/min.
10. Terminate the process after approximately 30 min or as soon as the liver stays irreversibly deformed through perfusion pressure (*see* **Note 5**).
11. Add ice-cold hepatocyte wash medium containing 10% FBS to inactivate collagenase activity and trypsin residues to prevent overdigestion of the cells (*see* **Note 6**).
12. Use forceps and scissors to comb the liver gently. Collect the cell suspension with suction bottle and then pour the suspension through the prepared funnel with four-layer gauze into centrifuge bottles to eliminate tissue debris (Fig. 6).
13. Centrifuge for 10 min with $50 \times g$ at 4 °C. Discard the supernatant; resuspend the cell pellet in hepatocyte wash medium and centrifuge for two more times. For third centrifugation, centrifuge at the same speed for 10 min (*see* **Note 7**).
14. After removal of the supernatant, weight the cell pellet. Suspend the cell pellet in hepatocyte culture medium.

$$\text{Calculation}: 1\ \text{g cell pellet} \approx 1.6 \times 10^8\ \text{cells}$$

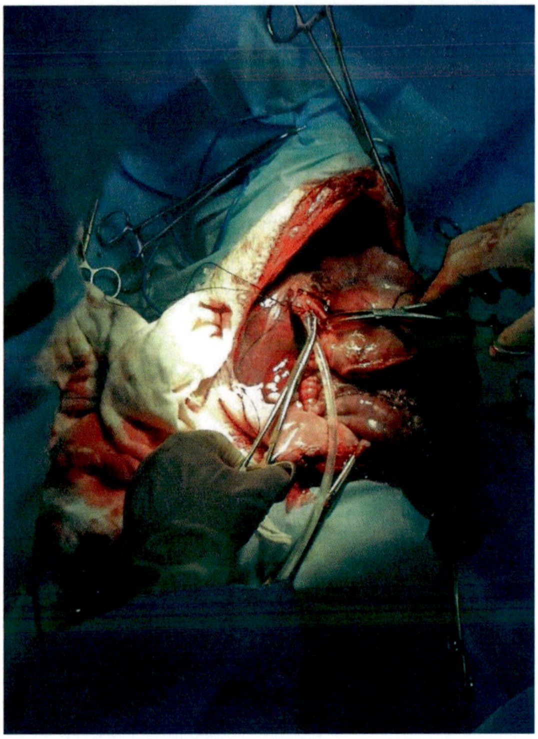

Fig. 4 Cannula through portal vein for perfusion

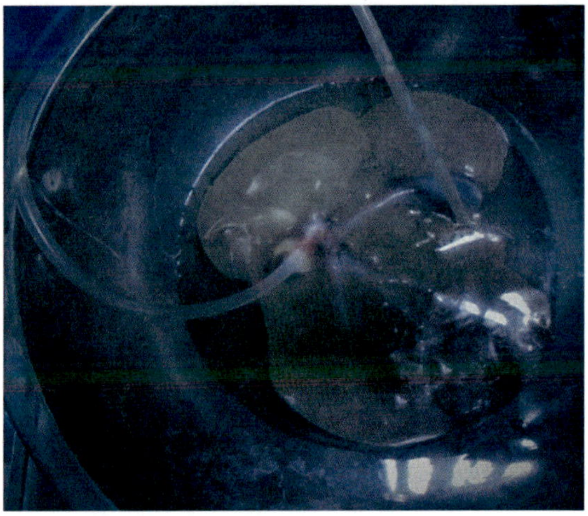

Fig. 5 Recycled perfusion for enzyme digestion

15. Take sample to estimate the vitality and cell number, using trypan blue exclusion using a hemocytometer (*see* **Note 8**).

Fig. 6 Isolated hepatocytes in centrifuge bottle

$$\text{Calculation : Total cell count} = \text{Counted cells} \times 10^4 \times \text{dilution} \times \text{total volume(mL)}$$

$$\text{Calculation : Viability(\%)} = \frac{\text{Viable cell count}}{\text{Total cell count}} \times 100\%$$

3.2 Resected Liver Tissue Isolation Procedure

1. A sterile work area is prepared when a surgical drape sheet is unwrapped.
2. Take the liver piece out and place it onto the plate. A favorable resection area can be cannulated into at least three vessels. Incise the liver tissue with a single cut with the scalpel to expose more vessels, if necessary (*see* **Note 9**).
3. Manually fix cannulas into the vessels in turn. Perfuse the liver using the same perfusion solution IA (2 L), IB (1 L), and II (1 L) at the speed until an adequate perfusion pressure has been reached (*see* **Note 4**).
4. Set up the recirculation tubing. Terminate the recycle perfusion when the liver is deformed (*see* **Note 5**).
5. Conduct the same procedures described above after **step 11**.

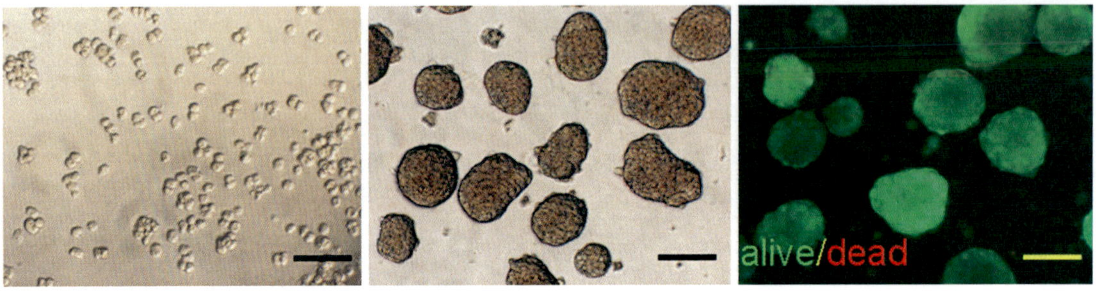

Fig. 7 Morphology of isolated porcine hepatocytes. (Left) fresh isolated hepatocytes; (Middle) rocker culture after 24 h; (right) viability FluoroQuench staining. Scale bar = 50 μm

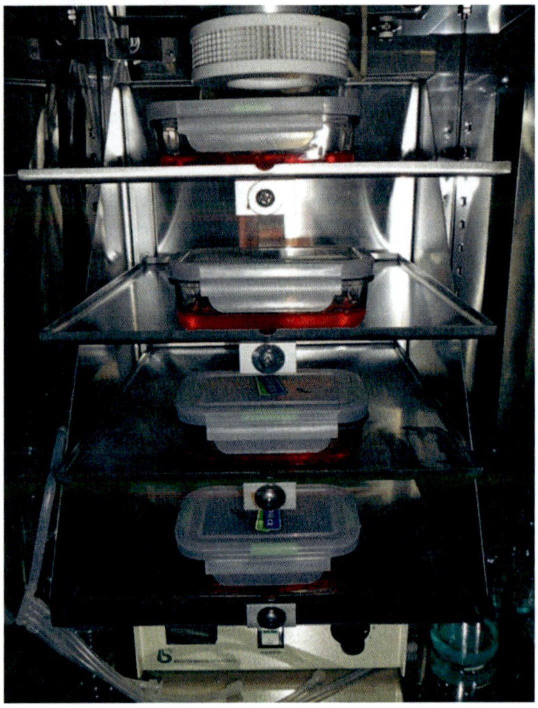

Fig. 8 Small-scale culture of hepatocyte spheroid formation

3.3 Hepatocyte Spheroid Culture for Liver Tissue Engineering

1. Suspend the harvested porcine hepatocytes in 1000 mL of hepatocyte culture medium at a density of 5 × 10^6 cells/mL, and inoculate into spheroid generation chamber. Spheroid chambers are incubated at 37 °C in a humidified atmosphere, supplied with 5% carbon dioxide gas mixture, and rocked continuously at 0.12 Hz for 24 h (Fig. 7) (*see* **Notes 10** and **11**).
2. For small-scale cultivation, suspend the harvested porcine hepatocytes in 20 mL of hepatocyte culture medium at a density of 1 × 10^6 cells/mL, and inoculate into a spheroid dish that is surface-siliconized with Sigmacote and incubate (Fig. 8) (*see* **Note 12**).

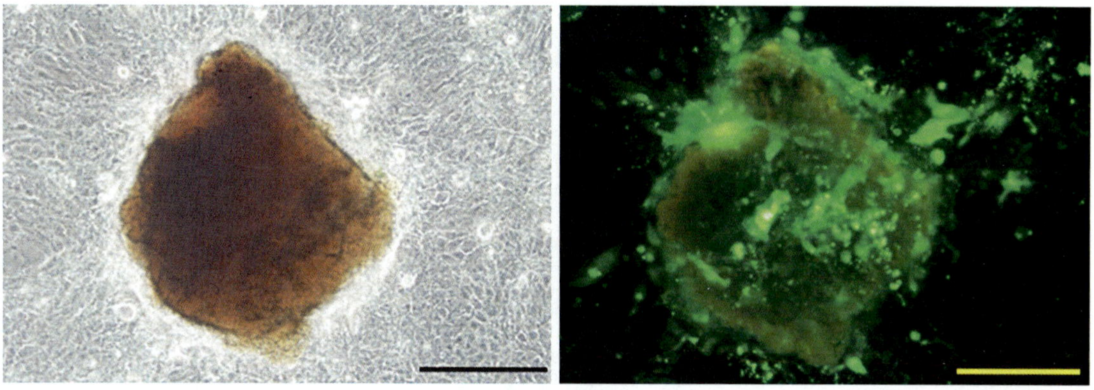

Fig. 9 Morphology of hepatic organoids cocultured hepatocytes with human umbilical vein endothelial cells. (Left) Light microscopy examination of hepatic organoids; (right) albumin expression of hepatic organoids

3. To form organoids, mix porcine hepatocytes and liver non-parenchymal cells at proper mixture ratio in hepatocyte culture medium immediately before inoculation into spheroid chamber or spheroid dish and cultivate then (Fig. 9).
4. For spheroid harvest, collect the hepatocyte spheroids from the chambers using a suction bottle, filter the suspension through the funnel to filter into centrifuge bottles, and then centrifuge at the speed of 50 × *g* at 4 °C. Discard the supernatant and suspend readying to use, or wash again if necessary.

3.4 Hepatic Bioengineered Tissue Construction

1. General anesthesia is administered to pigs (weight 6–10 kg).
2. A midline laparotomy from xiphoid to umbilicus is performed with adequate retraction and exposure.
3. Excise the liver and place it in a basin.
4. Ligate the hepatic artery, bile duct, and infrahepatic vena cava.
5. Cannulate the portal vein with silicon tube and tie; leave the suprahepatic vena cava as outflux.
6. Perfuse the liver as follows: ddH_2O for 1 h, 1% Triton X-100 for 3 h, ddH_2O for 3 h, 1% SDS for 3 h (recycle), ddH_2O for 3 h, 1% SDS for 3 h (recycle), 1% Triton X-100 for 3 h, ddH_2O for 3 h, PBS for 3 h. All the process should be performed at a speed of 200 mL/min (Fig. 10) (*see* **Notes 13–15**).
7. For scaffold heparinization, perfuse the liver with the heparinization perfusion solution (recycle) at a speed of 100 mL/min for 4 h.
8. For scaffold genipin crosslinking, perfuse the liver with the genipin solution for 12 h (recycle), followed by ddH_2O for 12 h (recycle) and PBS for 6 h, at a speed of 100 mL/min.
9. Cut the liver into disks of 1 × 1 × 1 cm in diameter and place into 24-well plates for lyophilization.

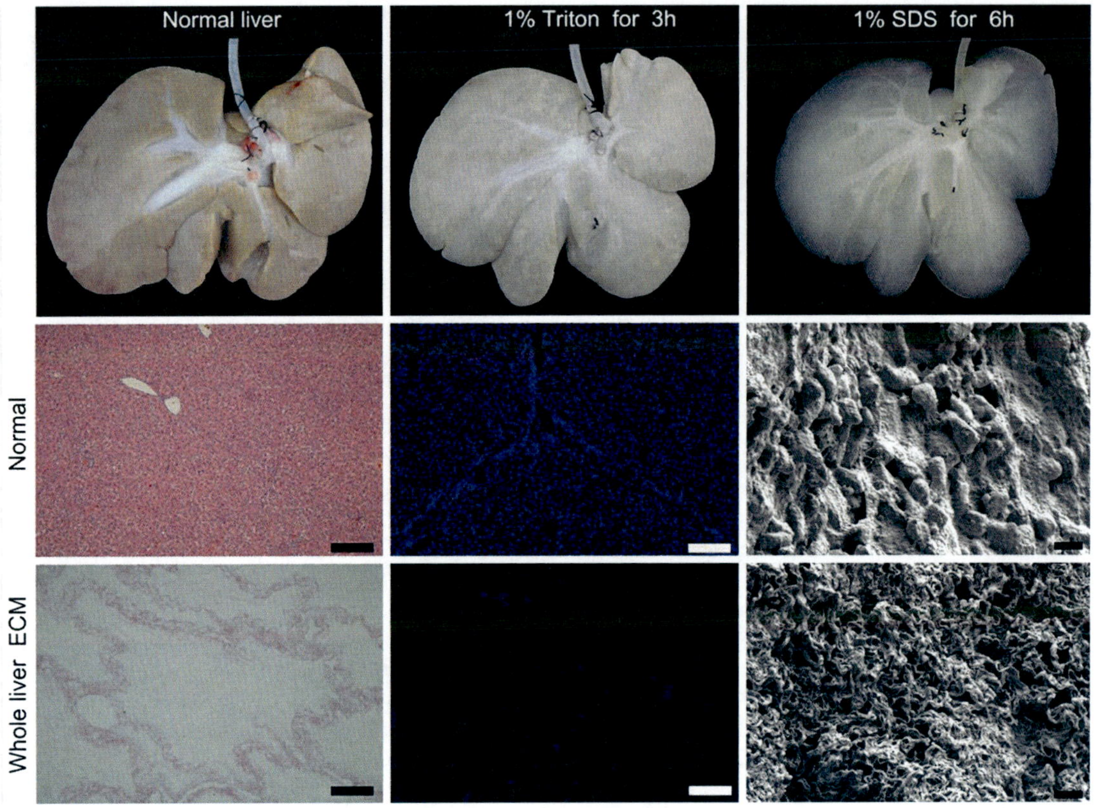

Fig. 10 Whole-organ porcine native and decellularzied liver. Representative images of (upper) macrograph; (lower left) H&E staining; (lower middle) DAPI staining and (lower right) SEM analysis of ultrastructure

10. Sterilize the scaffolds with gamma irradiation (1000 rad) for 1–2 h.
11. Prior to cell seeding on scaffolds, incubate the disks in hepatocyte culture medium at 37 °C to rehydrate overnight.
12. Pipette the hepatocyte spheroids in the center of decellularized liver scaffold disks using vacuum suction device (Fig. 11), and allow them to attach for 2 h, followed by the addition of medium, and then culture in ultralow attachment 6-well plates for 48 h (Fig. 12) (*see* **Notes 16** and **17**).
13. For whole bioengineered liver construction, sterilize the whole liver scaffolds with gamma irradiation and cannulate into the influx and outflux. Perfuse the liver in the hepatocyte perfusion recycle system with hepatocyte culture medium, and incubate at 37 °C in a sterile humidified atmosphere, supplied with 5% carbon dioxide gas mixture (Fig. 13). Inoculate the hepatocyte spheroids into the recycle system in multiple steps at 10–15 min intervals through a port (Fig. 14) (*see* **Note 18**).

Fig. 11 Suction device for cell seeding

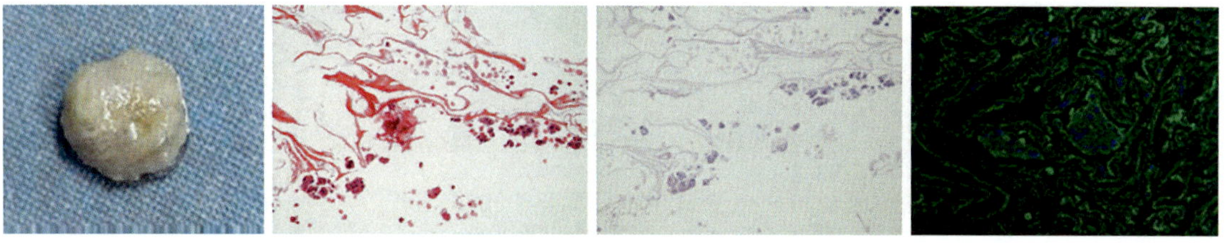

Fig. 12 Morphology of hepatic bioengineered tissue. (Left) Macrograph; (left middle) H&E staining; (right middle) Toluidine Blue staining and (right) albumin expression

$$\text{Calculation : Total cell number} = \text{Initial cell number} - \text{Cell number in perfusate}$$

4 Notes

1. High-protein diets (40%) are routinely given to donor pigs for 7 days before the hepatocytes harvests to induce secretion of albumin and other liver-specific proteins.

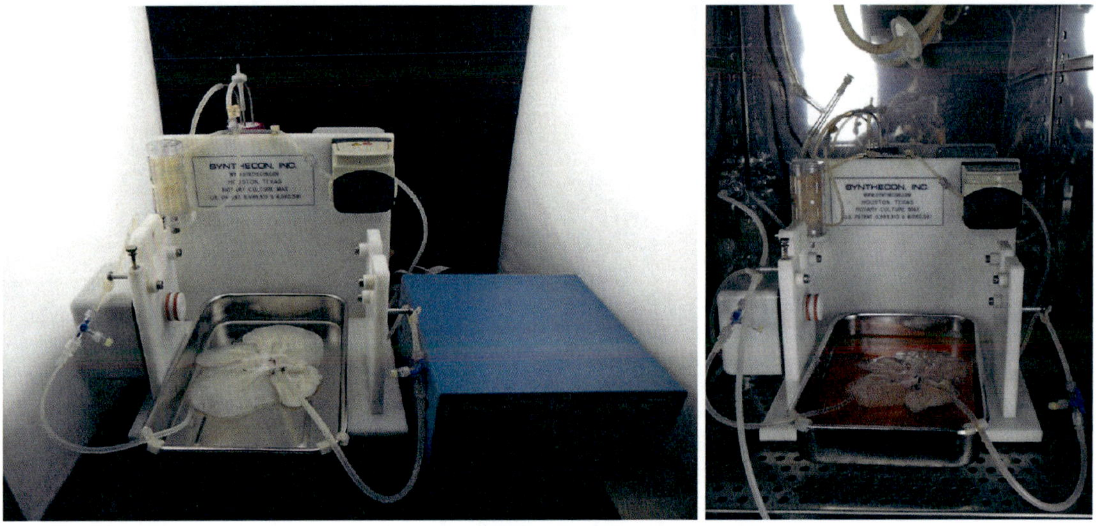

Fig. 13 Hepatocyte perfusion recycle system. (Left) Perfusion of the whole decellularized liver scaffold with the hepatocyte perfusion recycle system and (right) the hepatocyte perfusion recycle system in an incubator

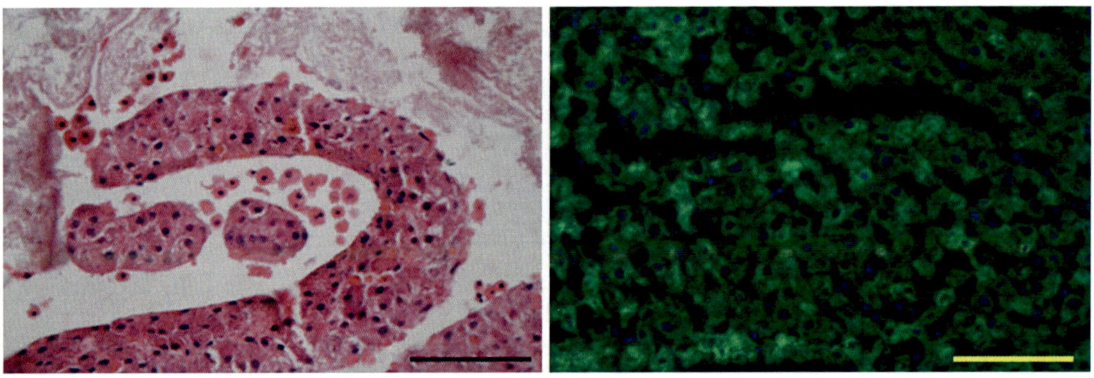

Fig. 14 Morphology of whole bioengineered liver. (Left) H&E staining and (right) albumin (green) expression

2. The volume of perfusion solution can expand according to the weight of the pigs.
3. There are many commercialized collagenases on the market to isolate liver cells, such as Roche Liberase research grades and Serva collagenase NB grades. The compositions vary from batch to batch, and subsequently its proteolytic activity strongly varies among different types. It is important to test before using routinely.
4. The collagenase is temperature sensitive; the activity of the collagenase can be greatly influenced. Keep the collagenase and perfusion solution II solution 2 cool before use. It is important to control the solution temperature using a water bath within the optimum temperature to obtain the optimal efficiency.

5. The enzyme perfusion time required for digestion is not predetermined. Perfusion process can terminate as soon as the liver stays irreversibly deformed through perfusion pressure or surgeons observe the evidence of adequate tissue digestion.
6. The hepatocytes are sensitive to concentration of sodium. Since the concentration of sodium can be influenced by temperature, pH of distilled water, pH calibration procedure, etc., it is recommended to adjust the sodium concentration to 140 mM with distilled water to get optimal outcome.
7. When purifying suspensions of hepatocytes, a separation of cells is recommended by density gradient centrifugation using a Percoll gradient.
8. To ensure persistent good viability during culture, at least 90% of the cells should exclude trypan blue on first examination.
9. It is possible to send the harvest hepatocytes or resected liver tissue to other laboratories. Hepatocytes in suspension can be cold stored in preservation solution for 24–72 h. It is recommended to flush all blood from the resected liver tissue, cannulate and perfuse the tissue with preservation solution, and then transport on ice.
10. Hepatocyte spheroids can both be maintained in serum-free and serum-containing culture medium. Serum-containing formulations are often based on Williams'-E Medium and are ideal for short-term cultures, up to 10 days. Serum-free formulations were developed for long-term cultures (several weeks) and require an adaptation period.
11. Hepatocyte spheroids in suspension show better outcomes when cold stored for transport.
12. Oxygen consumption can indicate cellular respiration of hepatocyte spheroids. Albumin synthesis is one of the classical assays for the study of liver-specific function. Assessing of ammonia degradation and ureagenesis are critical to evaluate the function of hepatocyte spheroids.
13. Single or multiple freeze-thaw cycles can be administrated before the decellularization process. The rapid thermal change brought on by repeated freezing and thawing results in an osmotic response and the formation of intracellular ice crystals, which accelerate cell lysis.
14. The rate and pressure of perfusion should depend upon the specific characteristics of the liver; excess pressure may lead to the thinning and damage of vascular walls.
15. Residual SDS in the ECM leads to cellular toxicity in the subsequent recellularization process. It is very important to make a final rinse at the end of decellularization.

16. It is easier to suspend the hepatocyte spheroids into 20–50 μL hepatocyte culture medium at high density and then pipette in the center of decellularized liver scaffold disks.
17. The decellularized liver scaffold degrades with the culture time. Modifications can slow this process down.
18. The required cell number for recellularization varies depending on scaffold volume, seeded cell type, and purpose of the recellularized liver.

References

1. Berry MN, Friend DS (1969) High-yield preparation of isolated rat liver parenchymal cells: a biochemical and fine structural study. J Cell Biol 43(43):506–520
2. Jiang QD, Li HP, Liu FJ et al (2013) Isolation and identification of bovine primary hepatocytes. Genet Mol Res 12(4):5186–5194. https://doi.org/10.4238/2013.October.30.3
3. Shulman M, Nahmias Y (2013) Long-term culture and coculture of primary rat and human hepatocytes. Methods Mol Biol 945:287–302. https://doi.org/10.1007/978-1-62703-125-7_17
4. Naik S, Trenkler D, Santangini H, Pan J, Jauregui HO (1996) Isolation and culture of porcine hepatocytes for artificial liver support. Cell Transplant 5(1):107–115
5. Li Y, Wang Y, Wu Q, Li L, Shi Y, Bu H, Bao J (2016) Comparison of methods for isolating primary hepatocytes from mini pigs. Xenotransplantation 23(5):414–420. https://doi.org/10.1111/xen.12259
6. Nelson LJ, Newsome PN, Howie AF et al (2000) An improved ex vivo method of primary porcine hepatocyte isolation for use in bioartificial liver systems. Eur J Gastroenterol Hepatol 12(8):923
7. Maruyama M, Totsugawa T, Kunieda T et al (2003) Hepatocyte isolation and transplantation in the pig. Cell Transplant 12(6):593–598
8. Mitaka T (1998) The current status of primary hepatocyte culture. Int J Exp Pathol 79 (6):393–409
9. Rush GF, Gorski JR, Ripple MG, Sowinski J, Bugelski P, Hewitt WR (1985) Organic hydroperoxide-induced lipid peroxidation and cell death in isolated hepatocytes. Toxicol Appl Pharmacol 78(3):473–483
10. Boess F, Kamber M, Romer S et al (2003) Gene expression in two hepatic cell lines, cultured primary hepatocytes, and liver slices compared to the in vivo liver gene expression in rats: possible implications for toxicogenomics use of in vitro systems. Toxicol Sci 73(2):386–402
11. Hickey RD, Lillegard JB, Fisher JE et al (2011) Efficient production of Fah-null heterozygote pigs by chimeric adeno-associated virus-mediated gene knockout and somatic cell nuclear transfer. Hepatology 54 (4):1351–1359. https://doi.org/10.1002/hep.24490
12. Hickey RD, Mao SA, Glorioso J et al (2014) Fumarylacetoacetate hydrolase deficient pigs are a novel large animal model of metabolic liver disease. Stem Cell Res 13(1):144–153. https://doi.org/10.1016/j.scr.2014.05.003
13. Lloyd TD, Orr S, Patel R et al (2004) Effect of patient, operative and isolation factors on subsequent yield and viability of human hepatocytes for research use. Cell Tissue Banking 5 (2):81–87
14. Bao J, Fisher JE, Lillegard JB et al (2013) Serum-free medium and mesenchymal stromal cells enhance functionality and stabilize integrity of rat hepatocyte spheroids. Cell Transplant 22(2):299–308. https://doi.org/10.3727/096368912X656054
15. Alexandre E, Cahn M, Abadieviollon C et al (2002) Influence of pre-, intra- and post-operative parameters of donor liver on the outcome of isolated human hepatocytes. Cell Tissue Banking 3(4):223–233
16. Puviani AC, Ottolenghi C, Tassinari B, Pazzi P, Morsiani E (1998) An update on high-yield hepatocyte isolation methods and on the potential clinical use of isolated liver cells [Review]. Comp Biochem Physiol A Mol Integr Physiol 121(2):99–109
17. Bartlett DC, Hodson J, Bhogal RH, Youster J, Newsome PN (2014) Combined use of N -acetylcysteine and Liberase improves the viability and metabolic function of human hepatocytes isolated from human liver. Cytotherapy 16(6):800–809

18. Solanas E, Sostres C, Serrablo A et al (2015) Effect of dimethyl sulfoxide and melatonin on the isolation of human primary hepatocytes. Cells Tissues Organs 200(5):2–13
19. Sellaro TL, Ranade A, Faulk DM et al (2010) Maintenance of human hepatocyte function in vitro by liver-derived extracellular matrix gels. Tissue Eng A 16(3):1075
20. Sérandour AL, Loyer P, Garnier D et al (2005) TNFalpha-mediated extracellular matrix remodeling is required for multiple division cycles in rat hepatocytes. Hepatology 41(3):478–486
21. Hammond JS, Gilbert TW, Howard D et al (2011) Scaffolds containing growth factors and extracellular matrix induce hepatocyte proliferation and cell migration in normal and regenerating rat liver. J Hepatol 54 (2):279–287
22. Fernández M, Semela D, Bruix J, Colle I, Pinzani M, Bosch J (2009) Angiogenesis in liver disease. J Hepatol 50(3):604–620
23. Lin P, Chan WC, Badylak SF, Bhatia SN (2004) Assessing porcine liver-derived biomatrix for hepatic tissue engineering. Tissue Eng 10(8):1046–1053
24. Crapo PM, Gilbert TW, Badylak SF (2011) An overview of tissue and whole organ decellularization processes. Biomaterials 32 (12):3233–3243
25. Yagi H, Fukumitsu K, Fukuda K et al (2013) Human-scale whole-organ bioengineering for liver transplantation: a regenerative medicine approach. Cell Transplant 22(2):231–242
26. Maghsoudlou P, Georgiades F, Smith H et al (2016) Optimization of liver decellularization maintains extracellular matrix micro-architecture and composition predisposing to effective cell seeding. PLoS One 11(5): e0155324
27. Wang Y, Bao J, Wu Q et al (2015) Method for perfusion decellularization of porcine whole liver and kidney for use as a scaffold for clinical-scale bioengineering engrafts. Xenotransplantation 22(1):48–61. https://doi.org/10.1111/xen.12141
28. Noishiki Y, Miyata T (2010) A simple method to heparinize biological materials. J Biomed Mater Res A 20(3):337–346
29. Murugesan S, Xie J, Linhardt RJ (2008) Immobilization of heparin: approaches and applications. Curr Top Med Chem 8 (2):80–100
30. Bao J, Shi Y, Sun H et al (2011) Construction of a portal implantable functional tissue-engineered liver using perfusion-decellularized matrix and hepatocytes in rats. Cell Transplant 20(5):753–766. https://doi.org/10.3727/096368910X536572
31. Bruinsma BG, Kim Y, Berendsen TA, Ozer S, Yarmush ML, Uygun BE (2015) Layer-by-layer heparinization of decellularized liver matrices to reduce thrombogenicity of tissue engineered grafts. J Clin Transl Res 1(1):48–56
32. Bao J, Wu Q, Sun J et al (2015) Hemocompatibility improvement of perfusion-decellularized clinical-scale liver scaffold through heparin immobilization. Sci Rep 5:10756. https://doi.org/10.1038/srep10756
33. Wang Y, Bao J, Wu X et al (2016) Genipin crosslinking reduced the immunogenicity of xenogeneic decellularized porcine whole-liver matrices through regulation of immune cell proliferation and polarization. Sci Rep 6:24779. https://doi.org/10.1038/srep24779
34. Badylak SF, Valentin JE, Ravindra AK, Mccabe GP, Stewartakers AM (2008) Macrophage phenotype as a determinant of biologic scaffold remodeling. Tissue Eng A 14(11):1835
35. Loh QL, Choong C (2013) Three-dimensional scaffolds for tissue engineering applications: role of porosity and pore size. Tissue Eng B Rev 19(6):485–502
36. Martino MM, Briquez PS, Güç E et al (2014) Growth factors engineered for super-affinity to the extracellular matrix enhance tissue healing. Science 343(6173):885–888
37. Mazza G, Rombouts K, Hall AR et al (2015) Decellularized human liver as a natural 3D-scaffold for liver bioengineering and transplantation. Sci Rep 5:13079
38. Sotogutierrez A, Zhang L, Medberry C et al (2011) A whole-organ regenerative medicine approach for liver replacement. Tissue Eng Part C Methods 17(6):677–686
39. Lin YQ, Wang LR, Wang JT et al (2015) New advances in liver decellularization and recellularization: innovative and critical technologies. Expert Rev Gastroenterol Hepatol 9(9):1–9

Chapter 19

Pig-to-Macaque Islet Xenotransplantation

Suzanne Bertera, Michael F. Knoll, Carmela A. Knoll, David K. C. Cooper, Massimo Trucco, and Rita Bottino

Abstract

The advancement toward a clinical application for porcine islets to cure diabetes in humans must include reproducible long-term successes in non-human primate (NHP) models. Many dedicated researchers around the world are continuing to work toward this goal. In this chapter, we describe procedures for islet isolation of pancreatic islets from adult and neonatal/fetal pigs. We further include procedures for the induction of diabetes in non-human primates and subsequent insulin therapy, islet transplantation, immunosuppression, and also the daily maintenance of xenotransplanted NHPs. The procedures that we outline in this chapter are ones that we have successfully utilized in pig-to-NHP islet transplantation models. However, where appropriate, alternative methods will also be identified.

Key words Islet isolation, Xenotransplantation, Pigs, Neonatal pigs, Non-human primates, Immunosuppression, Diabetes

1 Introduction

Beta cells are a subpopulation of pancreatic cells residing in the islets of Langerhans that produce the hormone insulin to help regulate glycemic balance in all animals including humans. Diabetes is caused when insulin is either not produced sufficiently (type 1 diabetes) or is used inefficiently (type 2 diabetes) so that normal blood glucose regulation cannot be maintained without intervention, most commonly with medication or exogenous insulin [1]. Pancreas transplantation from cadaveric donors is a clinically acceptable method of beta cell replacement that can restore glycemic control in recipients and has a physiologic response that provides greater therapeutic advantages over exogenous insulin. Whole pancreas transplantation, however, is a major surgery and is accompanied by frequent and often life-threatening complications [1, 2]. Alternatively to whole organ transplantation, islet cell transplantation from cadaveric donors is a relatively noninvasive procedure in which islet cells that comprise perhaps 1–2% of the make-up

Cristina Costa (ed.), *Xenotransplantation: Methods and Protocols*, Methods in Molecular Biology, vol. 2110,
https://doi.org/10.1007/978-1-0716-0255-3_19,

of the pancreas are separated from other pancreatic cells using sophisticated methods and then infused into the recipient's liver [3]. Successful outcomes following clinical islet allotransplantation are approaching the level of whole pancreas transplantation, without the associated fears of morbidity [2]. Insulin independence can often be achieved after islet transplantation but only maintained with difficulty as islets are lost over time [4]. Despite the potential return to exogenous insulin, islet transplantation can lead to lasting benefits for the islet recipient, both in quality of life and in limiting secondary complications associated with diabetes [5–7]. Because clinical islet allotransplantation requires lifelong immunosuppression in order to prevent the recipient from rejecting the donor cells, the procedure is typically reserved for patients who suffer from type 1 diabetes with severe potentially life-threatening hypoglycemia or in conjunction with kidney transplantation that already necessitates immunosuppression [6]. Another major caveat with islet transplantation is that it often requires more than one organ to provide the number of islets needed for transplantation. The scarcity of organ donors has led to the exploration of animals as alternative sources for islets. Islet xenotransplantation is an appealing alternative to the use of cadaveric human organs and has been well studied in animal models [8].

Adult pigs are an excellent choice as an islet source for a number of reasons. Humans already mass-produce pigs for commercial purposes, demonstrating a willingness to utilize them on a large scale. Tissues from pigs are already used for select medical purposes including heart valve replacement. Their physiology and metabolic requirements are similar to humans. Porcine insulin is only one amino acid different from human insulin and was commonly used before the advent of recombinant human insulin. Physical characteristics including organ size, gestational period, and litter size, as well as the ability to make them more compatible with the human immune system by altering certain immunomodulary genes, make them well suited to providing organs for donation. Pigs can also produce relatively high yields of islets suitable for transplantation. Multiple donors may still be required; however, they would be easily obtainable. The age and breed of the pig are important donor characteristics that should be carefully chosen in order to maximize islet yield. Several breeds including Large White, Landrace, and minipigs have been reported to provide good outcomes. Whatever the breed, females over 2 years of age that have produced several litters of piglets (so-called retired breeders) are preferred due to their ability to easily provide large numbers of good islets [9–11]. The porcine donor can be selected and bred especially for islet xenotransplantation to provide a specific pathogen-free (SPF) healthy islet source and allow for easy scheduling of patient surgeries. Moreover, pigs with genetic modifications can be generated to improve compatibility with humans. Knocking out of pig antigens that are targets of natural antibodies and transgenic expression of anticomplement and anticoagulation factors have been shown to enhance success of islet xenotransplantation in preclinical models [12, 13].

A pig pancreas, however, is morphologically different from a human pancreas. Porcine islets typically have a less compact shape than human islets and dissociate after isolation more easily than human islets making them more fragile and difficult to isolate. Retired breeders are the exception to this rule, producing more compact islets that are morphologically similar to human islets. While the technical features of the pig pancreas may make the procedure to isolate islets more challenging, large yields of good islets can still be obtained using well-defined methods. To date, preclinical and clinical studies using pig islets have shown great promise [12, 14].

Neonatal/fetal porcine islets are also regarded as a good source of replacement islets as they can be isolated relatively easier than adult pig islets using a well-established procedure [15, 16]. Additionally, they are hardier than islets isolated from adult pigs allowing them to more easily survive the stresses involved in islet isolation and transplantation. Like adult pig donors, neonatal/fetal pigs can also be obtained from genetically altered breeds to enhance compatibility with humans. The challenge to using neonatal/fetal pig donors is that they produce neonatal islet-like cell clusters (NICC) instead of mature islets. NICCs are immature cells that do not yet produce insulin. After isolation, they must be given additional time to mature in culture and will require even more time after transplantation before they are functional enough to reduce blood glucose to a normal level in recipients. However, once transplanted, NICC may increase in size and proliferate and facilitate differentiation of ductal cells, thus increasing the mass of insulin-producing beta cells [17]. Despite their delay in producing insulin, NICCs are often preferred over adult islets for the other advantages that they offer. They have been tested in NHPs [18] and clinically [19, 20] with positive outcomes. Obviously, the use of NICCs would require more donors; however, these could be produced on demand. On an organ-to-organ basis, the islet isolation process is less time-consuming than with adult pigs, which would help to compensate for the larger number of donors needed.

Pig islets have been extensively studied in rodent and non-human primate (NHP) models of diabetes. Preclinical pig-to-NHP studies have demonstrated tremendous success in controlling diabetes especially over the last 5 years [12]. Further studies with NHP are likely to be necessary to test immunosuppressive drugs that would be compatible with clinical islet xenotransplantation. When using non-human primates in transplantation research, many drugs must be administered and biological samples taken requiring daily manipulation of the animal. These must be done in the most humane method suited to the task. Daily chemical restraint is undesirable and constant use of a squeeze cage adds more stress to the animal in an already stressful situation. One alternative that obviates the use of other restriction devices is the tether system [21]. One end of the tether is attached to the cage on a swivel, while the other end is attached to a jacket worn by the animal. The animal is then surgically fitted with vascular lines that are threaded through the tether to the outside of the cage. This allows the animal to move freely while facilitating the delivery of most drugs and biological sampling without daily sedation or restraint. This method improves the ease in which drugs can be administered while at the same time reducing the stress on the animal.

In the upcoming sections, we will describe materials and methods for the isolation and transplantation of adult and neonatal porcine islets based on our own and other protocols in which we have successfully transplanted non-human primates with porcine islets. The following procedures involving animals have all been approved by the local IACUC and conform to the regulations established by the USDA for the safety and well-being of animals used in research.

2 Materials

2.1 Animals

1. Neonatal/fetal pigs. Most protocols typically use 3–7-day-old animals, but good results have been obtained in older and younger neonates. Several donors [4–8] are needed for one recipient (*see* **Note 1**).
2. Adult pigs. Age is one of the most important considerations for a good outcome in islet isolation. Because of the morphology and fragility of pig islets, the best age to use is 2 years or older. The second consideration is breed. Breeds that are reported to have good yields include Large White, Landrace, and minipigs and also pigs with genetic modifications. One adult pig donor will provide enough islets for transplant into one or two recipients (*see* **Note 2**).
3. Cynomolgus macaques (*Macaca fascicularis*) or rhesus macaques (*Macaca mulatta*), 2–5 years of age, are most commonly used (*see* **Note 3**).

2.2 Neonatal/Fetal ICC Isolation

1. Laminar flow biological cabinet (about 1.5–2 m wide).
2. 50 mL conical tubes (sterile).
3. Small stainless-steel surgical pan for cleaning pancreata.
4. Forceps and surgical scissors at least 15 cm long with long blades.

5. 500 μm nylon filter mesh and holder.
6. Sterile underpads.
7. Aspiration system.
8. Disposable pipettes (5, 10, 25 mL).
9. Collagenase type XI.
10. Non-tissue culture T-175 flasks (Sarstedt).
11. 37 °C water bath with thermometer.
12. 0.22 μm sterile filters.
13. University of Wisconsin (UW) or other similar organ preservation solution.
14. Ham's F-10 culture medium (for 1 L): 1 bottle Ham's F-10 powder, 1.2 g $NaHCO_3$, 1.22 g Nicotinamide, 0.7 g D-glucose, 0.15 g L-glutamine, 0.178 g $CaCl_2$ anhydrous, 5 g BSA Fraction 5 (endotoxin free), 1 mL IBMX (from 50 mM stock solution), 5 mL Penn/strep, up to 1 L H_2O.
15. Washing medium: 1 L HBSS, 0.35 g $NaHCO_3$, 2.38 g Hepes powder, 5 mL Penn/strep, 2.5 g bovine serum albumin (BSA) Fraction 5 (endotoxin-free).

2.3 Adult Pig Islet Isolation

1. Two laminar flow biological cabinets (about 1.5–2 m wide).
2. Large and small stainless-steel surgical pans.
3. Lab stand, base, and clamp.
4. Stainless-steel tubing coil.
5. Surgical pack consisting of large curved scissors, small scissors, large forceps, four to six hemostats.
6. Digital thermometer with dual readout to report the temperature of the water bath and inside the digestion chamber.
7. 500- and 1000 mL beakers (three each).
8. Two 1000 mL collecting bottles with caps.
9. Three ice buckets and one large plastic bin for ice.
10. Centrifuge with rotors for large 250 mL conical tubes.
11. 50 and 250 mL conical centrifuge tubes.
12. Racks for 250 conical tubes.
13. 5 and 10 mL pipettes.
14. Aspiration system.
15. 10 and 60 mL syringes.
16. 14G–24G catheters.
17. 2.0 silk suture with needle.
18. No. 10 disposable scalpels.
19. 30 mm Petri dishes.

20. Non-tissue culture T-175 flasks (Sarstedt).
21. 500 mL filter flasks (0.2 μm).
22. Temperature probe (Tyco).
23. Flexible silicone tubing, ID/OD 0.125/0.250 in.
24. Sterile underpads and drapes.
25. Peristaltic pumps.
26. Water bath with thermometer.
27. 750 mL Ricordi digestion chamber with stainless-steel screen and three to four siliconized glass marbles (1-cm diameter).
28. COBE 2991 Cell Processor.
29. COBE blood cell processing sets.
30. Polysucrose stock solution (Euro-Collins base) (Cellgro).
31. 1.108, 1.096, 1.037 polysucrose islet density gradients (Cellgro).
32. Collagenase: Serva NB1 or Roche Liberase DL or Vitacyte CIzyme with neutral proteases.
33. Culture media CMRL 1066 pH 7.4, 10% heat-inactivated porcine serum, 10 mM nicotinamide, 100 units/mL penicillin, 0.1 mg/mL streptomycin, 2 mM L-glutamine, store at 4 °C.
34. Dithizone solution (30 mg diphenylthiocarbazone), 9 mL of DMSO, 300 μL of 30% ammonium hydroxide. After 0.2 μm filtered, store at 20 °C.
35. 2% serum media pH 7.4, 300 mL of HBSS, 2% heat-inactivated porcine serum, 100 units/mL penicillin, 0.1 mg/mL streptomycin, 20 mM HEPES, store at 4 °C.
36. 5% serum collecting media, pH 7.4, store at 4 °C. RPMI 1640, 5% porcine serum, 100 units/mL penicillin, 0.1 mg/mL streptomycin.
37. Personal protective equipment (gowns, gloves, sterile gloves, masks, bonnets, shoe covers, eye protection).

2.4 Pig-to-Monkey Islet Xenotransplantation

1. Zanosar STZ (Sicor Pharmaceuticals, Irvine, CA).
2. Sterile drapes.
3. Tether system with three channel swivel apparatus.
4. Medical grade tubing (ID 1/16, OD 1/8, wall 1/32).
5. Micro-bore tubing (ID 0.050, OD 0.09 "wall 0.02").
6. 22G–26G catheters.
7. Primate jacket with riveted tether end plate.
8. Peristaltic pumps.
9. RIA and ELISA kits for porcine and human C-peptide, insulin, and glucagon.

10. Standard operating room set-up including anesthesia machine, surgical instruments, cautery machine, operating table, surgical lights, etc.
11. Immunosuppressive and supportive agents: mycophenolate mofetil (MMF), antithymocyte globulin (ATG), methylprednisolone, anti-CD154 mAb, prostacyclin, dextran sulfate, aspirin, ganciclovir, valganciclovir, famotidine, ketamine, isoflurane, ketoprofen, buprenorphine plus other medications recommended by veterinary staff.
12. 25% glucose solution and 10% arginine solution for metabolic tests.
13. Fast-acting (Humulin R) and medium-acting (Humulin N) insulin.

3 Methods

3.1 Neonatal/Fetal ICC Isolation

1. Prepare Ham's F-10 culture medium and filter sterilize. Before using, warm to room temperature.
2. Prepare washing medium and filter sterilize. Before using, warm to room temperature.
3. Keep neonatal pancreata on ice in UW solution (or other organ preservative) until ready to process.
4. Prepare the hood by lining it with sterile underpads to create a sterile field. Place sterile cleaning pan, surgical instruments, nylon mesh and holder, and 50 mL conical tubes in sterile field.
5. Prepare 50 mL (per pancreas) collagenase type XI in HBSS at a concentration of 1 mg/mL and filter sterilize. Keep on ice until needed.
6. Clean each pancreas of any extraneous tissue and cut into 3–5 pieces.
7. Place pieces into a 50 mL conical tube, one pancreas per tube with about 50 mL washing media (Fig. 1a) (*see* **Note 4**).
8. Aspirate washing media down to 15 mL.
9. With sterile scissors, cut pancreas in the 50 mL tube continuously for 1 min with rapid motions.
10. Bring washing media back up to 50 mL and let the tissue pieces gravity settle for 1 min.
11. Again, aspirate washing media down to 15 mL and repeat **steps 9** and **10** for a total of three times until tissue pieces are a few mm in size.
12. Bring washing media back up to 50 mL and centrifuge for 1 min at 220 × *g* (room temperature).

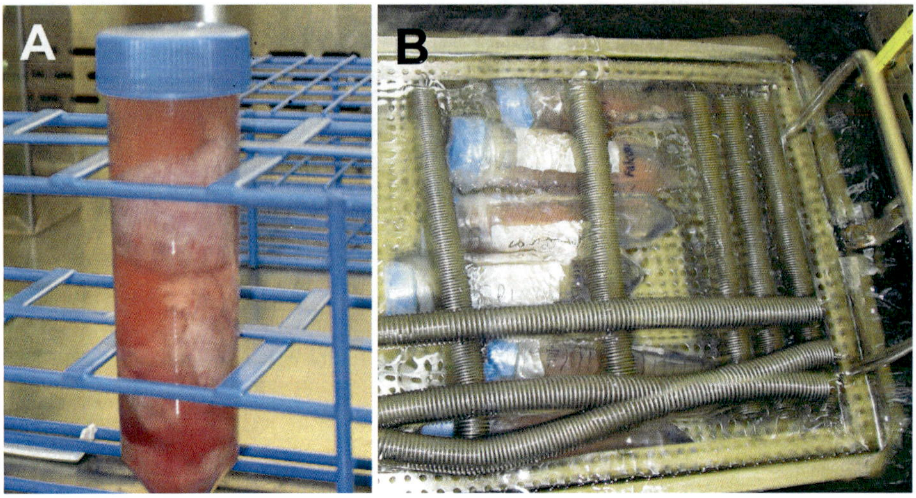

Fig. 1 Neonatal pancreas. (**a**) Neonatal pancreas in conical tube after cleaning and cut into pieces. (**b**) Conical tubes with chopped pancreatic tissue and collagenase (one pancreas per tube) placed in 37 °C shaking water bath. The tubes are held in place under the water with coiled springs

13. Aspirate HBSS completely from tube and add 25 mL freshly made collagenase (1 mg/mL).
14. Replace cap securely and wrap with parafilm.
15. Place horizontally in 37 °C water bath and shake at a medium pace for 7 min (Fig. 1b).
16. Remove from water bath and shake by hand rapidly for 1 min.
17. Check progress of the digestion by the size of the tissue particles (*see* **Note 5**).
18. Place tube back in the water bath and shake another 7 min (*see* **Note 6**).
19. Remove tube from water bath and again shake by hand rapidly for 1 min.
20. Pour digested tissue through the nylon filter mesh into a new 50 mL conical tube. Wash the digestion tube twice more and also pour through mesh into the new conical tube (Fig. 2a, b) (*see* **Note** 7).
21. Bring the total volume to 50 mL and centrifuge for 1 min at 250 × *g* (Fig. 2c).
22. Aspirate supernatant and resuspend pellet with HBSS washing medium. Bring volume up to 50 mL and centrifuge again. Repeat for a total of three washes.
23. Label two flasks for each pig pancreas.
24. After last wash, aspirate the medium and resuspend the pellet with 20 mL of Ham's F-10 medium.

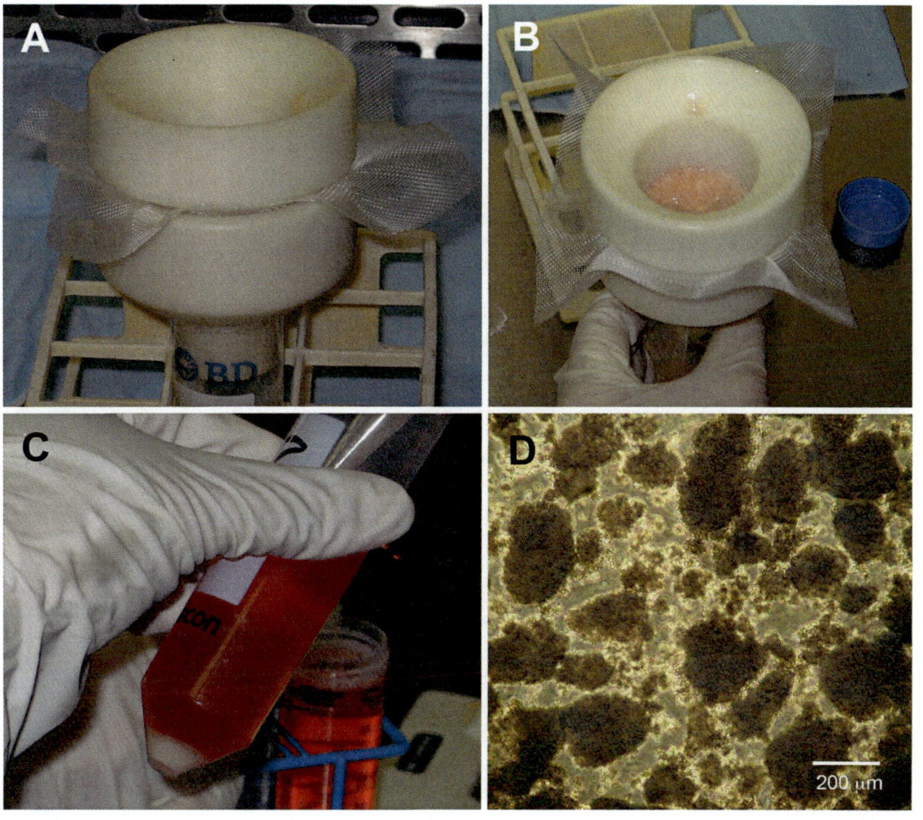

Fig. 2 NICC isolation. (**a**) Plastic device used to filter cell clusters from larger fragments of tissue. Nylon filter mesh is placed in the center of device. (**b**) Tissue is poured over the mesh and into a new conical tube after digestion and shaking to separate NICC from undigested tissue. (**c**) Packed cell pellet containing the NICC from one pancreas. (**d**) NICC in culture after isolation (×40 magnification)

25. Add 10 mL of resuspended NICC to each flask and bring the total volume of Ham's F-10 medium per flask to 70 mL.
26. Incubate the NICCs in a 37 °C water-jacketed incubator with 5% CO_2.
27. Refresh media on day 1 and every 2 days thereafter (*see* **Note 8**).

3.2 Determining Yield (NICC)

1. On days 7–9 pool together NICC in conical tube for counting. Counting is done either per pancreas or as a total of all pancreata if desired.
2. Resuspend NICC and remove a representative 0.5 mL of sample.
3. Under ×40 magnification, count each cluster that is equal to or larger than 50 μm (Fig. 2d) (*see* **Note 9**).
4. Multiply that number by the volume to ascertain the total number of NICC (*see* **Note 10**).

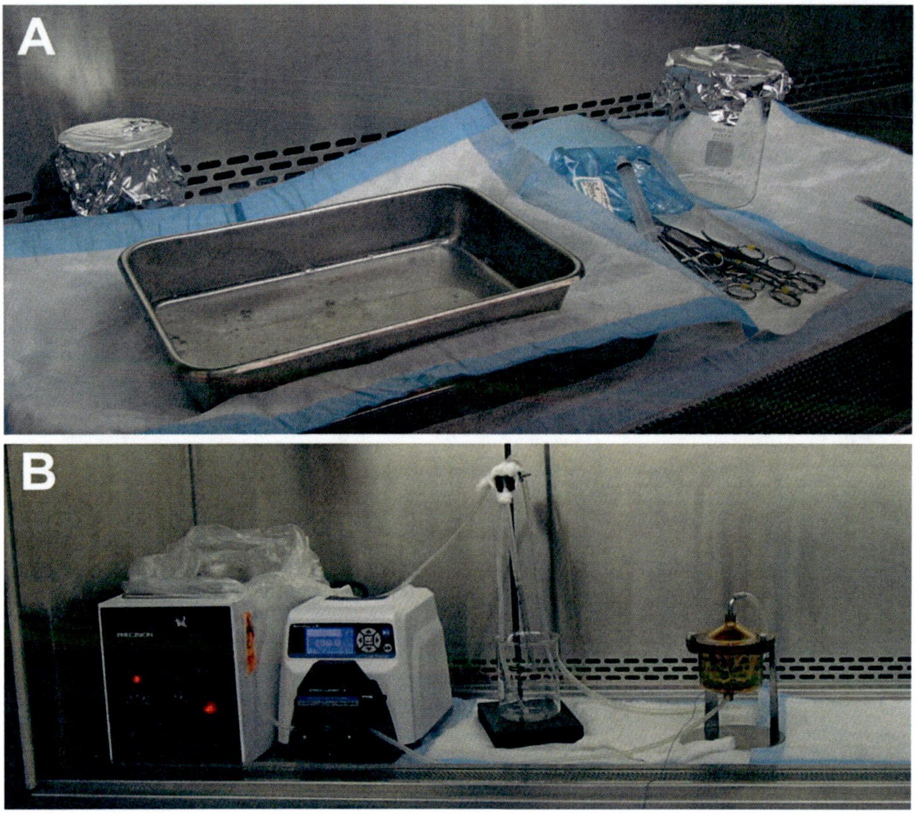

Fig. 3 Set-up for procedure. (**a**) Hood #1 shows cleaning pan on ice, surgical instruments and beakers in sterile field. (**b**) Hood #2 with (left to right) water bath, peristaltic pump, clamp stand with recirculating beaker and tubing, and Ricordi chamber

3.3 Adult Pig Islet Isolation

1. Keep pancreas on ice in UW solution (or other organ procurement fluid) until ready to process. Maintain sterility throughout the isolation process (*see* **Note 11**).
2. Prepare both hoods by lining them with sterile underpads to create a sterile field. Hood #1 is initially set up for cleaning the pancreas with a large (with ice) and a small stainless-steel surgical pan, surgical instruments for cleaning, 2–3 beakers (500 mL) for receiving trimmed non-pancreatic tissue and one larger beaker (1000 mL) with HBSS to rinse the pancreas. After cleaning, the same hood is then set up to receive digested tissue as it is removed from the Ricordi chamber. Islet gradient separation using the COBE 2991 is also performed in hood #1 (Fig. 3a).
3. Hood #2 is set up with the Ricordi chamber (with temperature probe), a peristaltic pump, water bath, stainless-steel coil, recirculating beaker, collecting bottle and digital thermometer for digestion and collection of tissue. Lab stand, base, and clamp are assembled and also placed in hood. The coil is attached to

the Ricordi chamber by recirculating tubing and placed in the water bath to maintain the proper temperature during tissue digestion. The digital thermometer is connected to the temperature probe from the Ricordi chamber and a separate probe from the water bath (Fig. 3b). Place 1 L bottle full with HBSS to warm in water bath.

4. Add 30 mL of Betadine to procurement fluid containing pancreas for 2–3 min. Rinse pancreas with cold HBSS in large beaker.
5. In large stainless-steel pan, place ice to cover the bottom and cover the ice with sterile drape or plastic sterile cover.
6. Place smaller sterile pan on top of sterile drape and fill with 300 mL cold 2% serum medium.
7. Place pancreas in smaller sterile pan and keep on ice during cleaning and cannulation. Inspect organ for cuts and make sure the intestine is intact (Fig. 4a) (*see* **Note 12**).
8. Remove the intestine carefully to avoid puncture (causing a possible contamination) and discard.
9. Clean the remainder of the pancreas of fat that is easily identified and separated (Fig. 4b) (*see* **Note 13**).
10. The cleaned pancreas is weighed and samples taken for histology (if desired).
11. Locate the main duct opening in the head of the pancreas. Insert a catheter of appropriate size and suture into place with a purse-string technique using 2-0 braided suture.
12. Locate the bridge of tissue that connects the head to the posterior lobe. Make a straight cut (close to the pancreas head) across the tissue that forms the bridge (Fig. 5a).
13. Examine the end of the tissue bridge and locate the duct for the posterior lobe. Insert a second catheter and suture into place.

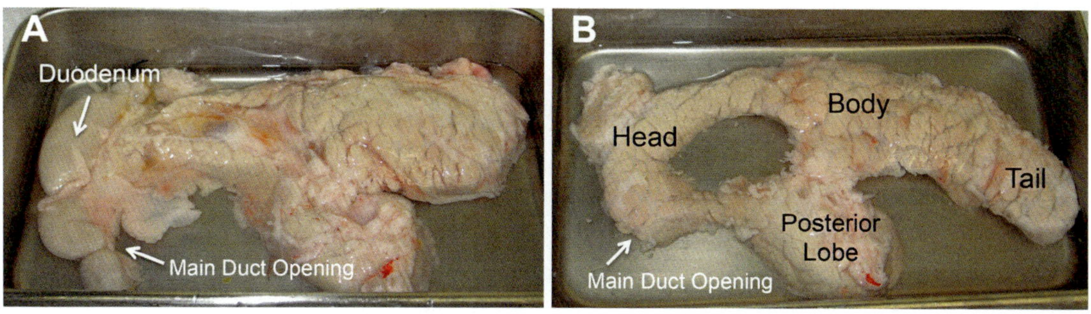

Fig. 4 Adult pig pancreas. (**a**) Pig pancreas before cleaning. The duodenum is connected to the head and adipose tissue is adherent. An arrow points to the position of the main pancreatic duct opening. (**b**) Pancreas after cleaning and removal of duodenum. Note characteristic shape and designation of sections of the porcine pancreas

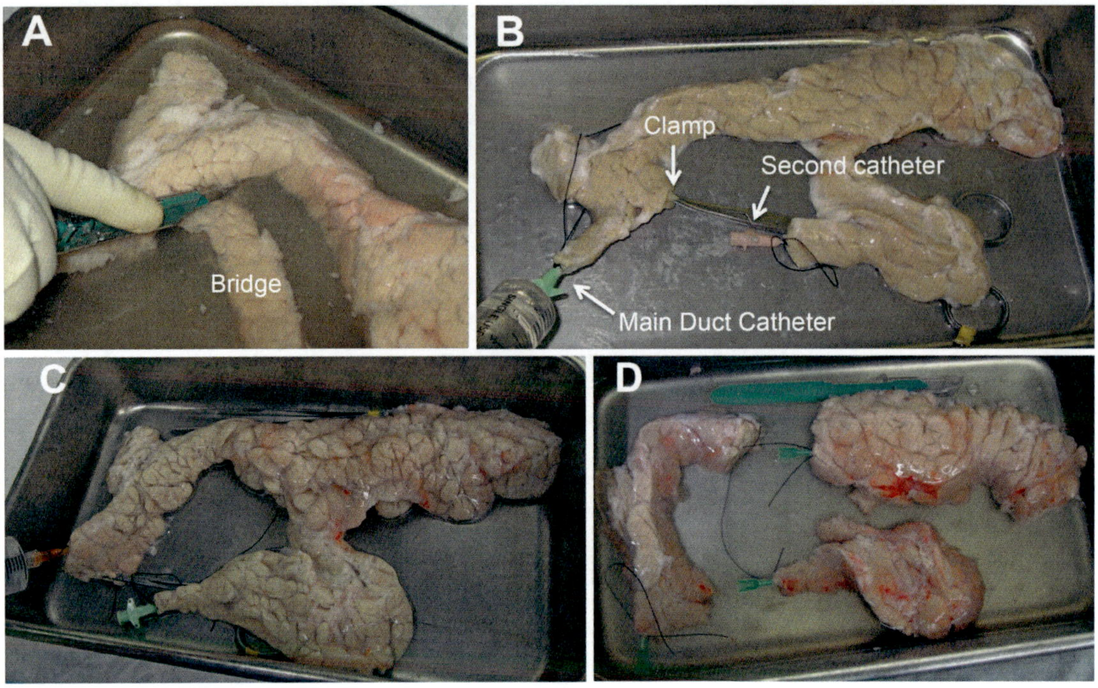

Fig. 5 Catheter placement. (**a**) Location of cut in the tissue bridge of the posterior lobe to place a second catheter. (**b**) Catheters placed and sewn into the main pancreatic duct and the duct to the posterior lobe. Clamp is placed to ligate the duct on the side near the head. (**c**) Distended pancreas after infusion of enzyme. (**d**) A third catheter may be placed in the body or tail if distention in these sections is inadequate

Clamp the duct on the other side of the incision (near the head) to prevent enzyme leakage (Fig. 5b) (*see* **Note 14**).

14. Mix together 7.5 Wunsch units collagenase and 0.13 mg protease per gram tissue weight. Keep on ice without agitation until collagenase is dissolved (*see* **Note 15**).
15. The collagenase/protease mixture is diluted in 500 mL of HBSS. The enzymes are either infused by syringe into the pancreatic duct at 4 °C or prewarmed to 24 °C. Our protocol uses prewarmed collagenase.
16. Remove the 2% serum media from the cleaning pan. Remove the pan and pancreas from the ice and place on sterile underpad in the hood.
17. Fill a 60 cm^3 syringe with the enzyme and connect it to the catheter in the pancreas head. Begin to infuse the collagenase with a slow steady pressure of about 20–30 mL/min. The pancreas should distend visibly without leaking. If leaks are present, clamp with hemostats (Fig. 5b) (*see* **Note 16**).
18. Refill the syringe and repeat with the second catheter to the posterior lobe. Alternate catheters until most of the collagenase

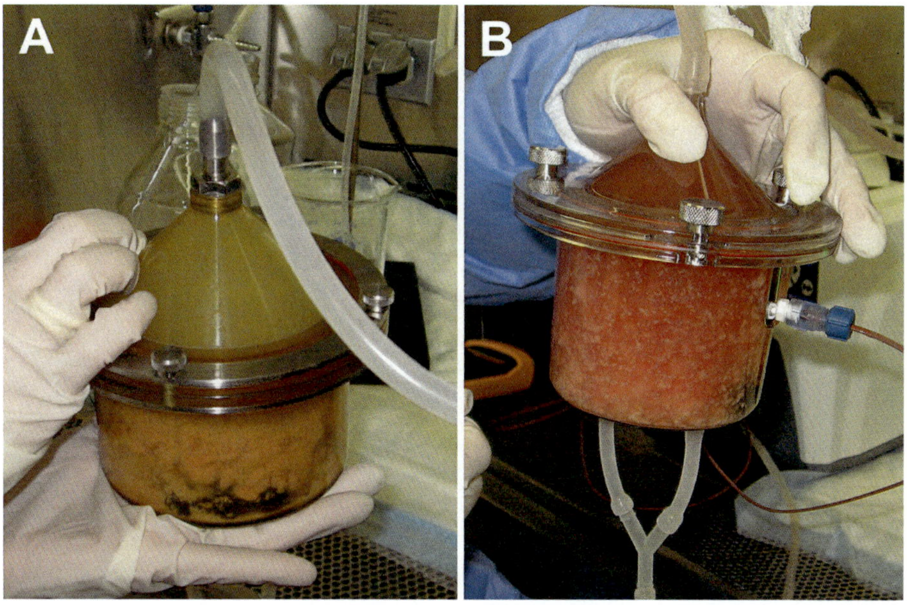

Fig. 6 (**a**) Pancreas in Ricordi chamber. Pancreas placed in Ricordi chamber with black marbles to aid in mechanical disruption of tissue. (**b**) Pancreatic tissue after digestion and disruption with marbles

has been infused and the pancreas is well distended. Infusion should be complete in 10–15 min (Fig. 5c) (*see* **Note 17**).

19. With infusion complete, remove catheters and cut away any gelatinous material surrounding the pancreas (*see* **Note 18**).
20. Cut the pancreas in 4–8 pieces. Transfer them sterilely to hood #2 and place them in the Ricordi chamber with 3–4 siliconized glass marbles. Fill the chamber with the collagenase and place the stainless-steel mesh filter over the chamber opening. The top is then screwed into place (Fig. 6a).
21. Any remaining enzyme is placed in the recirculation beaker and the chamber is allowed to completely fill using a peristaltic pump set to a speed of 80–90 mL/min. The direction of the flow is into the bottom of the chamber and out the top through the mesh filter. Place in-line stainless-steel coil in water bath.
22. Complete the circuit of flow into and out of the recirculation beaker by adding a small amount of the warm HBSS to the beaker. The HBSS will dilute the collagenase solution so only add enough to complete the fluid circuit that runs from the beaker, through the tubing and coil, into the Ricordi chamber, and back into the beaker.
23. The stainless-steel coil is used to maintain the desired temperature in the chamber (36 °C) as the enzyme travels through the circuit by placing the coil in and out of the water bath as

necessary. The temperature is monitored with the digital thermometer using the probes placed in the chamber and the water bath.

24. Once the circuit is complete, the chamber is shaken gently by rocking it back and forth at a slow speed. Pig islets are fragile and vigorous shaking will damage them. Shaking continues throughout the tissue collection (collection continues 30–90 min depending on the digestion and size of the pancreas).
25. As the collagenase activity continues, the marbles and shaking will mechanically disrupt the tissue (Fig. 6b). The digested tissue will begin to leave the chamber through the mesh filter. Samples are taken from the fluid to observe the progress of the digestion (*see* **Note 19**).
26. For samples, remove about 1 mL of fluid from the tubing coming out of the top of the Ricordi chamber. Let it flow from the tubing into a 35 mm Petri dish containing 100 μL dithizone. Dithizone stains endocrine tissue red within about 1 min. Dilute with an additional 2 mL PBS to the dish and examine under ×40 magnification.
27. Continue taking samples until at least 2–3 "free" islets are seen (stained red) without exocrine tissue (yellowish tissue) attached (Fig. 7a). If mostly trapped islets are seen (red-stained tissue surrounded by unstained tissue), do not start collecting tissue yet. More time is needed for the enzyme to break down the exocrine tissue (*see* **Note 20**).
28. After free islets are identified, begin collecting tissue into 1 L collecting bottles instead of letting it return to the recirculating beaker. As the tissue is collected, it is cooled to help inactivate the enzyme. Add the rest of the warm HBSS from **step 22** to the beaker to replace what is collected from the chamber.
29. From the collecting bottles, the tissue is transferred to 250 mL conical tubes and centrifuged for 3 min at 260 × *g*. The supernatant is discarded and the pellet is collected in a beaker containing 500 mL collecting medium. The beaker is placed on ice and will also receive the pellets from subsequent bottles until all the tissue from the digested pancreas has been collected.
30. When the warm HBSS has mostly been pumped through the system, continue collection with cold collecting medium. The collection is continued until most of the tissue has been digested and pumped out of the Ricordi chamber. A signal of when to stop is when the cell pellets become very small.
31. Stop collection by draining the recirculating beaker and chamber. The remaining undigested tissue in the chamber is

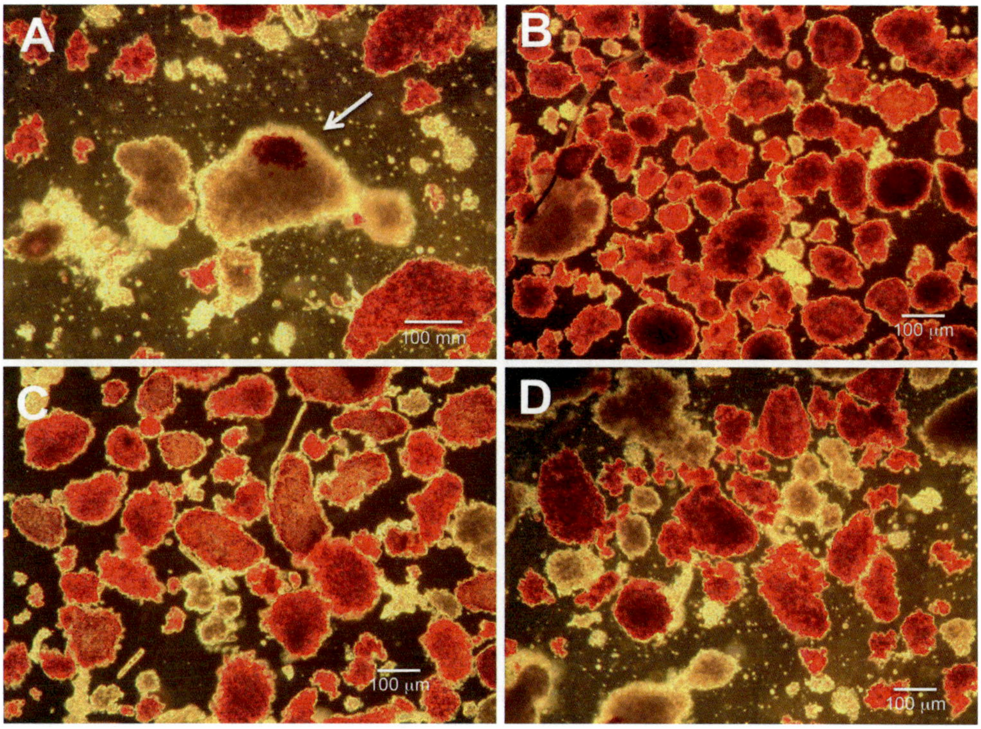

Fig. 7 Islet identification using dithizone. (**a**) Sample taken from Ricordi chamber to determine when to begin tissue collection with islets stained in red using dithizone. Sample shows several islets (red) that are free of exocrine (unstained) tissue signaling collection should begin. Arrow points to islet that is trapped in exocrine tissue. (**b**) Example of fraction with highest 70–95% purity after density gradient separation. (**c**, **d**) Samples from different tissue fractions after density gradient separation. These two samples have a similar purity between 30 and 70% and are usually pooled together

weighed and subtracted from the initial pancreas weight (*see* **Note 21**).

32. Resuspend the collected pellets and centrifuge as in **step 29**. Measure the amount of packed tissue and split into portions of 20–23 mL packed tissue (enough for 1 COBE separation run). Each portion is placed into a separate 250 mL conical tube. A large pancreas can yield enough tissue for 4–5 COBE runs.
33. The cells are resuspended in collecting medium and kept on ice until they are separated with the COBE 2991 cell processor using a density gradient.
34. For each COBE run, 20–23 mL of tissue is centrifuged, the supernatant removed and replaced with 225 mL polysucrose stock (Euro-Collins base) solution.
35. The COBE 2991 cell processor is prepared by loading it with a processing set consisting of a loading bag and inlet and outlet tubing (Fig. 8a).

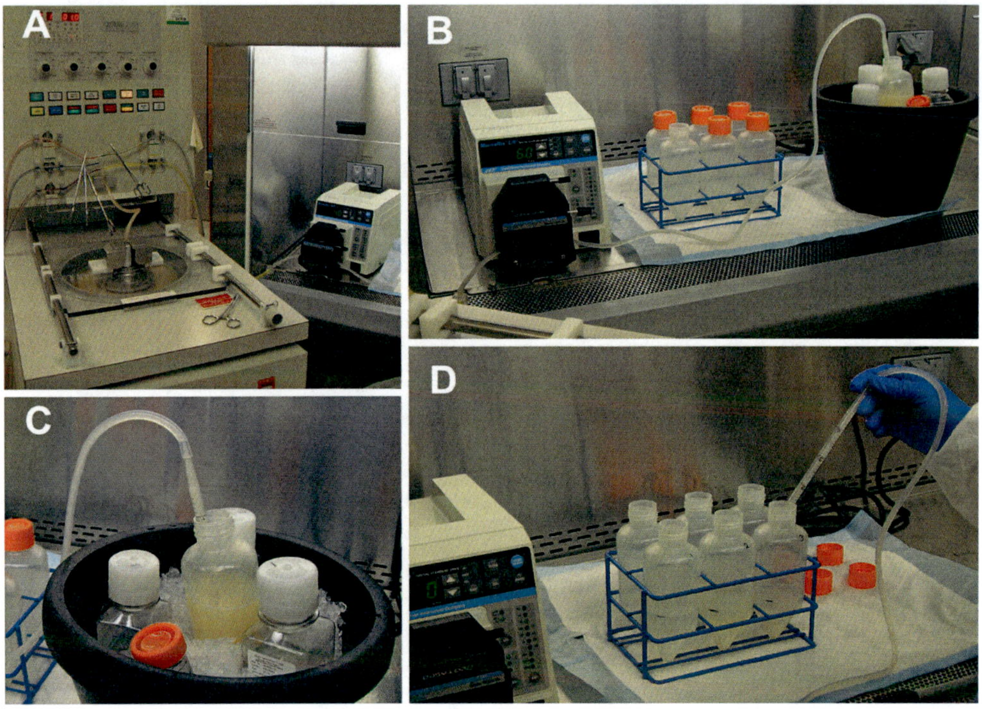

Fig. 8 Density gradient separation. (**a**) COBE 2991 cell processor fitted with a cell processing set consisting of a centrifuge loading bag to hold the cells and lines for loading and unloading gradients. (**b**) Hood set-up for running a density gradient to separate islets from acinar tissue. Left to right, peristaltic pump, 250 mL conical tubes to receive density fractions, ice bucket with density gradients, and collected pancreatic tissue. (**c**) Pancreatic tissue containing islets in 225 mL polysucrose being loaded into COBE bag through the inlet tubing to be followed by 75 mL/bottle of 1.108, 1.096, and 1.037 density gradients and finally 50 mL HBSS (in conical). (**d**) After 5 min of centrifugation with the gradients, the separated islet fractions are collected individually into conical tubes

36. The tubing is attached to a peristaltic pump and the digested pancreatic tissue is loaded into the loading bag (Fig. 8b, c). The COBE centrifuge is started and 75 mL/each of the following polysucrose islet density gradients are loaded at the rate of 40 mL/min in order from heaviest to lightest: 1.108, 1.096, 1.037 gradients. Finally load 50 mL of HBSS.
37. When loading is complete, stop the pump and release the feeder line. Engage the SuperOut option of the COBE to release any pressure in the system, and centrifuge the density gradient construct for 5 min.
38. Recover cells by setting the outflow rate to 100 and collecting the fractions in six 250 mL conical tubes labeled 1–6. Collect 100 mL in conical #1 and 75 mL each in the other five conicals. After the last conical is filled or the flow stops, the run is finished (Fig. 8D) (*see* **Note 22**).

39. Fill the conical tubes to 250 mL with collecting media and centrifuge 3 min at 310 × *g* at 4 °C.
40. Aspirate the supernatant leaving about 50 mL and resuspend again to 250 mL. Mix the tissue well and take a 1 mL representative sample from each conical to determine purity. Prepare sample as in **step 26** with dithizone and examine under ×40 magnification.
41. Samples with similar purity are pooled together. Typically two fractions are produced from the six original conical tubes, "Fraction A" with purity between 95 and 70% (Fig. 7b) and "Fraction B" with 70–30% (Fig. 7c, d) purity. Unless specifically needed, tissue with less than 30% purity is discarded.
42. The tubes are again centrifuged and supernatant discarded. Cell pellets with similar purity are combined and resuspended in 250 mL washing media.
43. Representative samples are taken for counting, the fractions are pelleted once more, and the pellet sizes are recorded. The islet-enriched tissue is split into flasks with 50–100 μL packed tissue and 50 mL culture media per flask.
44. Flasks are cultured at 24–27 °C and 5% CO_2. Media is changed at 24 hours after isolation and then every 2 days.

3.4 Determining Yield (Adult Porcine Islets)

1. Adult porcine islets are pooled together into a 250 mL conical tube based on purity. The tube is filled to 250 with mL of collecting (or culture) media.
2. A 0.5 mL representative sample is removed and placed into a 35 mm Petri dish with dithizone as in **step 26** (*see* **Note 23**).
3. Under ×40 magnification, count the islets by size as well as number. Equalizing the mass is necessary due to the varying sizes of the islets. Table 1 indicates the size categories and the correction to islet equivalent (IEQ).

3.5 Tethers and Indwelling Intravascular Catheters (See Note 24)

1. NHPs will be acclimated in the tethering cage while wearing the protective primate jacket for at least 7 days prior to tethering.
2. Following jacket acclimation, the animal will be connected to the metal tether cable and swivel (Fig. 9). This will take at least another 7 days until the animal is comfortable and eating normally.
3. After successful acclimation, a venous and arterial catheter will be surgically placed in the jugular vein and carotid artery in the right cervical area and subcutaneously tunneled from the neck to the right back. The catheters will be connected to the tubing and the swivel tether device at least 7 days before islet transplantation (*see* **Note 25**).

Table 1
Conversion to islet equivalents

Islet diameter range (μm)	Conversion to islet equivalent (IEQ)
50–100	$n/6.00$
100–150	$n/1.50$
150–200	$n \times 1.7$
200–250	$n \times 3.5$
250–300	$\mathrm{n} \times 6.3$
300–350	$n \times 10.4$
>350	$n \times 15.8$

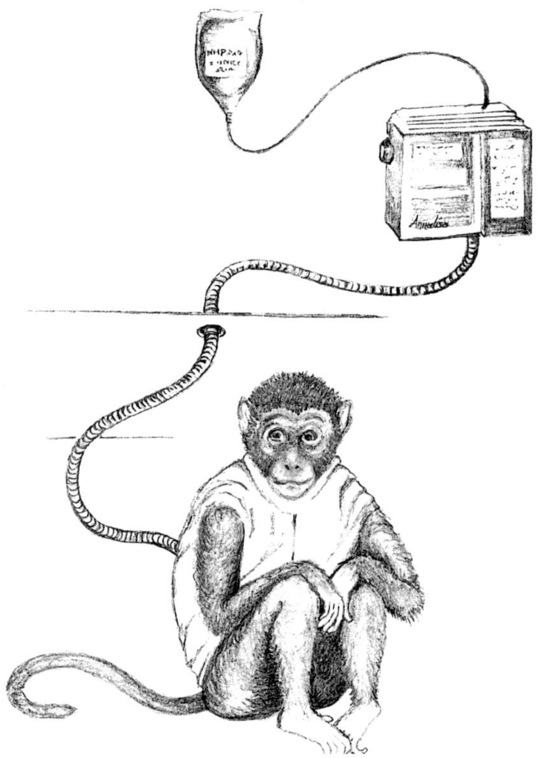

Fig. 9 Primate tether system. Illustration of primate fitted with jacket attached to tether that is connected to a swivel at the top of the animal's cage. The surgically placed intravascular lines are threaded through the tether and out of the cage and connected to a pump or other device to allow blood draws and medication delivery while allowing freedom of movement and lessening stress on the animal

4. To ensure that the lumen of the catheter remains patent, 2–4 mL/h of 0.9%-heparinized saline (2 IU/mL of heparin) is dripped through the catheter continuously. The lines are checked visually twice daily.
5. Minimally once per week, the animal is sedated and the wounds inspected and cleaned, skin in contact with the jacket checked for irritation and other maintenance as needed.
6. Bimonthly (or more frequently if needed) the swivel systems and jackets are changed (*see* **Note 26**).

3.6 Streptozotocin Diabetes Induction of NHPs

1. Allow several days for NHP to recover from catheterization surgery.
2. Fast animal overnight and prehydrate with 10 mL/kg/h of 0.9% sterile saline i.v. for 8–10 h before streptozotocin (Zanozar) administration.
3. Administer STZ at a rate of 125 mg/kg i.v. over a time period of 3–5 min [22].
4. For the next 48 h, frequent blood samples are taken for metabolic testing. Measures are taken to correct hypoglycemia or ketoacidosis as needed (*see* **Note 27**).
5. Diabetes is confirmed by two consecutive blood glucose readings of >350 mg/dL and primate C-peptide reduced by greater than 75% (*see* **Note 28**).

3.7 Maintenance of Diabetic Primates

1. Test blood glucose at least twice per day. Generally, test in the morning before the first meal of the day and after the evening meal.
2. Cynomolgus macaques have a normal blood glucose of 40–123 mg/dL. Insulin therapy is initiated to maintain a value in the range of 150–200.
3. An intermediate acting insulin (such as Humulin N) is given starting with a dose of 0.25 IU/kg twice a day and adjusted according to the needs of the animal (*see* **Note 29**).
4. An animal with indwelling vascular lines has approximately 10 μL of blood removed from the arterial catheter to test glucose levels using a glucometer (any commercial glucometer is permissible).
5. An animal without vascular lines must have blood drawn from the skin of the body. This is usually done with a lancing device on the arm, leg, or tail (*see* **Note 30**).
6. Inject insulin as necessary based on blood glucose level to maintain normoglycemia and avoid ketoacidosis. Injections are subcutaneous and may be administered also on the arm, leg, or tail.

3.8 Metabolic Testing (See Note 31)

3.8.1 Intravenous Glucose Tolerance Test (IVGTT)

1. Test is initiated the morning after food has been withheld overnight (approx. 15 h).
2. Blood glucose is measured with a glucometer at time point 0.
3. 0.5 g/kg glucose (25% solution) is injected i.v.
4. Subsequent blood glucose measurements are taken at 2, 5, 15, 30, 60, and 90 min post-injection.

3.8.2 Insulin Tolerance Test (ITT)

1. Animal fasting and preparation for ITT is the same as IVGTT.
2. 0.1 IU/kg fast-acting (Humulin R) insulin is injected i.v.
3. Collect 1.0 mL blood samples at 0, 15, and 30 min.
4. Test blood glucose with glucometer for each time point.
5. Centrifuge remaining blood sample.
6. Collect and freeze serum at −20 °C for glucagon testing by ELISA
7. After last blood sample is taken at 30 min, infuse a bolus of 5 mL of 50% dextrose to reverse hypoglycemia.

3.8.3 C-peptide Levels

1. During the IVGTT, an additional 0.6 mL blood samples are taken at time points 0, 5, 15, and 90 min for determination of monkey and/or porcine c-peptide levels.
2. Keep blood samples on ice until all samples are collected.
3. Centrifuge samples at 870 × g for 15 min to separate the serum.
4. Freeze serum samples at −20 °C until determining C-peptide levels with ultrasensitive radioimmunoassay kit.

3.8.4 Arginine Stimulation Test (AST)

1. Test is initiated the morning after food has been withheld overnight (approx. 15 h).
2. Blood glucose is measured with a glucometer at time point 0.
3. 70 mg/kg arginine (10% solution) is injected i.v.
4. Subsequent blood glucose measurements are taken at 2, 3, 4, and 5 minutes post-injection.

3.9 Immunosuppression of NHPs (See Table 2 and Note 32)

1. Starting on day −7 before transplant (Tx), recipient is given 100 mg/kg/day MMF and 81 mg/day aspirin, both by mouth.
2. Day −3 and −1 recipient is given 25 mg ATG i.v. over 7 h.
3. Day −1, 0, 3, 7, 11, and 15 and every other week subsequently recipient is given 25 mg/kg anti-CD154 mAb bolus i.v.
4. Day of transplant give the following: Prostacyclin 20 ng/kg/min starting 30 min before Tx for 3 h, methylprednisolone 10 mg/kg bolus, dextran sulfate 5 mg/kg bolus then 2 mg/kg/h for 6 h.

Table 2
Immunosuppression and supportive care schedule for xenotransplanted NHPs (*see* Note 32)

Medication	Dose	Route	Timetable	Target level
Mycophenolate mofetil (MMF)	100 mg/kg/day	p.o.	Start on day −7 and continue until endpoint	Trough level: 3–5 μg/mL
Aspirin	81 mg/day	p.o.	Day −7 to +7	
Antithymocyte globulin (ATG)	25 mg 5–25 mg	i.v. over 7 h	Day −3 Day −1	<500 CD3+ cells/μL
Anti-CD154 mAb	25 mg/kg	i.v. bolus	Day −1, 0, 3, 7, 11 and 15 then every other week until endpoint	
Prostacyclin	20 ng/kg/min	i.v.	3 h infusion starting 30 min before Tx	
Methylprednisolone	10 mg/kg	i.v. bolus	Day of Tx	
Dextran sulfate (a)	5 mg/kg	i.v. bolus	Day of Tx	Activated clotting time 150–190 s
Dextran sulfate (b)	2 mg/kg	i.v. over 6 h	Day of Tx	Activated clotting time 150–190 s
Ganciclovir	5 mg/kg/day	i.v. bolus	Day 0 to removal of vascular lines	
Famotidine (a)	0.25 mg/kg/day	i.v. bolus	Day 0 to removal of vascular lines	
Famotidine (b)	1 mg/kg/day	p.o.	After line removal until endpoint	
Valganciclovir	15 mg/kg BID	p.o.	After line removal until endpoint	

5. Day of transplant until removal of vascular lines, ganciclovir 5 mg/kg/day bolus and famotidine 0.25 mg/kg/day bolus.
6. Stop aspirin on day 7 after Tx.
7. Begin after line removal valganciclovir 15 mg/kg × 2 daily bolus and Famotidine 1 mg/kg/day bolus both by mouth.

3.10 Transplant Technique for Intraportal Infusion of Islets or NICC

1. Recipients are fasted approximately 15 h on the day of surgery.
2. Macaques are sedated with ketamine 10 mg/kg i.m. and prepared for surgery following standard operating room procedures.
3. Surgical anesthesia is achieved and maintained with ≤2% isoflurane and oxygen using endotracheal intubation and positive pressure mechanical ventilation.

4. To maintain stable heart rate during surgery, normal saline and dopamine (1–10 mg/kg) is given i.v.
5. The abdomen is surgically prepared by shaving, cleaning with a Betadine scrub, and placing sterile drapes to frame the surgical site.
6. A ventral midline laparotomy is performed.
7. The islets are suspended in 20 mL of sterile CMRL-1066 media supplemented with 4 mg/kg dextran sulfate and placed into a 60 mL syringe fitted with sterile i.v. tubing and connected to an intravenous catheter (*see* **Note 33**).
8. Islets are intraportally infused slowly through an intravenous catheter (22–26 gauge) via gravity drainage of the islets into the portal vein. The islet cell suspension is gently agitated throughout the 10–15 min infusion to prevent clumping.
9. The portal vein is closed with direct pressure for at least 5 min following catheter removal.
10. Ensure bleeding had stopped before closure.
11. The incision is closed in two layers with subcuticular stitches used to close the skin.
12. Before recovery from anesthesia, 2 mg/kg ketoprofen is given i.m. and the animal is returned to the jacket and tether. Catheters are inspected for patency and signs of infection (*see* **Note 34**).
13. Buprenorphine (0.01 mg/kg) is given postoperatively twice a day for 3 days in addition to the other medications listed in Table 2.

4 Notes

1. Pigs are considered neonatal if they are 30 days old or younger. The advantage in using neonatal islet-like cell clusters (NICC) is that they have the capability to proliferate. Conversely, because they are not fully formed islets, they require several weeks (in vitro and in vivo) to mature into insulin-secreting cell clusters.
2. Islets from juvenile pigs are known to be particularly fragile, and many researchers commonly report dissociation of the islets to single cells within the first 24 h. Other breeds of pigs including wild type are also suitable for islet isolation although the age is still a significant factor. The best type of pig for islet isolation of any breed is an older female that has had two or more litters (retired breeder). In our experience, islets from genetically engineered pigs with at least transgenic expression of human CD46 (provided by Revivicor, Inc. Blacksburg, VA) resulted in the best outcome in a pig-to-NHP model.

3. A factor that is considered when transplanting islets into NHP recipients is the recipient's body weight. Since the number of islets transplanted is based on weight, fewer islets will be needed for a smaller animal.
4. One person can typically manipulate two pancreata at one time. If there are more pancreata than people, keep them on ice and complete the procedure with the first two organs before starting on subsequent isolations. When processing islets, keep each pancreas separate in case of contamination.
5. The size of the tissue particles should become smaller as digestion proceeds. The optimal size should be equivalent to particles of sand. A few larger pieces may remain that will be removed during filtration.
6. Digestion may be faster or slower than average and actual time of digestion is determined by the size of the particles. The time may be shortened or lengthened to generate the correct particle size. Total digestion time should be approximately 10–14 min.
7. Alternatively to the plastic device shown here, a sterile funnel may be used to hold the nylon mesh. The tissue is poured through the mesh inside the funnel and collected in a conical tube.
8. NICCs are typically used for transplantation on days 7–9 after isolation.
9. NICCs are not established mature islets. Therefore, clusters are counted individually and not as islet equivalents (IEQ).
10. Counts can vary but average between 3000–5000 NICC per kg donor body weight.
11. The pancreas should be well flushed of blood. Blood interferes with the enzyme action and results in a poor yield of islets.
12. Any cuts in the pancreas will result in leakage when the enzyme is infused; therefore it is important to clamp any noticeable leaks. Leakage results in an inferior isolation.
13. Take care not to make cuts in the pancreas when cleaning it of fat and other tissue.
14. Frequently the pancreatic ductal system in pigs deviates slightly from organ to organ. There is a good possibility that the duct between the body and posterior lobe is not sufficiently developed for the collagenase to properly infuse the lobe from the main duct opening. Placing a second catheter diminishes the chance of an insufficient enzyme infusion.
15. Shaking collagenase results in decreased enzyme activity.
16. The pancreas must distend as much as possible to insure a good yield. If the enzyme is leaking out, there will be insufficient

pressure to push the enzyme into the small capillaries of the parenchyma and around the islets.

17. If the distal portion of the pancreas (tail) does not distend properly, the pancreas may be cut near the non-distended part and a third catheter placed to infuse enzyme directly into that portion (Fig. 5d).
18. This "jelly" needs to be removed as it can clog the filter in the Ricordi chamber and prevent the tissue from being collected.
19. Samples may be taken after shaking starts. A good practice is to take them every 2–3 min until free islets are seen.
20. Free islets have no exocrine tissue attached. Trapped islets still have exocrine tissue surrounding them. This is easily visualized with dithizone stain.
21. With a successful isolation, it is not unusual to have as much as a third of the pancreas undigested. This could be due to many things such as incomplete infusion or blood remaining in the tissue.
22. Most islets are found in collection tubes 2–5 with the more pure fractions found in 2 and 3.
23. If there are too many islets to count accurately, a smaller sample may be taken.
24. The tethering system described above represents the model that we have successfully utilized and which we believe provides utility, safety, and comfort for both the animals and their handlers. More recently, a novel alternative technique to tethering was described by Graham et al. that utilizes implantable vascular access ports [23].
25. A gastric line may also be surgically placed and used with this device.
26. Because of the high risk of infection, tether systems should only be used for the period from STZ induction to glucose stabilization after transplantation (approx. 2–8 weeks).
27. Dextrose is given to correct hypoglycemia; sodium bicarbonate is given for ketoacidosis.
28. If IVGTT or arginine stimulation test is performed, no increase in primate C-peptide should be seen.
29. Alternatively, a fast-acting insulin (Humulin R) may be given; however blood glucose should be checked more frequently (3–4 times daily).
30. Monkeys can often be trained through positive operant conditioning to present a body part to be pricked for blood glucose tests or for daily insulin injections. The less the animal is stressed through manipulation, the more accurate the blood glucose reading will be.

31. All tests are given at the inclusion of the animal into the protocol to provide a baseline value before STZ diabetes induction for then subsequently to follow and document the progression of the study.
32. The immunosuppression regimen outlined here represents our own experience [13]. Successful alternatives for immunosuppression of non-human primates exist that have been reviewed in Park et al. [12]. Shin et al. have demonstrated graft function for over 700 days using anti-CD154 mAb-based immunosuppression [24]. It is important to note, however, that none of the immunosuppression plans that have allowed for long-term graft function after pig islet xenotransplantation are clinically translatable due to their reliance on co-stimulation blockers that have been associated with adverse events in humans [25].
33. The number of adult islets transplanted using this protocol that resulted in long-term blood glucose normalization was 75,000–100,000 IEQ/kg recipient body weight [13]. The number of neonatal pig islets transplanted into macaques was approximately 50,000 [18, 26].
34. A veterinarian must be available at all times to oversee medical management of primates and recommend treatment in case of emergent conditions such as gastrointestinal upset, hemorrhage, etc.

Acknowledgments

The authors kindly wish to recognize and thank Anna Lisa Haughwout for her splendid illustration of the primate tether system.

References

1. Sutherland DE, Gruessner AC, Gruessner RW (1998) Pancreas transplantation: a review. Transplant Proc 30(5):1940–1943
2. Moassesfar S, Masharani U, Frassetto LA et al (2016) A comparative analysis of the safety, efficacy, and cost of islet versus pancreas transplantation in nonuremic patients with type 1 diabetes. Am J Transplant 16(2):518–526
3. Shapiro AM, Lakey JR, Ryan EA et al (2000) Islet transplantation in seven patients with type 1 diabetes mellitus using a glucocorticoid-free immunosuppressive regimen. N Engl J Med 343(4):230–238
4. Ryan EA, Paty BW, Senior PA et al (2005) Five-year follow-up after clinical islet transplantation. Diabetes 54(7):2060–2069
5. Foster ED, Bridges ND, Feurer ID et al (2018) Improved health-related quality of life in a phase 3 islet transplantation trial in type 1 diabetes complicated by severe hypoglycemia. Diabetes Care 41(5):1001–1008
6. Fiorina P, Shapiro AM, Ricordi C, Secchi A (2008) The clinical impact of islet transplantation. Am J Transplant 8(10):1990–1997
7. Thompson DM, Meloche M, Ao Z et al (2011) Reduced progression of diabetic microvascular complications with islet cell transplantation compared with intensive medical therapy. Transplantation 91(3):373–378
8. van der Windt DJ, Bottino R, Kumar G et al (2012) Clinical islet xenotransplantation: how close are we? Diabetes 61(12):3046–3055

9. Nagaraju S, Bottino R, Wijkstrom M, Trucco M, Cooper DKC (2015) Islet xenotransplantation: what is the optimal age of the islet-source pig? Xenotransplantation 22:7–19
10. Prabhakaran S, Hering BJ (2008) What strain of pig should be used? Xenotransplantation 15:83–86
11. Liu Z, Hu W, He T et al (2017) Pig-to primate islet xenotransplantation: past, present, and future. Cell Transplant 26(6):925–947
12. Park CG, Bottino R, Hawthorne WJ (2015) Current status of islet xenotransplantation. Int J Surg 23(Pt B):261–266
13. Bottino R, Wijkstrom M, van der Windt DJ et al (2014) Pig-to-monkey islet xenotransplantation using multi-transgenic pigs. Am J Transplant 14(10):2275–2287
14. van der Windt DJ, Bottino R, Casu A et al (2009) Long-term controlled normoglycemia in diabetic non-human primates after transplantation with hCD46 transgenic porcine islets. Am J Transplant 9(12):2716–2726
15. Korsgren O, Sandler S, Landström AS, Jansson L, Andersson A (1988) Large-scale production of fetal porcine pancreatic isletlike cell clusters. An experimental tool for studies of islet cell differentiation and xenotransplantation. Transplantation 45(3):509–514
16. Korbutt GS, Elliott JF, Ao Z, Smith DK, Warnock GL, Rajotte RV (1996) Large scale isolation, growth, and function of porcine neonatal islet cells. J Clin Invest 97(9):2119–2129
17. Trivedi N, Hollister-Lock J, Lopez-Avalos MD et al (2001) Increase in beta-cell mass in transplanted porcine neonatal pancreatic cell clusters is due to proliferation of beta-cells and differentiation of duct cells. Endocrinology 142(5):2115–2122
18. Cardona K, Korbutt GS, Milas Z et al (2006) Long-term survival of neonatal porcine islets in nonhuman primates by targeting costimulation pathways. Nat Med 12(3):304–306
19. Valdes-Gonzalez R, Rodriguez-Ventura AL, White DJG et al (2010) Long-term follow-up of patients with type 1 diabetes transplanted with neonatal pig islets. Clin Exp Immunol 162:537–542
20. Wang W, Mo Z, Ye B, Hu P, Liu S, Yi S (2011) A clinical trial of xenotransplantation of neonatal pig islets for diabetic patients. Zong Nan Da Xue Xue Bao Yi Xue Ban 36(12):1134–1140
21. McNamee GA Jr, Wannemacher RW Jr, Dinterman RE, Rozmiarek H, Montrey RD (1984) A surgical procedure and tethering system for chronic blood sampling, infusion, and temperature monitoring in caged nonhuman primates. Lab Anim Sci 34(3):303–307
22. Rood PP, Bottino R, Balamurugan AN et al (2006) Induction of diabetes in cynomolgus monkeys with high-dose streptozotocin: adverse effects and early responses. Pancreas 33(3):287–292
23. Graham ML, Rieke EF, Dunning M et al (2009) A novel alternative placement site and technique for totally implantable vascular access ports in non-human primates. J Med Primatol 38(3):204–212
24. Shin JS, Kim JM, Min BH et al (2018) Pre-clinical results in pig-to-non-human primate islet xenotransplantation using anti-CD40 antibody (2C10R4)-based immunosuppression. Xenotransplantation 25(1). https://doi.org/10.1111/xen.12356
25. Bottino R, Knoll MF, Graeme-Wilson J et al (2017) Safe use of anti-CD154 monoclonal antibody in pig islet xenotransplantation in monkeys. Xenotransplantation 24(1). https://doi.org/10.1111/xen.12283
26. Thompson P, Badell IR, Lowe M et al (2011) Islet xenotransplantation using gal-deficient neonatal donors improves engraftment and function. Am J Transplant 11(12):2593–2602

Chapter 20

Regulation of Clinical Xenotransplantation: A Reappraisal of the Legal, Ethical, and Social Aspects Involved

María Jorqui-Azofra

Abstract

Clinical trials of encapsulated pig islet transplantation are underway and are showing promising results. In addition, better immunosuppressant drug regimens and a wealth of new gene-edited pig varieties are raising new hopes in this field. Thus, consideration is also being given to the selection of patients for initial clinical trials of pig solid organ xenotransplantation. It seems likely that clinical trials of pig organ will be initiated within the next couple of years. For the moment, transplanting porcine islets to treat type 1 diabetes is generally viewed as the most likely to reach the clinical first. Further advances will hopefully make this approach a treatment choice for patients with this condition. Regulatory framework for this type of xenotransplantation is now available in several countries, and there is also a wider awareness of the importance of developing an internationally harmonized regulatory framework. For now, xenotransplantation is mainly conceived and regulated as a pharmaceutical product. The reframing of this emerging technology as cell therapy has redefined the normative landscape. Countries should be cautious about allowing xenotransplantation clinical trials to develop. But the original meaning of precaution, referred to safety measures in order to control risks of infection, is no longer seen as requiring restrictive limitations on fundamental rights. As with any new medicinal product, safety is certainly a crucial topic in research. Specific to xenotransplantation is the safety of porcine product. In this regard, regulatory frameworks should contain specific conditions about the safety of the source animals, of the xenotransplantation product, and of the manufacturing process. In turn, these frameworks should ensure that preclinical studies indicate safety and efficacy of the procedure and that risk management protocols are in place to identify, contain, and combat any outbreak of infection in a timely manner. The fragile balance between individual and collective rights and the tensions of globalization make necessary a coordinated international action to harmonize global practices in this field. Xenotransplantation clinical trials should be carried out in a context in which specific safety and ethical issues are addressed and in an environment in which specific practices that facilitate public engagement as a form of shared responsibility for regulatory decision-making are promoted as well.

Key words Xenotransplantation clinical trials, Individual rights, Informed consent, Public health, Infectious disease, Globalization, Public participation, Risk-benefit evaluation, Advanced therapy medicinal products

Cristina Costa (ed.), *Xenotransplantation: Methods and Protocols*, Methods in Molecular Biology, vol. 2110,
https://doi.org/10.1007/978-1-0716-0255-3_20,

1 Introduction

Xenos is Greek for "foreign" or "strange," meaning that xenotransplantation is, by definition, cross-species transplantation or the transplantation of living cells, tissues, or organs from one species to another, such as from pigs to humans. By contrast, the term allotransplantation refers to the transplantation between individuals of the same species—normally between human beings matched by tissue typing. Autotransplantation is the transplantation from and to the same individual, for example, the relocation of the skin from the thigh to the arm in order to treat burns.

As a result, xenotransplantation has come to be defined as animal-to-human transplantation and, more specifically, as any procedure that involves the transplantation, implantation, or infusion into a human recipient of either live tissues or organs retrieved from animals or human body fluids, cells, tissues, or organs that have undergone ex vivo contact with live nonhuman animal cells, tissues, or organs (*see* **Note 1**).

Therefore, xenotransplantation is defined in line with international terminology to include a range of procedures that involve the use of animal living products in human therapies (*see* **Note 2**). For several reasons (social, ethical, physiological, financial, and logistical), all current efforts toward xenotransplantation focus on the use of pigs as potential source animal. The reasons include similar physiology, size, low costs, and the ability to be cloned and easily genetically modified (*see* **Note 3**).

The chance to reframe xenotransplantation was offered by the encouraging preliminary results of clinical trials involving cellular transplants, such as porcine islets. These therapies would be widely applicable to diseases involving large populations, such as diabetes disease [1]. In this regard, islet xenotransplantation is viewed as the most advanced form of xenotransplant, and initial clinical trials are underway in diabetic patients [2–6].

Cell-based medicinal products (CBMPs) are subject to specific legal framework, which apply also to xenotransplantation products—these include live cells, tissues, or organs used in xenotransplantation (as defined above). Acellular products like heart valves and decellularized cornea are not considered a cell therapy product, but rather a medical device with its own regulation (*see* **Note 4**). This study is mainly focused on cell therapy products because these are expected to reach the market earlier than xenogeneic organs (e.g., pancreatic islets, hepatocytes, and cellularized cornea) [7]. However, it is important to note that considerable progress has been made in the experimental laboratory to improve organ xenograft survival in several pig-to-nonhuman primate systems [6, 8]. Thus, the pace of progress in this area suggests also that clinical trials may no longer be a distant prospect.

Notwithstanding this, xenotransplantation is now mainly conceived of and regulated as a pharmaceutical product. This creates an attractive market for it and means also that the impact of potential risks has resized and reduced. As we will see, the safety issues are still central to the development of xenogeneic CBMPs which are regulated by European regulatory authorities as advanced therapy medicinal products (ATMPs). Broadly defined, ATMPs are medicines for human use that are based on genes, tissues, or cells. They can be classified into three main types: gene therapy medicines, somatic cell therapy medicines, and tissue-engineered medicines. In addition, some ATMPs may contain one or more medical devices as an integral part of the medicine, which are referred to as combined ATMPs (e.g., cells plus an encapsulation tool) (*see* **Note 5**). This classification is used in Europe. By contrast, in the USA, it is used as cellular and gene therapy products [7]. In the USA, the Food and Drug Administration's (FDA's) definition for ATMPs is broader (*see* **Note 6**). In this regard, xenogeneic CBMPs are regulated by European regulatory agencies as ATMPs, in the USA as cell and gene therapy products. The recommendations included in the respective xenotransplantation guidelines add to these regulations and directives [7].

In Europe, three directives relevant to xenotransplantation are currently in place, making it possible for Member States to have the corresponding provisions on this field: Directive 2001/18/EC of the European Parliament and of the Council of 12 March 2001 on the Deliberate Release into the Environment of Genetically Modified Organisms and Repealing Council Directive 90/220/EEC; Directive 2001/20/EC of the European Parliament and of the Council of 4 April 2001 on the Approximation of the Laws, Regulations and Administrative Provisions of the Member States Relating to the Implementation of Good Clinical Practice in the Conduct of Clinical Trials on Medicinal Products for Human Use, replaced by Regulation (EU) 536/2014 of the European Parliament and of the Council of 16 April 2014 on Clinical Trials on Medicinal Products for Human Use, and Repealing Directive 2001/20/EC; and Directive 2001/83/EC of the European Parliament and of the Council of 6 November 2001 on the Community Code Relating to Medicinal Products for Human Use (*see* **Note 7**).

The general regulatory framework for cells and tissues are based on two main documents: Directive 2004/23/EC on setting standards of quality and safety for the donation, procurement, testing, processing, preservation, storage, and distribution of human tissues and cells and Regulation (EC) No 1394/2007 on advanced therapy medicinal products and amending Directive 2001/83/EC and Regulation (EC) No 726/2004, which standardized the approval for the production and placing on market of gene therapy, somatic cells therapy, and tissue-engineered products (*see* **Note 8**). The marketing authorization of these medicines is carried out through the centralized procedure requested by the EMA, in which a specialized committee for advanced therapies (CAT) was created for assessing the quality, safety, and efficacy of ATMPs and following scientific developments in this field. It was established in accordance with Regulation (EC) No 1394/2007 on ATMPs as a multidisciplinary committee, gathering some of the best available experts in Europe (*see* **Note 9**).

Therefore, the regulatory framework for ATMPs is based on the standardized procedure established by Regulation 1394/2007 and the specific decisions adopted by CAT in each concrete case (*see* **Note 10**).

Given the special safety requirements needed, this regulatory framework on ATMPs establishes specific rules, including the traceability of donors, recipients, and products so as to prevent possible negative public health events (*see* **Note 11**).

It would not be possible to engage in an ethico-legal assessment of xenotransplantation without an understanding of the benefits it promises to mankind, the burdens it may place on humans, the likely chance of its success, and whether other ways of providing its promised benefits are reasonably possible. Thus there is a need for any cautious reflection to follow the most important scientific advances in this field very closely. A number of ethico-legal aspects exist, and these need to be appropriately addressed so as to move this biomedical technology forward in a manner that optimizes its potential benefits to society and reasonably minimize any risks (*see* **Note 12**).

The ethical and legal considerations include some that are applied to any experimental therapy, and others that are unique to xenotransplantation, such as those related with the need of identifying the steps that should be followed to have a reasonable or favorable risk-benefit ratio and challenges in developing an appropriate informed consent process. This chapter will focus primarily on reviewing regulatory framework addressing xenotransplantation, taking into consideration the recent advances in this field in order to reexamine the criteria and circumstances where clinical trials may be medically, ethically, and legally justified.

In the light of current guidelines, regulations, and literature in xenotransplantation, the discussion below reviews the specific conditions that should be met to ensure that initiation of clinical xenotransplantation trials is ethical and legally acceptable. These trials should be approved, monitored, and reviewed by an institutional review board and a national data, safety, and monitoring board to assure safe and ethically sound trial conduct and to protect the rights of prospective participants and the public at large. The conflicting normative aspects involved in this field will be considered, particularly those related with individual rights and the collective dimension of public health and safety. The regulatory approaches developed at European and international level to address some of these difficulties will also be discussed. Finally, some recommendations for ensuring the initiation of clinical xenotransplantation trials within an appropriate ethical and regulatory framework will be underlined.

2 A Framework for Risk–Benefit Evaluations in Xenotransplantation

2.1 Favorable Risk–Benefit Ratio

According to the most relevant literature on the subject, to be ethical, the research developed with xenotransplantation product should be valuable, meaning that it embodies a therapeutic option that could lead to improvements in the quality and length of human life. Certainly, the great promise of xenotransplantation is that it may result in important clinical benefit to diseased recipients. The interest in it is in direct response to a worldwide shortage of human organs, tissues, and cells. The supply is insufficient to treat all patients who are waiting for kidney, liver, islet cells, pancreas, heart, or lung transplantation, which provides a great incentive to xenotransplantation research [9].

As has been mentioned, recent developments in this field are focused on tissue replacements and cell grafting. It is true that record survival of primates with pig organs raises hope. In this regard, there has been an upsurge of interest in this area in recent years. This resurgence can be attributed to a combination of factors: the improvement in efficacy in several preclinical models, the rapid development of genome-editing technology, and the specter of porcine endogenous retrovirus (PERV), which has receded significantly [6]. Notwithstanding this, researchers are still discovering new mechanisms of immune rejection and debating which genetic changes are best at forestalling them [10]. For now, cell therapies are showing promising results and have been recently tested in some clinical trials. Of particular current clinical interest is porcine cell pancreatic islet transplantation for the treatment of type 1 diabetes (*see* **Note 13**). It is hoped that the great progress achieved in this field may ultimately allow islet xenotransplantation to be performed without chronic immunosuppression and hence extend its applicability to the broader populations of diabetic patients [11–13]. Positive safety and efficacy outcomes of clinical trials have already been achieved in this field (*see* **Note 14**). Porcine cell pancreatic islet transplantation, along with the potential to treat diabetes, may also reduce or negate the demand on human islet cells which are limited in availability. Porcine islets have the potential to provide a therapeutic benefit without having the same supply limitations. Therefore, the research developed with these xenogeneic porcine islet products is justified from a social value perspective, since it offers the potential for new approaches for metabolically unstable type 1 diabetic patients (*see* **Note 15**).

Indeed, the promising results obtained in this subgroup of diabetics may offer the possibility of extending the indication to a broader population of diabetic patients who will benefit from tight metabolic control without experiencing hypoglycemia [4].

Although the abovementioned and other xenotransplantation products hold much promise and some success has been achieved, further preclinical studies are now underway to confirm, among other issues, certain safety parameters. To this end, investigators and sponsors planning to conduct clinical trials using xenotransplantation products, as well as the relevant institutional review boards (IRBs) and research ethics committees involved, should determine, and ask for, a favorable risk-benefit ratio. It requires methodological evaluation as to whether xenotransplantation clinical trials are scientifically sound and are capable of generating a relatively high expectation of benefit to the recipient and ultimately to society. It requires empirical judgment about how rigorous and relevant the preclinical scientific data are with respect to this expectation and concerns the possibility of exposing recipients and the society at large to unknown and potentially serious risks of infections. Normative evaluation is also required for these ethical and regulatory conditions and should be satisfied before proceeding with any clinical xenotransplantation trial. The discussion below will consider these evaluations to ensure that xenotransplantation procedures are undertaken in a context in which specific safety, efficacy, and ethical issues are addressed, in particular those associated with minimizing the potential infectious disease risks. To this end, we will take into account the regulatory landscape in Europe, the relevant initiatives recently launched at the international level, and some of the most important steps taken by the Food and Drug Administration (FDA) and the European Medicines Agency (EMA) with the publication of some key guidance documents.

2.1.1 Principle-Based Actions for the Conduct of Clinical Xenotransplantation Trials

For any ethically conducted trial, risks to individual subjects and to society must be minimized. Important efforts have been carried out by both national and international bodies to develop guidelines for minimizing the infectious risks of xenotransplantation (*see* **Note 16**). The sections below outline the regulatory framework ensuring that a comprehensive evaluation of the xenotransplantation products is performed and any potential safety concerns addressed before a clinical xenotransplantation trial is allowed to proceed. This framework contains specific issues concerning the source animals, how these animals are housed, fed, and tested for the presence of infectious agents. It also ensures preclinical studies indicating safety and the efficacy of the procedure are sufficiently described. This framework includes also a set of risk management mechanisms that are applicable to the individual recipient and to public safety.

Selective Breeding and Screening of the Source Animals

There is a widespread agreement that qualification and control of the source animals is of high importance in mitigating the risk of introducing new infections from animals to humans. In this respect, there are specific recommendations for herd and source animal origin (e.g., closed herds or colonies housed in a well-controlled, pathogen-free environment with high standards of animal welfare); for genetically modified animals (e.g., genetic engineering animals must be fully characterized and confirmation of the nature of the inserted or deleted gene must be given); for animal husbandry (e.g., procedures should be developed to identify and prevent incidents that negatively affect the health of the herd or colony or that could negatively impact on the barrier facility or the specific pathogen-free status of the herd); for veterinary control (e.g., protocols for monitoring the herd for disease and infectious agents should exist); for quarantine (e.g., once individual source animals have been chosen, they are put under quarantine for a defined period of time to allow completion of screening procedures); for animal facilities (e.g., animal facilities should be operated in such a way as to minimize the exposure of the animals to infectious agents and to prevent cross-contamination both among animals and animals and humans); for transportation (e.g., barriers equivalent to, or better than, those in place at the facility should be maintained during transit to avoid source animal contamination); for screening and detection of infectious agents in source animals (e.g., periodic testing for adventitious agents, adequate and validated diagnostic assays, and methodologies for surveillance of known infectious agents from the source animal should be available prior to initiating clinical trials); for obtaining and maintaining archive samples from the source animals (e.g., a protocol for archiving samples should be established and validated to ensure traceability); and for herd records (e.g., all animal records should be archived for a period at least equal to that of the archived samples) (*see* **Note 17**).

Manufacturing

Procedures must be in place not only to prevent cross-contamination of harvested and/or processed xenotransplantation products but also to ensure its safety and quality. Final product testing for xenotransplantation products is an additional required element which includes testing for safety, identity, purity, and potency of the product. Final product testing for xenotransplantation products can be challenging because the products include living cells (*see* **Note 18**).

Preclinical Testing

The need for securing benefit over harm should be also based on rigorous preclinical studies using the most relevant animal models (*see* **Note 19**). As recommended by international organizations and most regulatory authorities, investigators must provide clear justification for the trial, including adequate preclinical data on safety and efficacy, usually from nonhuman primate testing (*see* **Note 20**). These studies should take into account the risks to the research subjects and to society imposed by the corresponding trial. The criteria by which preclinical data should be considered appropriate to move forward with clinical studies have been intensely debated. In this regard, it is important that investigators and/or sponsors planning to carry out clinical trials using xenotransplantation products make use of the various regulations and guidance documents published by regulatory agencies, such as FDA and EMA (*see* **Note 21**).

Clinical Trials

Clinical xenotransplantation trials should be designed to ensure that prospective recipients and society are not exposed to an unacceptable risk. This concept requires consideration of the risk both to trial participants and to the public in general. In addition, it requires considerations of efficacy, safety, and individual and general benefits, since the concept of acceptability must be informed by the balance between risks and potential benefits. Thus, the risk of significant recipient disease or contagion should appear to be small and may be acceptable when weighed against the potential benefits to both transplant candidates and society at large of expanded access to life-saving transplants [14]. An IRB with relevant expert representation must establish, among other issues, the following: whether there is adequate scientific evidence from preclinical studies to allow a trial in humans to proceed; whether the proposed trial will provide a clinical benefit and valuable information with regard to safety and efficacy; whether there is adequate information on the safety and toxicity of the materials to be used in a trial in humans; whether the investigators have the qualifications to conduct the proposed trial; and whether an appropriate risk assessment has been conducted by them, ensuring that infectious disease risks are extremely low to justify proceeding with the proposed trial (*see* **Note 22**).

In questioning this latter aspect, we must take into account the abovementioned criteria (e.g., data on safety of the source animal, safety of the xenotransplantation product, safety of the manufacturing process, adequate preclinical data). In addition to these criteria, other safety measures, such as the monitoring procedures for xenotransplant recipients, and, where deemed appropriate, their close contacts and healthcare professionals, are crucial for managing the potential infectious disease risk to the public. This should also include the development of procedures for archiving their specimens.

The risk management mechanisms applicable to the individual recipient and to public safety can be summarized as follows: patient selection—the clinical development of xenotransplantation products should involve initially patients with serious life-threatening disease for whom adequately safe and effective alternative therapies are not available or where there is a potential for a clinically significant improvement with increased quality of life, following the procedure—(*see* **Note 23**); counseling regarding behavior modification and other issues associated with risk of infection should be provided to the patient and made available, where deemed appropriate, to the patient's family and other close contacts, and that such counseling should continue to be available (*see* **Note 24**); informed consent process for recipients and an education and counseling process for them should be carried out; a detailed plan for screening, treating, and control of unexpected infectious diseases should be in place; procedures to ensure the appropriate care and follow-up of the recipients and the necessary safety measures during the initial period after treatment must be undertaken; a system to follow-up close contacts of the recipients may be required when relevant, that is, the monitoring of these later should not be required unless data exist to demonstrate that the recipient is infected [5]; archiving of patient plasma, tissue specimens, and medical records, which should contain all relevant information for both safety and maintenance of efficacy profiling.

One of the most challenging aspects of the recommended safety measures by a xenotransplantation protocol is the requirement of long-term monitoring of the prospective recipients [15, 16]. The monitoring responsibilities expected and required of subjects are difficult due to problems with predicting and controlling their behavior and the potential for violating their fundamental rights and freedoms, such as the individual's right to withdraw according to what is endorsed by some regulatory bodies and by some authors. It may be possible that allowing a recipient to drop out the monitoring program would handicap the scientific community's ability to track and treat any potential disease, thus subjecting the public to risk of infection. In what follows, we explore some problems raised by "compelling" lifelong monitoring, indicating that while active compliance is crucial, significant obstacles hinder obtaining such compliance.

2.1.2 Main Issues Raised by the Lifelong Monitoring of Xenograft Recipients

Lifelong monitoring has been considered as a crucial activity for managing the risk to the public of a potential infectious disease (*see* **Note 25**). The effectiveness and efficiency of public health action is directly related to the quality—and for many purposes the quantity—of relevant surveillance data in order to provide a foundation for a rapid response to emerging infectious disease. Indeed, the objective of surveillance and monitoring would be to ensure that there is prompt recognition, investigation, and management of any suspected or confirmed infections which arise in xenotransplant recipients, their close contacts, or in healthcare professionals. Additional benefits to be gained from this safety measure would be prompt feedback of information to all of those concerned in xenotransplantation and the provision of information which is appropriate to evaluate and review the system. Therefore, the best hope for recipient monitoring in order to be able to implement development and implementation of treatments such as vaccines to combat infectious agents lies in carrying out routine physical evaluations and collecting and archiving tissue and/or body fluid specimens obtained from each recipient (*see* **Note 26**). To this end, active compliance with long-term surveillance is crucial.

The risk of transmission of infection in the course of treatment with a xenogeneic medicinal product requires that possible conditioning factors in the prospective recipient's subsequent life—as well as those safety measures that may intrude on their family and intimate contacts—be explained during the informed consent process. Prospective recipients should be also informed about the importance of complying with the corresponding surveillance responsibilities. Among these are, for instance, those related with the need of regular postclinical checkups and testing schedule, informing about possible changes of address/contact numbers, timely reporting of all unexplained illnesses, following specific behavioral guidelines with respect to exchanges of body fluids with intimate contacts, education of family members and intimate contacts about their need to take precautions associated with infectious disease risks, possible isolation and quarantine, travel movements, etc. (*see* **Note 27**).

Undoubtedly, the subjection to rigorous postoperational monitoring by the medical staff has an effect, to a greater or lesser extent, on the freedom and privacy of prospective recipients and, where deemed appropriate, their close contacts. This requirement emphasizes the difference between xenotransplantation and other forms of medical intervention. It is unusual for any patient to be required to agree to lifelong and highly intrusive monitoring, but it is even more unusual that contacts should effectively be subject to intrusion into private matters, because of treatment being carried out on a third party [17]. In this regard, all effort should be also made to adequately inform prospective recipients that transmission of infectious agents cannot be excluded. That is, before the subject provides his/her consent, he/she must receive the information about the convenience for that person, where appropriate, to convey that information for family members and intimate contacts (*see* **Note 28**).

Therefore, the stringency of the surveillance responsibilities required of xenotransplant recipients and, when appropriate, their close contacts indicates an unusually high level of restriction on their capacity to make life choices. However, in terms of consent, no clinical xenotransplantation trial can proceed unless a valid agreement is obtained from the prospective recipients that they will comply with the surveillance program. This requires, as outlined above, that the full extent of the program be explained before they are asked to consent and that they have the commitment to comply. Notwithstanding this, prospective xenotransplant recipients, as is common in all clinical trials, should be granted they can withdraw from the trial at any stage (*see* **Note 29**). It is certainly a key principle of the Declaration of Helsinki that the potential subject is free at any stage to withdraw from a research program (*see* **Note 30**).

Unlike the position held by certain bodies and regulatory documents, this principle should be also applicable in this field. The xenotransplant recipient who gives an agreement to conform to the surveillance and monitoring program and then subsequently withdraws could certainly become a risk to himself/herself and to the public. However, the requirement for lifelong monitoring does not undermine the individual xenotransplant recipients' right to withdraw their consent, despite what is stated in some regulatory documents and by some authors in this respect (*see* **Note 31**).

Reasons that conflict with reviewing the recipient's consent as an irrevocable agreement include the importance of protecting the recipient's right to freely decide on decisions that affect their own health and personal status. It is true that xenotransplant recipients might risk the health of the entire community by refusing to comply with the surveillance and monitoring program. However, from a legal and ethical perspective, it may be only acceptable to withdraw respect of a xenotransplant recipient's autonomy and fundamental rights (such as the right to physical integrity), in the face of circumstances where there is wide agreement that respecting it potentially undermines the health and safety of others [18]. Progress in xenotransplantation shows that the transfer of an infectious microorganism is either less likely than previously thought or readily manageable. This emerging technology, in particular, xenogeneic cell-based medicinal products, is no longer perceived with unknown and so unpredictable risks; quite the contrary, it is associated with well-known hazards. The main legitimizing debate on safety is focused on traceability, that is, keeping track of all elements involved, from subjects to processes to products. Given the much lower risk involved in this field and the illegitimacy of constrains imposed by an irrevocable and prolonged surveillance, this requirement must be reconsidered. A public authority may intervene, as we will see, in accordance with national law and in conformity with the principles of necessity and proportionality depending on the concrete circumstances in which public health protection so requires.

The clinical use of these products (ATMPs) in humans "may be associated with specific risks to the patient and to third parties. These risks are determined by various risk factors, which are related to the quality, biological activity and application of the ATMP. Since ATMPs are very diverse in nature (...), a flexible approach to address and evaluate potential risks associated with the clinical use of ATMPs is described in the «Risk-based approach»" (*see* **Note 32**). Given the specific safety requirements needed, the Regulation authorizes strict rules, including the complete traceability of donors, recipients, tissues, and products in order to prevent potentially negative public health events.

As has been indicated, not only products but also clinical trials with all ATMPs are regulated together. In this regard, safety measures earlier confined to xenotransplantation have now been extended to all clinical trials with ATMPs. According to guidelines in which the standards for all clinical trials with allogeneic and xenogeneic cells are set, *informed consent from the subject of the trial is required so as to legitimately perform the follow-up*. In these guidelines on good clinical practice specific to ATMPs, the subject has to be properly informed of, and agree to, a wide range of issues (*see* **Note 33**). Concerning the right to withdraw from a trial, it is stated that "Where a subject is withdrawn from a trial at their own request or based on a decision of the investigator the follow up should be maintained, *subject to the consent of the subject*" (cursive is own) (*see* **Note 34**).

Therefore, subjects involved in ATMPs trials should be informed about the importance of following the essential responsibilities concerning with traceability, including provisions of for subject data protection and confidentiality, follow-up before and following the trials, communicating risks to close contacts and offspring where they could be at risk, etc. That is, *xenotransplantation safety measures are still the rule*, *but* unlike the compulsory measures originally established by the Council of Europe Recommendation Rec (2003) 10 on xenotransplantation, *the right of subjects to make their own choice or decide on research participation has to remain the core prerequisite for enrolling any person in a clinical trial. And all researchers are bound to follow this patient's right and fundamental principle of medical ethics.* Otherwise, xenotransplantation would be difficult of being ethically and legally legitimated.

Consequently, if it can be argued that the surveillance responsibilities can be presented as important to follow, it seems difficult to see how compliance could be fully enforced from and ethical and legal perspective. In this regard, the Council of Europe Recommendation states that "If after the xenotransplantation has been carried out, the recipient or his or her close personal contacts refuse to comply with the constraints associated with xenotransplantation, public authorities should intervene and take appropriate measures, *where public health protection so requires*, *in conformity with principles of necessity and proportionality*. Depending on the circumstances and in accordance with the procedures provided for by national law, such measures might include registration, compulsory medical follow-up and sampling" (cursive is own, *see* **Note 35**). In what follows, we discuss how the principles of necessity and proportionality may be interpreted in these situations in which it is not only a matter of respecting the free, individual decision of the xenotransplant recipient but also of balancing individual autonomy against collective public health interests which can be of greater priority.

As some authors affirm, the generally applicable public health law provisions in most countries are inadequate to enforce long-term surveillance when the xenotransplant recipients are asymptomatic and the nature and communicability of possible pathogens are undetermined [19, 20]. Certainly, there is no evidence to date establishing the existence of an imminent, substantial, and highly fatal infectious disease risk associated with xenotransplantation, the ground for declaring a public health emergency and for imposing certain mandatory requirements to noncompliant recipients. It is true that the viability of a surveillance program depends upon its ability to collect epidemiological data in the absence of such compelling evidence, regardless of whether the recipients appear to be asymptomatic or not [20]. Indeed, the potential risks of infections cannot be completely ruled out, since undetectable organisms can remain in a latent state within the source animal and recipient for indefinite time [19, 20].

It is also true that public health authorities retain the authority to take appropriate measures in those specific situations where it is deemed necessary to intervene to prevent and protect society from the spread of infectious diseases. The coercive measures in other public health contexts are based on the rationale that the subject presents a sufficient risk to society to justify elevating the public interests over the individual ones. The same argument could be applied where xenotransplant recipients might risk the health of the entire community and the requirement, where deemed appropriate, of adopting measures in order to prevent the propagation and the risk of pandemics. According to the concrete circumstances, these measures could adopt a coactive expression. The significance of such measures would make it necessary to dispose of an adequate legal framework in order to conciliate the rapidity and effectiveness of the response with respect and the minimal restriction to the fundamental rights of the affected persons. In general, national regulations that would be relevant in case the competent authorities had to prevent the spread and the risk of epidemics might be those which include the following measures: measures of recognition, treatment, hospitalization, or control when there are *rational* or *evident signs* that allow to suppose the existence of hazard for the public health due to the specific health situation of a person or group of persons or the specific health conditions in which an activity is developed.

Those measures that may be imposed to noncompliant recipients should respond to the *principles of necessity* and *proportionality*. Concerning the first one, it is true that long-term monitoring requirements may compromise xenotransplant recipients' fundamental rights. As outlined above, a detailed plan for screening, treating, and control of unexpected infectious disease should be undertaken to avert and prevent a potential xenogeneic pandemic. Requiring that xenotransplant recipients comply with surveillance seems reasonable. Despite its relevant purpose of assessing rigorously the risks and benefits of the xenotransplant in a scientifically valid manner, the subject has the right to withdraw from the trial, and the follow-up should be maintained subject to his/her consent. Notwithstanding this, the least restrictive measures should be taken once there are evident signs that allow to suppose the existence of a xenogeneic disease. For instance, coercive isolation and quarantine should only be used as a "last resort" (*see* **Note 36**). These measures would require the highest level of evidentiary justification. Coercive social distancing is only justified if it results in net benefits to society as a whole. According to what it is established in the European Convention on Human Rights (Article 5.1.e), "detention" must be a "lawful" detention. In this respect, the Guide on Article 5 of this European Convention "Right to liberty and security," states that "The essential criteria when assessing the "lawfulness" of the detention of a person "for the prevention of the spreading of infectious disease" are whether the spreading of the infectious disease is dangerous to public health or safety and whether detention of the person infected is the *last resort* in order to prevent the spreading of the disease, *because less severe measures have been considered and found to be insufficient to safeguard the public interest*" (cursive is own). "When these criteria are no longer fulfilled, the basis for the deprivation of liberty ceases to exist" (*see* **Note 37**). From this perspective, it may be argued that the detention of noncompliant but asymptomatic xenotransplant recipients, when they have withdrawn from a xenotransplantation clinical trial and the follow-up at their own request, is unjustified and unconstitutional by virtue of being disproportionately intrusive. In the event that the recipient presents the various symptoms that may indicate an infectious syndrome, hospital admission and isolation may be required until the nature of the process is defined more accurately. These measures should be based on the corresponding clinical protocol, the level of perceived risk for the community, and in the respective local regulations.

On the other hand, measures taken should be linked, as far as possible, with the respect of xenotransplant recipients' fundamental rights. For instance, those measures related with the need of collecting and archiving tissue and/or body fluid specimens for epidemiological surveillance purposes should be respectful with the individuals' rights included in the data protection normative. And, as the risks of transmitting new infectious diseases into the population are reassessed and minimized, the number and time frames of recipients' surveillance responsibilities should be revised accordingly.

In addition to comply with *necessity* requirements, it should be determined whether the measures taken are *proportionate*. The restriction of a right is proportional if it is balanced because more benefits for the general interest are derived from it than damages against other goods or values in conflict. In this case, xenotransplant recipient's rights—such as the right to physical integrity on which is based informed consent—have to be confronted with the serious risk for public health derived from an uncontrolled transmission of pathogens or the risk of introducing new infectious diseases into the general population, a dramatic threat that, as has been said, can legitimate, on concrete circumstances, certain constraints on personal autonomy. Therefore, reasons of general interest may justify the limitations on certain fundamental rights and freedoms, authorizing medical interventions in persons suspected of suffering diseases that involve the specific risk of affecting public health. The rule of proportionality is the one that should weigh those limitations both in its intensity and in the time of application.

Therefore, as any other emerging technology, xenotransplantation requires a safe but ethical and legal way to proceed. As has been mentioned, the impact of xenotransplantation on xenotransplant recipients requires an adequate evaluation of the extent to which the potential benefits of xenotransplantation might outweigh the risks and harms. In this regard, another important issue that should be taken into account when this risk-benefit analysis is undertaken makes reference to the criteria for patient selection. As outlined before, clinical xenotransplantation trials should involve patients with serious or life-threatening disease for whom adequately safe and effective alternative therapies are not available and for whom receiving a xenotransplantation product would be both medically and ethically justified. One wonders, in this respect, whether researchers should accept the responsibility for excluding prospective xenotransplant recipients from the trial when they are unlikely to abide by critically important surveillance responsibilities (*see* **Note 38**).

2.1.3 Public/Citizens' Participation on Decision-Making in Xenotransplantation

According to some authors, the individual-centered nature of traditional consent fails to take into account the potential risks of transmitting an infectious disease to third parties and to society at large [19]. All these issues have been extensively discussed elsewhere and are not our concern here (*see* **Note 39**). Instead, we want to focus on another important issue that has been introduced into the consent process: public/citizens' participation in decision-making in this controversial technology.

At the beginning of the twenty-first century, two important initiatives of public consultation have been launched by the Canadian (2000) (*see* **Note 40**) and the Australian governments (2001) (*see* **Note 41**). Both countries understood the public health risks posed by xenotransplantation as serious enough to require their citizens' involvement in a decision (*see* **Note 42**). More recently New Zealand engaged some fractions of the populations—including the Maori—in a similar exercise in connection with xenocell products (*see* **Note 43**), while the EU institutions have been widely using web-based consultations to legitimize the placing in the market of new pharmaceuticals such as cell therapy, gene therapy, and tissue engineering products (*see* **Note 44**).

However, as far as the ethical aspects are involved, some reflections can be presented here. There is an ethical argument that, because clinical xenotransplantation trials have the potential to impact a wider population beyond the individual participants, a form of collective consent should be also required. Certainly, there is a widespread agreement that where risks are collective, "publics" should not only be educated about the risk but should also be given an opportunity to participate actively and meaningfully in the decision about whether and under what conditions they are exposed to the risk [21, 22]. The most diverse groups are included in the rubric of "publics" (*see* **Note 45**). Citizens must be involved in all stages of discussion on the issues involved in xenotransplantation—such as the potential risks that this biomedical technology poses to society as a whole, the feasibility and significance of the potential benefits, the fragile balance between individual and collective rights in matters of health, allocation of resources, the ethics of animal research, etc.—and have their voices and perspectives incorporated into decision-making (*see* **Note 46**).

To recap, different public voices, in and outside of official arenas, should be included in a deliberative and open process with the aim of not merely gauging public attitudes to xenotransplantation, but having the right to participate in decision-making. To this end, the way of framing citizens' engagement would require not only diverse open forms of consultations but also a broader range of participatory activities—including learning processes, deliberations, interactions between scientists and citizens, etc.—with the aim of contributing to an "opening" of decision processes on xenotransplantation to a broader range of publics, knowledge, and values (*see* **Note 47**).

2.1.4 Toward a Worldwide Harmonization of the Regulatory Framework of Xenotransplantation

Today there are different regulatory approaches to xenotransplantation, and the potential public health risk associated with this technology makes this uncoordinated approach problematic. In the USA and Europe, the response to regulatory inconsistency has been made necessary, as has been seen, to develop and standardize regulations and guidelines for the clinical use of xenotransplantation products.

The different approaches to xenotransplantation are, in general, a reaction to the potential risk of infectious disease. Public consultations in some countries, such is the case in Canada, resulted in recommendations not to proceed with xenotransplantation clinical trials involving humans until the risks were better understood and could be better managed. However, when such countries decide not to proceed with clinical xenotransplantation trials, they have no assurance that they will be protected from that risks they do not wish to accept. Public health risks of xenotransplantation have the potential to transcend national boundaries. That is to say, xenotransplantation is of local, national, and international concern since, as evidenced by other human infections of animal origin, viruses can quickly and effectively transverse geopolitical borders.

Therefore, regardless of a State's position on xenotransplantation, an adequate legal framework for monitoring global application of this biomedical technology would be necessary, in particular, or at least, public health safety measures, taking account of xenotransplantation's expansion into the medical tourism industry. The issue of "xenotourism," which is the travel of prospective xenotransplant candidates to countries where xenotransplantation procedures are less stringently regulated, has raised critical questions for both monitoring of xenotransplant recipients and mechanisms to protect public health. There need to be mechanisms of identifying and monitoring xenotourists and xenotransplant recipients in general. Therefore, xenotourism is another issue linked to public safety that requires all nations to bear their fair share of responsibility regarding the control of infectious disease risks [23]. In the absence of internationally agreed regulations and monitoring procedures, the most assiduous safety measures of any nation, or group of nations, are likely to be ineffective or unsuccessful. This problem arises because of a constantly mobile population. Prospective xenotransplant candidates could receive a xenograft in one country, which may or may not have appropriate regulatory and monitoring process, and later return to their home country where public health authorities lack awareness of the procedure or enter another one without ever having to state that they are the recipient of a xenograft.

This problem is highly complex and requires, as noted earlier, international agreement on regulations covering xenotransplantation. It may be appropriate to establish xenotransplantation alerts in a similar manner to a notifiable infectious disease so that physicians who may see such patients are obliged to register them and to inform local or national authorities.

To recap, xenotourism calls for global standardization of xenotransplantation [24]. It is necessary to have a specific public health surveillance and response protocol in place for this technology in the event of novel diseases arising. A commitment from each government to ensure this protocol is in place should be a condition for approving clinical xenotransplantation trials.

Globalization clearly presents important difficulties for crafting consensus on regulatory policy in xenotransplantation. It becomes increasingly apparent that both national and international regulatory responses are required. In what follows, we review what kind of measures should be addressed to protect individual and public interests at the State level while acknowledging the challenges presented by globalization.

International Oversight and National Regulatory Frameworks

Clinical xenotransplantation trials should be conducted only in countries where regulations and adequate resources empower a national health authority to effectively regulate these trials. This regulation needs to be influenced and guided by the international community to harmonize global practices. The global threat of infectious diseases forces States to cooperate with each other so as to develop global governance, which involves certainly the interaction of States, international organizations, and non-state actors to shape values, policies, and rules. To this end, important steps have been taken toward the harmonization of xenotransplantation procedures and policies. For instance, in 2001 the World Health Organization (WHO)—that was one of the first international organizations to deal with xenotransplantation—released a guidance document to facilitate consideration and implementation of a strategy for international cooperation and coordination on xenogeneic infection /disease surveillance and response (*see* **Note 48**). In 2004, the World Health Assembly (WHA) adopted Resolution 57.18 (*see* **Note 49**), which was regarded as an important step toward minimizing infectious risks associated with xenotransplantation and a commitment of more than a hundred countries to the development of adequate, harmonized regulatory procedures [25]. The resolution ensures accountability and traceability, harmonization of practices, and ethics review, which are included as part of practice requirements. It also promotes international cooperation and collaboration to face xenotourism. In 2008, the WHO, in collaboration with the Chinese Ministry of Health, the University of Central-South China, and the International Xenotransplantation Association (IXA), conducted a global consultation on clinical xenotransplantation trials (*see* **Note 50**). The consultation with experts representing sovereign States resulted in the Changsha Communiqué (*see* **Note 51**), in which principles relating to xenotransplantation were listed, followed by recommendations to (1) the WHO, (2) Member States of the WHA, and (3) investigators and proposers of xenotransplant procedures and trials. These recommendations are intended to provide guidance that, if followed, will enable safe and efficient clinical trials to be carried out [26].

Certainly, it is of high importance that international collaboration to prevent unregulated xenotransplantation and to coordinate xenotransplantation vigilance, surveillance, and response to infections is encouraged. The role played by the WHO and other international agencies is crucial in this respect. For instance, some regulatory agencies, such as the FDA and the EMA, have been active participants in open dialogues with these international organizations and with regulatory bodies of individual countries, concerning the regulatory requirements for clinical xenotransplantation trials.

The FDA is the sole agency responsible for regulating clinical xenotransplantation trials in the USA. It evaluates the risks and benefits of these trials and determines whether the clinical trials may proceed. This agency has published diverse guidance documents with recommendations for sponsors of clinical xenotransplantation trials (*see* **Note 52**). These documents address several key conditions—such as those related with qualification and control of the source animals, manufacturing, preclinical, and several safety issues raised by xenotransplantation—in order to enable the initiation of clinical trials under the highest scientific, ethical, and regulatory standards (*see* **Note 53**).

The EMA has published also key guidance documents for xenogeneic cell-based therapies, such as the "Guideline on Xenogeneic Cell-based Medicinal Products" (2009). This guideline is an annex to the guideline on human cell-based medicinal products and deals specifically with requirements unique to xenogeneic specificities. It provides general principles to be taken into account for the development and assessment of xenogeneic cell-based products without prejudice to medical practice or national legislation, which may be applicable. This guideline addresses issues such as sourcing and testing of animals, manufacture, quality control, nonclinical and clinical development, surveillance, animal health, and welfare in the sourcing materials. It also contains requirements for the development and marketing authorization of xenogeneic cell-based medicinal products for human use. Particularly, this guideline is intended for products entering the marketing authorization procedure. However, the principles laid down in the guideline should be considered by applicants entering into clinical trials (*see* **Note 54**). As has been said before, the recommendations included in these xenotransplantation guidelines add to the respective regulations and directives mentioned (xenogeneic CBMPs regulated by European regulatory agencies as ATMPs and in the USA as cell and gene therapy products).

Undoubtedly, it is crucial a coordinated international effort, such as that initiated by WHO and other international agencies, to harmonize xenotransplantation procedures in accordance with the best available scientific data and the highest ethical and regulatory standards. However, this effort may not be sufficiently strong to make a difference in terms of compliance with desirable legal standards. Governmental regulatory bodies have the sovereign power to interpret those principles on conducting clinical xenotransplantation trials that do not necessarily translate into adequate protection for individual rights and public health. International cooperation on xenotransplantation that is effective will need to consider ways of combining legally binding measures with soft law provisions, agreements, and/or statements of principle, at local, national, and international level (*see* **Note 55**).

Therefore, because of the community risk, clinical xenotransplantation trials need to be effectively regulated. Regulation should have a legal basis with powers to ban unregulated procedures and enforce compliance with regulatory requirements for (1) herd and source animal origin, history and pathogen-specific status, animal husbandry practices, qualifications for the facilities, and high standards of animal welfare (*see* **Note 56**); (2) safety of the manufacturing process; and (3) risk management protocols; all the risks should be appropriately managed under the regulatory framework. To this end, several key conditions should be also considered in detail: the establishment of a register, xenotransplant recipients must be registered in an appropriate database while ensuring that patient privacy is protected (*see* **Note 57**); long-term storage of animal and patient samples, pre- and post-treatment, as well as records (*see* **Note 58**); and a system in place for vigilance and surveillance with appropriate plans to identify and respond to any indication of xenotransplantation-related infection in a timely manner.

As noted earlier, effective monitoring and surveillance will require cooperation at local, national, and international level—e.g., monitoring of patients and trial participants will require that patients and clinical teams are aware of the importance of reporting any problems to local or national authorities (*see* **Note 59**). Compliance with a requirement to submit to lifelong monitoring should not be enforceable. It is not ethical and legally justifiable to require patients to waive some of their fundamental rights. As noted earlier, when an individual fails to voluntarily take required precautions against the possibility of infecting others, legal sanctions and/or the use of force may be only justified based on necessity and proportionality requirements where public health protection so requires. States have the power to limit personal freedoms and rights when public health or security is at stake. At national level, health authorities, as outlined above, will be responsible for implementing a surveillance system as part of their own regulatory framework, including a patient register. Indeed, a national register and biological sample archive are also among the key provisions to protecting public health from the risks of novel infections. They provide the basis for traceability in the event that a novel infection is detected. Therefore, there should be rigorous analysis of trial outcomes, and all recipients should be registered in an appropriate database while their confidentiality should be protected (*see* **Note 60**).

Another problem for current regulation in this regard is the difficulty of ensuring adherence to the (costly) monitoring measures, above all in developing nations that would not have an appropriate national surveillance system. Moreover, when we take into account the resources to combat a new epidemic, risks are not equally shared. Developed countries have far greater resources to put in the fight against a new virus than developing nations. The inequity between developed and developing nations with respect to the risks and benefits associated with xenotransplantation is another social concern that has been also highlighted [27–29]. As stated by the Changsha Communiqué, consideration should be given from the beginning to future equitable access to this therapy.

On the other hand, there are some xenotransplantation experiments that are currently being conducted in countries with no regulatory framework or without appropriate regulatory procedures to deal with this technology and its potential negative outcomes. "Xeno-havens" are believed to amplify global contagion due to a perceived lack of regulatory oversight, scientific and ethical standards, and patient monitoring [24]. To this end, as has been outlined by the IXA: "Without organized international cooperation, the best efforts at minimizing these risks in countries with appropriate regulatory oversight may be thwarted by the free travel of individuals undergoing unmonitored xenotransplantation in countries lacking such regulation" [16].

There is currently an agreement that the results of trials that breach xenotransplantation guidelines or the ethical and regulatory standards regarded as the global standard to which all countries must adhere should not be accepted for publication in high-ranked peer-reviewed journals. There is certainly an international inventory designed to document clinical xenotransplantation trials (*see* **Note 61**); however, there is no international or national requirement to document or share such information. In this regard and despite these efforts of international cooperation and standardization, there is currently no international sanction for noncompliance to international xenotransplantation guidelines or, for that matter, an authority to impose sanction (*see* **Note 62**).

The Review Process of Clinical Xenotransplantation Trials as a Multidisciplinary Exercise

Investigators and proposers of clinical xenotransplantation trials should provide all the information required by the regulatory authorities to assess the benefits and risks and determine how these latter can be minimized. As noted earlier, certain requirements—such as those related with the safety of the source animals, of the xenotransplantation product, and of the manufacturing process, appropriate preclinical studies that provide support concerning the safety and efficacy, risk management protocols, etc.—may prove the capacity to demonstrate a favorable risk-benefit ratio or to convince, for instance, that the risk of introducing new infectious diseases into the general population is extremely low. This implies that trials should be approved, monitored, and reviewed by an IRB and a national safety monitoring board, with representation of all required expertise in order to (1) assure validity and scientific value of the trial, a scientific assessment should be carried out to establish, among other issues whether the trial will provide clinical benefit, whether there is adequate scientific evidence from experimental studies, whether the investigators, in addition to have the corresponding qualifications and experience, have conducted appropriate risk assessment of their proposed procedures, patient selection, etc. and (2) ensure the ethical conduct of the clinical research and the protection of the current and prospective participants rights, an ethical assessment should be undertaken to review to what extent the proposed trial meets ethical requirements, including adequacy of the information and informed consent process, justification for research, minimization of harm, distributive justice, social and cultural considerations, etc.

Xenotransplantation is no longer seen as a source of unique risks. In this regard, safety measures earlier confined to this emerging technology have now been extended to all clinical trials with ATMPs. The compulsory measures originally established by regulatory documents have been rethought. Therefore, informed consent from the subject of the xenotransplantation clinical trial is now required in order to legitimately perform the follow-up. Notwithstanding this, as has been seen, coercive powers should be exercised on the basis of an evident hazard to public health. That is, compulsory physical examination or isolation entails a reasonable suspicion that the person is contagious or could pose harm to others. The measures adopted by the national authorities must be appropriate to prevent the threat. In this regard, they must strive to ensure that there is a reasonable fit between the coercive measures imposed on individuals and the public health benefit that they seek to achieve.

As noted earlier, another important issue that have been introduced into the consent process is public/citizens participation in decision-making in xenotransplantation. There should be government commitments to support and develop different forms of collaboration within the civil society and between civil society organizations and the international scientific community about this biomedical technology [30]. A lot of effort should be made to create a more knowledgeable and more responsible citizenship, to favor an open dialogue between scientists and citizens, and to make policy-making more transparent and accessible (*see* **Note 63**). Not only have advancements been made in defining risks resulting from known pathogens, but also data generated over the past years suggests that the risks of human infection resulting from xenotransplantation are less likely and readily manageable. All these efforts may apply to public health protection, as it has been demonstrated that forms of participatory management of infectious diseases may work better than practices based merely on control and risk assessment. To this end, States should promote public awareness of xenotransplantation practices and their risks. As it has been suggested, the next step in this respect should go through a more complex meaning of public participation. The shift that is taking place goes from mere passive approaches to develop consensus and risk acceptance to a shared accountability for risk control and regulatory decision-making.

Therefore, assessment of clinical xenotransplantation trials or treatments with xenogeneic medicinal products is certainly a multidisciplinary exercise, requiring expertise for characterization, manufacturing and control, and nonclinical, clinical, ethical, legal, cultural, and social aspects. All these areas should be represented in the corresponding committees involved in the evaluation process of the proposed trials.

3 Conclusions

In questioning the attainability of a favorable risk-benefit ratio, we must take into account, on the one hand, that potential benefits of xenotransplantation are both significant and achievable, as some positive safety and efficacy outcomes of islet xenotransplantation trials are showing, and organ xenotransplants are promising to offer. On the other hand, we must take note of the significant progress that has been made in recognizing and excluding infection risks. Not only have advancements been made in defining risks resulting from known pathogens, but also data generated over the past years suggests that the risks of human infection resulting from xenotransplantation are less likely and readily manageable. More so, the risk of the spread of infection to contacts of the xenograft recipient also appears to be low. Indeed, it is encouraging that, to date, porcine endogenous retroviruses (PERVs) have not been associated with any current, past, or latent infection in xenografted patients [5, 8, 14, 31]. It is possible that newer molecular techniques including the generation of PERV knockout swine using CRISPR-Cas9 technology could limit or completely exclude PERV transmission [6, 8, 32, 33]. Furthermore, several drugs are now available that would be effective in treating any adverse clinical syndrome that resulted from their transfer [33]. It should not be left unsaid that some types of xenotransplantation products, such as cell-based xenotransplantation products, imply a significantly smaller risk of virus transmission than xenotransplants of vascularized organs [5]. Cells can be best screened for a spectrum of infectious agents in advance. Furthermore, xenogeneic cell transplant barriers to immunology, such as encapsulation techniques, may control viral transmission as well [19, 34]. Therefore, it seems closer the possibility of excluding predictable factors of the infectious risk beforehand, which is of major concern in this biotechnology.

International recommendations and guidelines for well-monitored global application of xenotransplantation practices and policies have been set in place in order to exclude this potential risk of transferring animal viruses to transplant recipients and the wider community. And, although this risk cannot be completely eliminated, there is widespread agreement that it is sufficiently low and can be managed appropriately via herd selection, documented health screening strategies, rigorous preclinical safety, and efficacy data using the most relevant animal models, in addition to those set of risk management components that are applicable to the individual recipient (*see* **Note 64**).

Therefore, under such internationally agreed regulations, only clinical xenotransplantation trial proposals that are deemed safe, provide a significant and attainable possibility of therapeutic success, and have protocols that ensure the highest ethical and regulatory standards for both human recipients and animal participants should be allowed to proceed. If these precautions are taken, under the international supervision, the potential risk of cross-species infection may be considered extremely low and manageable and, thus, outweighed by the benefits of this biomedical technology. The question is, as outlined earlier, how to make the international normative level more effective through legally binding measures in this field, since inadequacies in national regulations may produce undesired local effects and have also global implications, in terms of safety (*see* **Note 65**). Of equal importance is to open deliberative processes to a broader range of the publics taking into account views regarding morals, religions, and preferences. In each country, practices should be promoted for engaging with publics within the scientific community as a form of shared responsibility for regulatory decision-making.

4 Notes

1. This definition replicates the definition set out in Directive 2003/63/EC of 25 June 2003 amending Directive 2001/83/EC of the European Parliament and of the Council on the Community Code Relating to Medicinal Products for Human Use, at Annex I, Part IV "Specific statement on xeno-transplantation medicinal products." This definition is widely accepted and has been adopted by the Changsha Communiqué, for which the World Health Organization served as

the Secretariat. Available at: http://www.who.int/transplantation/xeno/ChangshaCommunique.pdf?ua=1 (last visited 20 September 2018).

2. World Health Organization (WHO) (2001), WHO *Guidance on Xenogeneic Infection* /*Disease Surveillance and Response: A Strategy for International Cooperation and Coordination*, WHO, Department of Communicable Disease Surveillance and Response, WHO/CDS/CSP/EPH/ 2001–2002. Available at http://apps.who.int/iris/bitstream/handle/10665/68888/WHO_CDS_CSR_EPH_2001.2.pdf;jsessionid=21D1539699DDDDAC156F0EF479DF9F1E?sequence=1 (last visited 20 September 2018), pp. 1–2; Council of Europe (2003), *Report on the State of the Art in the Field of Xenotransplantation*, Council of Europe: Strasbourg. The full report is available at http://www.coe.int/t/dg3/healthbioethic/activities/06_xenotransplantation_en/XENO(2003)1_SAR.pdf (last visited 21 September 2018), page 16; *See* also the Recommendation informed by the State of the Art Report, Recommendation Rec (2003) 10 of the Council of Ministers to Member States on Xenotransplantation, Article 3; The United States Food and Drug Administration (FDA) Department of Health and Human Services, Public Health Services, Guideline on Infectious Disease Issues in Xenotransplantation, January 19 (2001). Available at http://www.fda.gov/OHRMS/DOCKETS/98fr/960311g2.pdf (last visited 22 September 2011); US Department of Health and Human Services. Secretary's Advisory Committee on Xenotransplantation (SACX) (2004), *Draft Report on the State of the Science in Xenotransplantation*, page i. Available at http://www.nelsonerlick.com/PDF/NIH%20Report%20on%20State%20of%20Xenotransplantation%202005.pdf (last visited 25 September 2018); FDA. Center for Biologics Evaluation and Research. Updated December 2016. *Guidance for Industry: Source Animal, Product, Preclinical, and Clinical Issues Concerning the Use of Xenotransplantation Products in Humans.* Available at https://www.fda.gov/downloads/biologicsbloodvaccines/guidancecomplianceregulatoryinformation/guidances/xenotransplantation/ucm533036.pdf (last visited 25 September 2018), pp. 1 and 6.

3. US Department of Health and Human Services. Secretary's Advisory Committee on Xenotransplantation (SACX) (2004), *Draft Report on the State of the Science in Xenotransplantation*, p. 6. Available at http://www.nelsonerlick.com/PDF/NIH%20Report%20on%20State%20of%20Xenotransplantation%202005.pdf (last visited 25 September 2018).

4. Medical Devices Containing Materials Derived From Animal Sources (Except for In Vitro Diagnostic Devices)—Draft Guidance for Industry and Food and Drug Administration Staff; January 23, 2014. This document, when finalized, will supersede "Medical Devices Containing Materials Derived From Animal Sources (Except for In Vitro Diagnostic Devices)" issued November 6, 1998. http://www.fda.gov/downloads/MedicalDevices/DeviceRegulationandGuidance/GuidanceDocuments/UCM381491.pdf (last visited 23 September 2018); WHO, medical device – full definition, http://www.who.int/medical_devices/full_deffinition/en/ (last visited 20 September 2018). In the USA, medical devices are overseen by the FDA, https://www.fda.gov/MedicalDevices/UCM2005076 (last visited 27 September 2018). In Europe is welcomed the adoption of the proposal for two new Regulations on medical devices which establish a modernized and more robust EU legislative framework to ensure better protection of public health and patient safety; *see* https://ec.europa.eu/growth/sectors/medical-devices/regulatory-framework_en (last visited 29 September 2018).
5. ATMP in Europe are defined by Regulation (EC) no 1394/2007 of the European Parliament and of the Council of 13 November 2007 on advanced therapy medicinal products and amending Directive 2001/83/EC and Regulation (EC) No 726/2004, (so-called ATMPs Regulation), Article 2 [1](a). EU Medicinal Products Directive 2001/83/EC also applies to such products. Commission Directive 2003/63/EC of 25 June 2003 amending Directive 2001/83/EC of the European Parliament and of the Council on the Community Code Relating to Medicinal Products for Human Use, Annex I, Part IV deals with ATMPs and concerns specific requirements for gene therapy medicinal products (using human autologous or allogeneic system, or xenogeneic system) and Cell Therapy Medicinal Products both of human or animal origin and xenogeneic transplantation medicinal products. The definition of an ATMP is viewed slightly differently by regulatory bodies. For example, in Europe an ATMP is defined as being gene therapy medicinal products, somatic cell therapy medicinal products, a tissue-engineered product, or a combined ATMP. See EMA 21 May 2015 EMA/CAT/600280/2010 rev.1 Committee for Advanced Therapies (CAT) Reflection paper on classification of advanced therapy medicinal products. Available at: https://www.ema.europa.eu/documents/scientific-guideline/reflection-paper-classification-advanced-therapy-medicinal-products_en-0.pdf (last visited 27 September 2018).

6. FDA's Office of Cellular, Tissue and Gene Therapies defines them as cellular therapy products (that) include cellular immunotherapies and other types of both autologous and allogeneic cells for certain therapeutic indications, including adult and embryonic stem cells. Human gene therapy refers to products that introduce genetic material into a person's DNA to replace faulty or missing genetic material, thus treating a disease or abnormal medical condition. https://www.fda.gov/biologicsbloodvaccines/cellulargenetherapyproducts/default.htm (last visited 25 September 2018).
7. Directive 2001/18 EC introduced common procedures and safety requirements for the approval of GMOs, such as engineered pigs for xenotransplantation; Directive 2001/20/EC on clinical trials established some specific and restrictive provisions on xenogeneic cell-based therapies and GMOs; and Directive 2001/83/EC included rules about the approval of trials involving animal cells and tissues.
8. The Regulation has been complemented by Directive 2009/120/EC of 14 September 2009 amending Directive 2001/83/EC of the European Parliament and of the Council on the Community Code Relating to Medicinal Products for Human Use as Regards Advanced Therapy Medicinal Products.
9. More information about the role of the CAT is available at: https://www.ema.europa.eu/committees/committee-advanced-therapies-cat
10. Given that ATMPs are very diverse, it is necessary a flexible approach to evaluate the potential risks. In this regard, see CAT-EMA Guideline on the risk-based approach according to annex I, part IV of Directive 2001/83/EC applied to Advanced therapy medicinal products: available at: https://www.ema.europa.eu/documents/scientific-guideline/guideline-risk-based-approach-according-annex-i-part-iv-directive-2001/83/ec-applied-advanced-therapy-medicinal-products_en.pdf (last visited 30 September 2018).
11. In this regard, the European Commission published Guidelines in 2009 on good clinical practice specific to ATMPs, which set the standards for all clinical trials with allogeneic and xenogeneic cells. See more detailed information in ref. 1. Guidelines available at: https://ec.europa.eu/health/sites/health/files/files/eudralex/vol-10/2009_11_03_guideline.pdf
12. This emerging technology gives rise to a host of ethical and legal issues ranging from the use and treatment of animals and its potential impact on individuals and the wider society. The problems concerning the use and treatment of animals used for scientific purposes merit further ethical analysis but exceed the limits of this study.

13. Porcine islet can replace malfunctioning human pancreatic cells unable to produce sufficient amounts of insulin, a hallmark of type 1 diabetes.
14. For instance, clinical xenotransplantation trials are currently underway in New Zealand, one of which regards islet xenotransplantation in diabetic patients. In this study, promoted by Diatranz Otsuka Ltd., encapsulated porcine islet cells were transplanted in patients affected by type 1 diabetes. Similar clinical trials of pig islet xenotransplantation were conducted in Buenos Aires. From the evidence provided by these trials "the balance of potential risk to potential benefit favours continuing with these endeavours, provided similar safety measures are employed": see more detailed information in ref. 4. Other authors state that "In New Zealand and Argentina, encapsulated neonatal porcine islets have been transplanted into nonimmunosuppressed T1DM patients but with minimal detectable function to date (. . .). Ongoing studies are required to validate such approaches, and the further application of CRISP-Cas9 to humanize the porcine genome will generate more promise, albeit additional ethical challenges": *see* refs. 5, 13, 35–37.
15. Socially valuable studies should collect data that are methodologically sound and address an important and unresolved scientific and clinical problem.
16. These guidelines apply to the procurement and screening of source animals, the clinical and preclinical testing of xenotransplantation products, and the post-xenotransplant monitoring recipients. See, for instance, Food and Drug Administration (FDA). Xenotransplantation Guidance Documents: FDA (2016) Source Animal, Product, Preclinical, and Clinical Issues Concerning the Use of Xenotransplantation Products in Humans; Guidance for Industry CBER; FDA (2001) PHS Guideline on Infectious Disease Issues in Xenotransplantation, January 19, 2001; (last visited 21 September 2018); European Medicines Agency (EMA, former EMEA). EMEA. Points to Consider Xenogeneic Cell Therapy Medicinal Products (2003). Available at: http://www.emea.europa.eu/docs/en_GB/document_library/Scientific_guideline/2009/09/WC500003893.pdf (last visited 21 September 2018); EMEA. Concept Paper on the Revision of the Points to Consider on Xenogeneic Cell Therapy Medicinal Products (2007). Available at: http://www.ema.europa.eu/docs/en_GB/document_library/Scientific_guideline/2009/09/WC500003811.pdf (last visited 21 September 2018); EMEA. Guideline on Xenogeneic Cell-based Medicinal Products (2009). Available at: http://www.emea.europa.eu/docs/en_GB/document_library/Scientific_guideline/2009/12/WC500016936.pdf (last visited 21 September 2018);

Organization for Economic Co-operation and Development (OECD). Xenotransplantation: International policy issues (1999). World Health Organization (WHO). Guidance on xenogeneic infection/disease surveillance and response: A strategy for international cooperation and coordination. Geneva (2001). Available at: http://whqlibdoc.who.int/hq/2001/WHO_CDS_CSR_EPH_2001.2.pdf (last visited 21 September 2018); WHO. Report of the 57th World Health Assembly. Human Organ and tissue transplantation (WHA 57.18) (2004). Geneva. Available at: http://apps.who.int/gb/ebwha/pdf_files/WHA57/A57_R18-en.pdf (last visited 21 September 2018); First WHO Global Consultation on Regulatory Requirements for Xenotransplantation Clinical Trials. Changsha, China, 19–21 November 2008. The Changsha Communiqué. Available at: http://www.who.int/transplantation/xeno/ChangshaCommunique.pdf; Second WHO Global Consultation on Regulatory Requirements for Xenotransplantation Clinical Trials October 17–19 2011, WHO, Geneva, Switzerland. Available at: http://www.who.int/transplantation/xeno/report2nd_global_consultation_xtx.pdf (last visited 21 September 2018).

17. "Although testing of the final xenotransplantation product for infectious agents is crucial, appropriate control of animal sources and husbandry provides important additional assurance for the safety of such products by controlling infections by both known and potentially even unknown agents. Therefore, the specific information supplied by the sponsor, regarding animal husbandry including housing, feeding, veterinary care, and drug and biologic treatment of source animal herds and individual source animals, will be crucial for FDA evaluation of the potential for safe use of cells, tissues, or organs from such source animals." See more detailed information with respect to source animal characterization in FDA (2016) Source Animal, Product, Preclinical, and Clinical Issues Concerning the Use of Xenotransplantation Products in Humans; Guidance for Industry, https://www.fda.gov/downloads/biologicsbloodvaccines/guidancecomplianceregulatoryinformation/guidances/xenotransplantation/ucm533036.pdf, pp. 7–21. This guideline describes the breeding and maintenance of donor animals in dedicated facilities with a so-called designated pathogen-free status. In addition to infectious pathogens that can affect the health status of the herd, animals should not contain microorganisms that have the potential of cross-species transmission, also called xenozoonosis. These pathogens are exogenous and can be removed from the donor herd by specific breeding technologies.

18. The regulatory requirements are much broader for cell-based medicinal products (CBMPs) than for organs, with respect to the manufacturing process. For further guidance concerning

characterization of xenotransplantation products and the manufacturing and process-related good manufacturing practice (GMP) considerations for harvest and processing of those products, see FDA (2016) Source Animal, Product, Preclinical, and Clinical Issues Concerning the Use of Xenotransplantation Products in Humans; Guidance for Industry, pp. 21–35. See also refs. 34, 38.

19. International Xenotransplantation Association has updated its original "Consensus Statement on Conditions for Undertaking Clinical Trials of Porcine Islet Products in Type 1 Diabetes." In this regard, Chapter 2a, previously Chapter 2, suggests several important revisions regarding the generation of suitable source pigs from the perspective of the prevention of xenozooneses. The newly added Chapter 2b discusses conditions for the use of genetically modified source pigs in clinical islet xenotransplantation (*see* refs. 39, 40. See also ref. 41 in which it is revisited the important topic of preclinical efficacy and safety data required to justify a clinical trial.

20. The World Health Organization (WHO). First WHO Global Consultation on Regulatory Requirements for Xenotransplantation Clinical Trials. The Changsha Communiqué. Changsha, China, 19–21 November 2008. It is essential that for any clinical xenograft trial, "the study design must be based on peer-reviewed objective evidence from primates studies that efficacy is likely relative to the best available alternative"; "In view of the possibility of transferring a porcine microorganism to the patient, and perhaps to the community at large, a clinical trial should only be undertaken if there is clear evidence of success in a pre-clinical non-human primate model": *see* ref. 41.

21. Generalized information on the kind of preclinical studies that should be done can be found in FDA (2016) Source Animal, Product, Preclinical, and Clinical Issues Concerning the Use of Xenotransplantation Products in Humans; Guidance for Industry. CBER. As stated in this guidance: "Preclinical studies are particularly valuable for gaining insight into safety issues which cannot be evaluated in human recipients for ethical or practical reasons. Consequently, you should design strong preclinical safety programs and also consult the FDA "Guidance for Industry: Preclinical Assessment of Investigational Cellular and Gene Therapy Products." November 2013; The European Medicines Agency's scientific guidelines on cell therapy and tissue engineering help medicine developers prepare marketing authorization applications for human medicines. https://www.ema.europa.eu/human-regulatory/research-development/scientific-guidelines/multidisciplinary/multidisciplinary-cell-therapy-tissue-engineering. In this regard, see, in particular, Guideline on Xenogeneic Cell-based Medicinal Products

(2009). Available at: http://www.emea.europa.eu/docs/en_GB/document_library/Scientific_guideline/2009/12/WC500016936.pdf. This document addresses the scientific requirements for medicinal products that contain viable animal cells or tissues as an active substance (xenogeneic cell-based medicinal products). It provides general principles for the development and assessment of xenogeneic cell-based products without prejudice to medical practice or national legislation, which may be applicable. See also guideline on safety and efficacy follow-up and risk management on Advanced Therapy Medicinal Products (25 January 2018) EMEA/149995/2008 rev. 1 https://www.ema.europa.eu/documents/scientific-guideline/draft-guideline-safety-efficacy-follow-risk-management-advanced-therapy-medicinal-products-revision_en.pdf; Guideline on the quality, non-clinical and clinical aspects of gene therapy medicinal product (22 March 2018) EMA/-CAT/80183/2014 https://www.ema.europa.eu/documents/scientific-guideline/guideline-quality-non-clinical-clinical-aspects-gene-therapy-medicinal-products_en.pdf (last visited 21 October 2018). European regulators have responded to the scientific advances in ATMPs by initiating the development of several new guidelines specific to these complex products. See Guidelines Relevant for Advanced Therapy Medicinal Products. EMA website. http://www.ema.europa.eu/ema/index.jsp?curl=pages/regulation/general/general_content_000298.jsp&mid=WC0b01ac05800862bd (last visited 21 October 2018).

22. In EU Committee for Advanced Therapies (CAT) was established by EMA to offer high-level expertise to assess the quality, safety, and efficacy of ATMPs. In the USA, regulatory product development considerations for ATMPs are quite similar in their basic concepts to those in the EU. However, there are different processes for clinical trial approval and green light to commence the study. Institutional Biosafety Committees (IBCs) and/or Institutional Review Boards (IRBs) are responsible for recommending protocols involving gene transfer to be submitted to the National Institutes for Health (NHI) for Recombinant DNA Advisory Committee (RAC) review and public consultation meeting. *See* ref. 42.

23. To see which criteria are considered in relation to the selection of appropriate participants for early-phase clinical trials of xenogeneic islet cell products in T1D, *see* ref. 43. Other papers have discussed the selection of patients for initial clinical trials for solid organ xenotransplantation; *see* ref. 44. *See also* ref. 33, in which potential conditions for which initial clinical trials of pig organ xenotransplantation may be justified are also described.

24. The terms "where deemed appropriate" means that there are circumstances where disclosure of personal information outweighs the benefits of individual privacy. Disclosure of patient's health status (for instance, concerning a specific communicable disease) to a third party may be justified in exceptional circumstances with compelling reasons, for example, partner pregnancy or unprotected sexual contact. If at all possible, the patient must be given the opportunity to consent to the disclosure first. Unless there is a clearly identified individual at risk, there is unlikely a duty to disclose confidential information. As stated by FDA (2016) Source Animal, Product, Preclinical, and Clinical Issues Concerning the Use of Xenotransplantation Products in Humans; Guidance for Industry, section X.J. "Intimate contacts of xenotransplantation product recipients include persons who have engaged repeatedly in activities that could result in intimate exchange of body fluids, including blood or saliva, with a xenotransplantation product recipient."
25. See EMEA (2009) Guideline on xenogeneic cell-based medicinal products. EMEA/CHMP/CPWP/83508/2009. The requirement for surveillance and monitoring is endorsed by the Council of Europe Recommendation (2003) 10 of the Committee of Ministers to Member States on Xenotransplantation (art. 13); FDA (2001) PHS on Infectious Disease Issues in Xenotransplantation; FDA (2016) Source Animal, Product, Preclinical, and Clinical Issues Concerning the Use of Xenotransplantation Products in Humans; Guidance for Industry, section X.G.
26. See EMA (2009) Guideline on xenogeneic cell-based medicinal products. EMEA/CHMP/CPWP/83508/2009; FDA (2001) PHS Guideline on infectious disease issues in xenotransplantation; FDA (2016) Source Animal, Product, Preclinical, and Clinical Issues Concerning the Use of Xenotransplantation Products in Humans; Guidance for Industry.
27. In addition, prospective xenotransplant recipients should be also informed that exclusion criteria for blood, tissues, ova, sperm and organ donation apply to them. Female recipients should be also advised not to breast feed. FDA (2016) Source Animal, Product, Preclinical, and Clinical Issues Concerning the Use of Xenotransplantation Products in Humans; Guidance for Industry; EMEA (2009) Guideline on xenogeneic cell-based medicinal products. EMEA/CHMP/CPWP/83508/2009.
28. See EMEA (2009) Guideline on xenogeneic cell-based medicinal products. EMEA/CHMP/CPWP/83508/2009. According to this guideline: "It may be, in rare instances, necessary to

collect blood samples of close contacts (e.g., family) prior to the procedure and to store them for retrospective testing." However, if such circumstances exist, then previous written consent must be obtained from them and after being properly informed. Coercive measures may be exercised by the national authorities on the basis of an evident hazard to public health.

29. See Regulation (EU) No 536/2014 of the European Parliament and of the Council of 16 April 2014 on clinical trials on medicinal products for human use and repealing Directive 2001/20/EC Text with EEA relevance. Under Chap. V "Protection of subjects and informed consent," Article 28.3 states that "*Any subject*, or, where the subject is not able to give informed consent, his or her legally designated representative, *may, without any resulting detriment and without having to provide any justification, withdraw from the clinical trial at any time by revoking his or her informed consent.* Without prejudice to Directive 95/46/EC, the withdrawal of the informed consent shall not affect the activities already carried out and the use of data obtained based on informed consent before its withdrawal" (cursive is own). See also Article 29 "Informed consent" whose paragraph 2 states that "Information given to the subject or, where the subject is not able to give informed consent, his or her legally designated representative for the purposes of obtaining his or her informed consent shall: (a) enable the subject or his or her legally designated representative to understand: (ii) *the subject's rights and guarantees regarding his or her protection, in particular his or her right to refuse to participate and the right to withdraw from the clinical trial at any time without any resulting detriment and without having to provide any justification*" (art. 29.2.a.ii) (cursive is own).

30. World Medical Association Declaration of Helsinki Ethical Principles for Medical Research Involving Human Subjects (2013). Principle 26 states that "In medical research involving human subjects capable of giving informed consent, each potential subject must be adequately informed of the aims, methods, sources of funding, any possible conflicts of interest, institutional affiliations of the researcher, the anticipated benefits and potential risks of the study and the discomfort it may entail, post-study provisions and any other relevant aspects of the study. *The potential subject must be informed of the right to refuse to participate in the study or to withdraw consent to participate at any time without reprisal.* Special attention should be given to the specific information needs of individual potential subjects as well as to the methods used to deliver the information" (cursive is own). See also Principle 31: "The physician must fully inform the patient which aspects of their

care are related to the research. The refusal of a patient to participate in a study or the patient's decision to withdraw from the study must never adversely affect the patient-physician relationship."

31. The Council of Europe considered the risks associated with xenotransplantation as unknown and unpredictable. In this respect, it was regarded necessary that some fundamental rights be waived. Council of Europe Recommendation Rec (2003) 10, of the Committee of Ministers to Member States on xenotransplantation, Article 21; European Commission Health & Consumer protection directorate-general *Opinion on the state of the art concerning xenotransplantation* adopted by the Scientific Committee on Medicinal Products and Medical Devices on 1st. October 2001, see paragraph 12, page 11: " Some of the measures that may need to be taken in a surveillance system may have legal implications as they could be in violation of the Declaration of Helsinki and other guidelines for research on human subjects. In the recent EC directive for conduct of clinical trials, the right of a subject to withdraw from a clinical trial is explicitly stated (EC 2001), and this could be a problem for prolonged surveillance in xenotransplantation clinical trials. As xenotransplantation has implications for public health, it may be that certain rights may have to be modified in such a way that surveillance can be continued. *Patients (and others?) could therefore have to agree to waive some of their human rights*" (cursive is own) Available at: http://ec.europa.eu/health/ph_risk/committees/scmp/documents/out38_en.pdf (last visited 21 October 2018). Some authors consider that "The potential but unquantified possibility of xenogeneic disease has resulted in a compelling argument for long-term, potentially life-long surveillance." From this perspective, it is argued that "the legal logic that undergird those laws (conventional public health laws) can also justify mandatory long-term surveillance of xenotransplant recipients." *See* ref. 45.

32. See Guideline on the risk-based approach according to annex I, part IV of Directive 2001/83/EC applied to advanced therapy medicinal products, 11 February 2013 EMA/CAT/CPWP/686637/2011, page 3. Available at: https://www.ema.europa.eu/documents/scientific-guideline/guideline-risk-based-approach-according-annex-i-part-iv-directive-2001/83/ec-applied-advanced-therapy-medicinal-products_en.pdf (last visited 23 October 2018).

33. Detailed guidelines on good clinical practice specific to advanced therapy medicinal products, Brussels, 03/12/2009. *See* section 11 paragraph 49. h). Available at https://ec.europa.eu/health/sites/health/files/files/eudralex/vol-10/2009_11_03_guideline.pdf (last visited July 2018).

34. Detailed guidelines on good clinical practice specific to advanced therapy medicinal products, Brussels, 03/12/2009. *See* section 8.2 paragraph 45.
35. Council of Europe Recommendation Rec (2003) 10, of the Committee of Ministers to Member States on xenotransplantation, Article 21.
36. Public health law may under some circumstances authorize quarantine of persons who have had high-risk infectious disease exposures or who are, in fact, infected. "Currently under the public health legislation contained within the Health Act 1956 (in New Zealand), the power to intervene when necessary regarding a communicable disease sits with the Medical Officer of Health. Regulatory powers exist to isolate individuals if required." The National Health Committee in New Zealand considers that the regulatory capacity currently provided within the abovementioned Health Act is sufficient to respond to an infectious disease. See NHC (2008) Advice on Living Cell Technologies Application for Xenotransplantation Clinical Trials in New Zealand. Available at: http://www.moh.govt.nz/notebook/nbbooks.nsf/0/8C460BAF8494AF90CC2579A5006C9CE6/$file/nhc-living-cell-technologies-report-oct082.pdf (last visited 20 October 2018).
37. See the Guide on Article 5 of the European Convention on Human Rights, "Right to liberty and security": https://www.echr.coe.int/Documents/Guide_Art_5_ENG.pdf (last visited 30 October 2018). From this viewpoint, it may be considered that xenotransplant recipients have to be isolated and quarantined in case that they show to be harboring a xenozoonotic infection, and this must be well justified by the circumstances of the specific case, that is, where public health protection so requires. Fundamental rights and freedoms should only be restricted to the extent necessary to most effectively reduce risks to public health. Public health laws should include procedural rights to protect the interests of individuals subject to compulsory measures. It is certainly difficult to imagine a less restrictive means to protect the public health than periodic testing of body fluids. However, in the absence of evident hazard or actual epidemic caused by xenogeneic infection, conventional public health law is insufficient to justify certain mandatory measures. Public health authorities must ensure that laws authorizing medical interventions without consent are never used to discriminate against vulnerable individuals. *See* refs. 18, 20, 46, 47.
38. As is stated by the World Health Organization (WHO). First WHO Global Consultation on…, op. cit., 2008, The Changsha Communiqué: "Investigators should select trial participants for

whom there is no adequately effective alternative therapy available and who understand the risks and consequences of the procedure, including the need for compliance with life-long follow up and who are motivated to modify their behavior accordingly." *See also* Second WHO Global Consultation on Regulatory Requirements for Xenotransplantation Clinical Trials. October 17–192,011, WHO, Geneva, Switzerland in which this statement is included. http://www.who.int/transplantation/xeno/report2nd_global_consultation_xtx.pdf See also FDA (2016) Source Animal, Product, Preclinical, and Clinical Issues Concerning the Use of Xenotransplantation Products in Humans; Guidance for Industry, section X.D: "(. . .) You should also consider the patient's ability to comply with public health measures as stated in the protocol, including long-term monitoring." *See also* ref. 43.

39. US Department of Health and Human Services Secretary's Advisory Committee on Xenotransplantation (SACX). Informed consent in clinical research involving xenotransplantation, Draft June 2004. Available at: http://www.med.nagoya-u.ac.jp/tx-immunology/xeno/ECE3E929-31FC-41FB-A7DB-E6B311A05C65_files/SACX%20informed%20consent.pdf (last visited 21 September 2018). *See* refs. 27, 30, 47, 48.
40. Canadian Public Health Association (CPHA) (2001), Animal-to-human transplantation: Should Canada proceed? A public consultation on Xenotransplantation. Available at: https://www.cpha.ca/sites/default/files/uploads/resources/xeno/execsum_e.pdf (last visited 20 September 2018).
41. NHMRC (National Health and Medical Research Council) Xenotransplantation Working Party XWP (2003) Animal-to-human transplantation research: How should Australia proceed? Response to the 2002 public consultation on Xenotransplantation on Draft Guidelines and Discussion Paper on Xenotransplantation. Public consultation 2003/4 NHMRC, Australian Government, Commonwealth of Australia.
42. *See* ref. 49, where the author compares and contrasts the Canadian and Australian responses on xenotransplantation to consider what lessons can be learned about the regulation of a complex and controversial biotechnology.
43. NHC (National Health Committee) National Health Committee's Advice on Living Cell Technologies Application for Xenotransplantation Clinical Trials in New Zealand, 2008. Available at: http://www.moh.govt.nz/notebook/nbbooks.nsf/0/8C460BAF8494AF90CC2579A5006C9CE6/$file/nhc-living-cell-technologies-report-oct082.pdf (last visited 25 September 2018). In this participatory model endorsed by

New Zealand, the Maori population was consulted about xenotransplantation as their different cultural values mattered in technoscientific innovation.

44. A targeted stakeholder consultation took place on the Regulation on ATMPs. See http://ec.europa.eu/health/human-use/advanced-therapies/developments/index_en.htm. As has been said, xenogeneic CBMPs are regulated by European regulatory agencies as ATMPs.

45. According to Michael M., two broad rhetorical categories of public are identified: Publics-in-General (PiGs) and Publics-in-Particular (PiPs). "The PiG can be regarded as an undifferentiated whole that is distinguished from science that is itself characterized globally in terms of some key dimension." "Publics-in-particular (PiP) can be broadly defined as those publics that have an identifiable stake in particular scientific or technological issues or controversies. "(...), PiPs can be associated with specific scientific projects, programs of research or technoscientific enterprises, are attached to recognizable "interests," and enact particular alliances with other actors." In general terms, this two-category distinction is based on organized and non-organized interests, respectively. *See* ref. 50. *See also* ref. 51 where the authors discern four major types of constructions of the public which they label as follows: the "general public" (constructed mainly through opinion polls which form a regular element in many consultation processes referring to issues of human genetics and other technoscientific issues); the "pure public" (described in terms of deliberative procedures utilized such as citizen juries, citizen conferences and particularly youth conferences); the "affected public" (e. g., patients); and the "partisan public" (it consists of organizations—not of individuals—who hold strong collective opinions on the issue or have a particular interest. Interviewees usually referred to such groups as stakeholders or lobbyists).

46. The engagement of the public is highlighted by the Council of Europe: "In accordance with the principles stated in Article 28 of the Convention on Human Rights and Biomedicine, member states should take active steps to ensure that the fundamental questions raised by xenotransplantation are the subject of appropriate public discussion particularly in light of relevant medical, psychological, cultural, ethical, legal, social and economic implications": Recommendation (2003) 10 of the Committee of Ministers to Member States on Xenotransplantation (Article 30).

47. As stated by Sobbrio P and Jorqui M: "Both xenotransplantation scientific communities and governmental institutions, (...), should be aware that the feasibility of an international

regulatory framework will require not just a rhetorical and shallow reference to the public(s), but a deep and shared sense of engagement and responsibility from people all over the world," *see* ref. 52.

48. WHO Guidance on Xenogeneic Infection/Disease Surveillance and Response: A Strategy for International Cooperation and Coordination, Geneva, Switzerland, 2001, Ref: WHO/CDS/CSR/EPH/2001.2. Available at: http://whqlibdoc.who.int/hq/2001/WHO_CDS_CSR_EPH_2001.2.pdf (Accessed September 2018).

49. Resolution WHA. 57.18. Human organ and tissue transplantation. In Fifty-seventh World Health Assembly, Geneva, 22 May 2004. Available at: http://www.who.int/ethics/en/A57_R18-en.pdf (Accessed July 2010).

50. Most of the participants were also members of the IXA. Five representatives of WHO participated as well as members of regulatory authorities of different countries, transplant physicians and surgeons, ethicists, and representatives of organ donation organizations. It is stated that effective regulation by individual governments is an essential condition to legitimately proceed with xenotransplantation. And it is added that "the regulatory system should be transparent, must include scientific and ethical assessment and should involve the public" (Principle 4). *See* ref. 26.

51. First WHO Global Consultation on Regulatory Requirements for Xenotransplantation Clinical Trials. Changsha, China, 19–21 November 2008. The Changsha Communiqué. Available at: http://www.who.int/transplantation/xeno/ChangshaCommunique.pdf (last visited 25 July 2018).

52. The FDA makes important efforts to face the challenge of regulating xenotransplantation while safeguarding public health, publishing guidance documents, collaborating with international organizations and other regulatory agencies, and promoting public engagement (*see* ref. 53).

53. In the US the FDA, for instance, issued in 2003 a Guidance for Industry covering the source of the animal, the product, and preclinical and clinical issues, updated December 2016: FDA (2016) Source Animal, Product, Preclinical, and Clinical Issues Concerning the Use of Xenotransplantation Products in Humans; Guidance for Industry. Other guidance documents, that have been mentioned earlier, are, for instance, the following ones: Guidance for Industry: Public Health Issues Posed by the Use of Non-Human Primate Xenografts in Humans (1999). Available at: http://www.fda.gov/downloads/BiologicsBloodVaccines/GuidanceComplianceRegulatoryInformation/Guidances/Xenotransplantation/ucm092866.pdf

(last visited 20 May 2018; Draft Guidance for Industry: Precautionary Measures to Reduce the Possible Risk of Transmission of Zoonoses by Blood and Blood Products from Xenotransplantation Product Recipients and Their Intimate Contacts (2002). Available at: http://www.fda.gov/downloads/BiologicsBloodVaccines/GuidanceComplianceRegulatoryInformation/Guidances/Blood/ucm080375.pdf (last visited 20 May 2018); PHS on Infectious Disease Issues in Xenotransplantation (2001). Available at: http://www.fda.gov/downloads/BiologicsBloodVaccines/GuidanceComplianceRegulatoryInformation/Guidances/Xenotransplantation/UCM092858.pdf last visited 20 May 2018).

54. EMEA (2009) Guideline on xenogeneic cell-based medicinal products. EMEA/CHMP/CPWP/83508/2009. Available at: http://www.emea.europa.eu/docs/en_GB/document_library/Scientific_guideline/2009/12/WC500016936.pdf (last visited 21 June 2018). Other guidance documents that have been published by this agency before this latter are the following ones: EMEA (2003) Points to consider on xenogeneic cell therapy medicinal products. CPMP/1199/02. Available at: http://www.emea.europa.eu/docs/en_GB/document_library/Scientific_guideline/2009/09/WC500003893.pdf (last visited 21 June 2018); EMEA (2007) Concept paper on the Revision of the Points to Consider on xenogeneic therapy medicinal products. EMEA/CHMP/165085/2007. Available at: http://www.ema.europa.eu/docs/en_GB/document_library/Scientific_guideline/2009/09/WC500003811.pdf (last visited 21 June 2018).

55. As has been argued: "An appropriate regulatory framework should be viewed of as an integrated multilevel system in which different normative sources may complement each other and in which different social components may be usefully activated" (*see* ref. 54).

56. Investigators and proposers of clinical trials using xenotransplantation products "must ensure that source animals are bred for the purpose and as safe as possible, using a closed colony of consistently known specific pathogen-free animals housed in a well-controlled pathogen-free environment with high levels of biosecurity" The Changsha Communiqué (2008).

57. "If a Member State has the capacity to regulate xenotransplantation and believes xenotransplantation should be carried out, it should ensure there is an effective national registry and regulatory process in place." In some countries, it is requested a central and independent register. For instance, in New Zealand: NHC (2008) Advice on Living Cell Technologies Application for Xenotransplantation Clinical Trials in

New Zealand. Available at: http://www.nhc.health.govt.nz/sites/www.nhc.health.govt.nz/files/documents/publications/nhc-living-cell-technologies-report-oct082.pdf at p. 16 (last visited 21 September 2018). In Australia, there is no central register, and long-term monitoring of patients rests with the corresponding institutions, such as hospitals, rather than a central agency. See NHMRC (2009) Discussion paper. Xenotransplantation: A review of the parameters, risks, and benefits. Available at: http://www.nhmrc.gov.au/_files_nhmrc/file/about/committees/expert/gtrap/nhmrc_xeno_discussion_paper_website.pdf at p. 22 (last visited 21 September 2018).

58. "Investigators must ensure storage of appropriate pre- and post-procedure specimens and maintain both the specimens and records in accordance with national regulatory guidelines." The Changsha Communiqué (2008). EMA states that the archiving strategy should be adequate for the intended use of these types of xenotransplantation products. Records should be kept for 30 years. EMEA. Guideline on Xenogeneic Cell-based Medicinal Products (2009). Available at: http://www.emea.europa.eu/docs/en_GB/document_library/Scientific_guideline/2009/12/WC500016936.pdf at p. 8 (last visited 25 September 2018). FDA recommends that all tissues and animal records should be archived for a minimum of 50 years. FDA (2016) Source Animal, Product, Preclinical, and Clinical Issues Concerning the Use of Xenotransplantation Products in Humans. Available at: https://www.fda.gov/downloads/biologicsbloodvaccines/guidancecomplianceregulatoryinformation/guidances/xenotransplantation/ucm533036.pdf (last visited 27 July 2018).

59. Each country should develop their own protective measures, as part of their own national regulatory framework. Any physician seeing a patient who has undergone a xenotransplantation procedure in another country should have the responsibility to report that procedure to their own national public health authorities, respecting the privacy and confidentiality regulations of their own country. Any physician performing a xenotransplantation procedure on a patient returning to another country should, in turn, report that procedure to their own national authority, which should report to the home country of the recipient. As recommended by SACX: "Public health authorities should maintain good communication with physicians and other health care providers who are likely to serve as the first line of defense against the spread of potential pathogens detected in xenotransplantation recipients" (*see* ref. 30).

60. NHMRC (2009) Discussion paper. Xenotransplantation: A review of the parameters, risks, and benefits. Available at: http://www.nhmrc.gov.au/_files_nhmrc/file/about/committees/expert/gtrap/nhmrc_xeno_discussion_paper_website.pdf at p. 20 (last visited 30 July 2018).
61. "The WHO is actively encouraging national health authorities to contribute information on the clinical xenotransplantation activities that are taking place in their countries. In parallel, reports from IXA members or any person aware of such practices may be a complementary source of information for this international inventory of xenotransplantation practices in humans" (*see* ref. 55).
62. "Member States should review their laws to determine whether they have adequate authority to regulate xenotransplantation, ban unregulated xenotransplantation and provide appropriate sanction for failure to comply": the Changsha Communiqué.
63. As is stated by the Changsha Communiqué "The regulatory system should be transparent, must include scientific and ethical assessment and should involve the public."
64. Adequate pathogen screening, animal selection, and the establishment of microbiological, genetic, and quality controls are increasing safety and efficacy of xenotransplantation clinical trials. "The risk for patients related to the potential transmission of porcine pathogens to humans is not zero, but it is no longer considered to be a reason to prohibit clinical trials. This phase transition in research and development has already been passed for islet transplantation (see the clinical trials at LCT and others mentioned above)," *see* ref. 36. A number of factors on which the first clinical trials of pig organ or cell transplantation will depend are described in the last pages of the paper included in ref. 33. According to some authors, the timing of the first clinical trials of pig organ or cell transplantation will depend on a number of factors. Some of these factors are described at the end of the paper included in ref. 33.
65. Despite the efforts of global standardization, there are really no effective mechanisms of global regulation for managing global threats, "let alone to enforce regulations over the sovereignty of scientific decision-making within the nation-state" (*see* ref. 24).

Acknowledgments

Financial Support and Sponsorship: This work has been supported by the research project DER2016-80138-R "Disability, Chronic Disease and Accessibility to Rights," 2017–2019.

References

1. Tallacchini M (2017) Technologies and EU law: the evolution of regulatory approaches and governance. In: Cremona M (ed) New technologies and EU law. Oxford University Press, Oxford, pp 9–37
2. Liu Z, Hu W, He T et al (2017) Pig-to-primate islet xenotransplantation: past, present, and future. Cell Transplant 26(6):925–947
3. Samy KP, Martin B, Turgeon NA, Kirk AD (2014) Islet cell xenotransplantation: a serious look towards the clinic. Xenotransplantation 21(3):221–229
4. Cozzi E, Tönjes RR, Gianello P et al (2016) First update of the International Xenotransplantation Association consensus statement on conditions for undertaking clinical trials of porcine islet products in type 1 diabetes—Chapter 1: update on national regulatory frameworks pertinent to clinical islet xenotransplantation. Xenotransplantation 23 (1):14–24
5. Denner J, Tönjes RR, Takeuchi Y et al (2016) First update of the International Xenotransplantation Association consensus statement on conditions for undertaking clinical trials of porcine islet products in type 1 diabetes—Chapter 5: recipient monitoring and response plan for preventing disease transmission. Xenotransplantation 23(1):53–59
6. Cowan PJ, Tector AJ (2017) The resurgence of xenotransplantation. Am J Transplant 17:2531–2536
7. Schuurman H-J (2015) Regulatory aspects of clinical xenotransplantation. Int J Surg 23:312–321
8. Cooper DKC, Pierson RN 3rd, Hering BJ et al (2017) Regulation of clinical xenotransplantation—time for a reappraisal. Transplantation 101(8):1766–1769
9. Cooper DKC (2015) The case for xenotransplantation. Clin Transpl 29(4):288–293
10. Servick K (2017) Xenotransplant advances may prompt human trials. Science 357(6358):1338
11. Bruni A, Gala-Lopez B, Pepper AR, Abualhassan NS, Shapiro AJ (2014) Islet cell transplantation for the treatment of type 1 diabetes: recent advances and future challenges. Diabetes Metab Syndr Obes 23(7):211–223
12. Mourad NI, Gianello PR (2017) Xenoislets: porcine pancreatic islets for the treatment of type I diabetes. Curr Opin Organ Transplant 22(6):529–534
13. Pepper AR, Bruni A, Shapiro AMJ (2018) Clinical islet transplantation: is the future finally now? Curr Opin Organ Transplant 23 (4):428–439
14. Pierson RN III (2009) Current status of xenotransplantation. JAMA 301(9):967–969
15. Sykes M, d'Apice A, Sandrin M (2003) Position paper of the Ethics Committee of the International Xenotransplantation Association. Xenotransplantation 10(3):194–203
16. Sykes M, Sandrin M, Cozzi E (2004) International cooperation on xenotransplantation. Nat Med 10(2):119
17. McLean SAM, Williamson L (2005) Xenotransplantation: law and ethics. Ashgate, Farnham, p 216
18. Rothblatt M (2004) Your Life or Mine: how geoethics can resolve the conflict between public and private interests in xenotransplantation. Ashgate, Aldershot
19. Ravelingien A (2007) Xenotransplantation and the harm principle: factoring out foreseen risk. J Evol Technol 16(1):127–149
20. Florencio PS, Ramanathan ED (2004) Legal enforcement of xenotransplantation public health safeguards. J Law Med Ethics 32:117–123
21. Bach FH, Ivinson A (2002) A shrewd and ethical approach to xenotransplantation. Trends Biotechnol 20(3):129–131
22. Clark MA (1999) This little piggy went to market: the xenotransplantation and xenozoonosis debate. J Law Med Ethics 27:137–152
23. Hall CM (2011) Biosecurity, tourism and mobility: institutional arrangements for managing tourism-related biological invasions. J Policy Res Tour Leisure Events 3(3):256–280
24. Cook PS, Kendall G, Michael M, Brown N (2011) The textures of globalization: biopolitics and the closure of xenotourism. New Genet Soc 30(1):101–114
25. Sykes M (2005) Commentary: World Health Assembly resolution 57.18 on xenotransplantation. Transplantation 79(6):636–637
26. Cooper DK (2009) Global consultation on regulatory requirements for xenotransplantation clinical trials. Invited commentary. Xenotransplantation 16(2):58–60
27. Gold ER, Adams WA (2002) Reconciling private benefit and public risk in biotechnology: xenotransplantation as a case study in consent. Health Law J 10:31–75
28. Sparrow R (2009) Xenotransplantation, consent and international justice. Dev World Bioeth 9(3):119–127

29. Cook PS (2007) Informed consent and human rights: some regulatory challenges of xenotransplantation. Social Alternatives 26 (4):29–34
30. Cooper DKC (2005) Draft reports for public comment from the U.S. Secretary's Advisory Committee on Xenotransplantation. Xenotransplantation 12:255–257
31. Denner J (2018) Why was PERV not transmitted during preclinical and clinical xenotransplantation trials and after inoculation of animals? Retrovirology 15(28):1–9
32. Yang L, Güell M, Niu D et al (2015) Genome-wide inactivation of porcine endogenous retroviruses (PERVs). Science 350 (6264):1101–1104
33. Cooper DKC, Gaston R, Eckhoff D et al (2017) Xenotransplantation—the current status and prospects. Br Med Bull 125(1):5–14
34. Arcidiacono JA, Evdokimov E, Lee MH et al (2010) Regulation of xenogeneic porcine pancreatic islets. Xenotransplantation 17 (5):329–337
35. Wynyard S, Nathu D, Garkavenko O et al (2014) Microbiological safety of the first clinical pig islet xenotransplantation trial in New Zealand. Xenotransplantation 21 (4):309–323
36. Yung GLP, Rieben R, Bühler L et al (2017) Xenotransplantation: where do we stand in 2016? Swiss Med Wkly 147:w14403
37. Cooper DKC, Matsumoto S, Abalovich A et al (2016) Progress in clinical encapsulated islet xenotransplantation. Transplantation 100 (11):2301–2308
38. Rayat GR, Gazda LS, Hawthorne WJ et al (2016) First update of the International Xenotransplantation Association consensus statement on conditions for undertaking clinical trials of porcine islet products in type 1 diabetes—Chapter 3: Porcine islet product manufacturing and release testing criteria. Xenotransplantation 23(1):38–45
39. Spizzo T, Denner J, Gazda L et al (2016) First update of the International Xenotransplantation Association consensus statement on conditions for undertaking clinical trials of porcine islet products in type 1 diabetes—Chapter 2a: source pigs— preventing xenozoonoses. Xenotransplantation 23(1):25–31
40. Cowan PJ, Ayares D, Wolf E, Cooper DK (2016) First update of the International Xenotransplantation Association consensus statement on conditions for undertaking clinical trials of porcine islet products in type 1 diabetes—Chapter 2b: genetically modified source pigs. Xenotransplantation 23(1):32–37
41. Cooper DKC, Bottino R, Gianello P et al (2016) First update of the International Xenotransplantation Association consensus statement on conditions for undertaking clinical trials of porcine islet products in type 1 diabetes—Chapter 4: pre-clinical efficacy and complication data required to justify a clinical trial. Xenotransplantation 23(1):46–52
42. Hurley P, Jurmeister S, Parsley K (2016) Advanced therapy medicinal products—an evolving regulatory landscape. file:///C:/Users/maria.jorqui.DYC/Downloads/Regulatory-Focus-201702%20(4).pdf (last visited 21 September 2018). This article was developed from a presentation at RAPS Regulatory Convergence in September 2016 in San Jose, California, US. It was the opening presentation in the "Global Developments in Cell and Gene Therapy" session
43. Hering BJ, O'Connell PJ (2016) First update of the International Xenotransplantation Association consensus statement on conditions for undertaking clinical trials porcine islet products in type 1 diabetes—Chapter 6: patient selection for pilot clinical trials of islet xenotransplantation. Xenotransplantation 23(1):60–76
44. Cooper DKC, Wijkstrom M, Hariharan S et al (2017) Selection of patients for initial clinical trials of solid organ xenotransplantation. Transplantation 101:1551–1558
45. Spillman MA, Sade RM (2007) Clinical trials of xenotransplantation: waiver of the right to withdraw from a clinical trial should be required. J Law Med Ethics 35:265–272:270
46. Selgelid MJ (2009) Pandethics. Public Health 123(3):255–259
47. Barker JH, Polcrack L (2001) Respect for persons, informed consent and the assessment of infectious disease risks in xenotransplantation. Med Health Care Philos 4(1):53–70
48. Bowman DM (2004) Bioethical and legal perspectives on xenotransplantation. Monash Bioeth Rev 23(3):16–29
49. Cheng M (2015) Islet Xeno/transplantation and the risk of contagion: local responses from Canada and Australia to an emerging global technoscience. Life Sci Soc Policy 11:12
50. Michael M (2009) Publics performing publics: of PiGs, PiPs and politics. Public Understand Sci 18(5):617–631
51. Braun K, Schultz S (2010) '... a certain amount of engineering involved': constructing the public in participatory governance arrangements. Public Understand Sci 19(4):403–419
52. Sobbrio P, Jorqui M (2014) An overview of the role of society and risk in xenotransplantation. Xenotransplantation 21:523–532

53. Bloom E (2007) National policies for xenotransplantation in USA. Xenotransplantation 14:345–346
54. Tallacchini M (2008) Defining an appropriate ethical, social and regulatory framework for clinical xenotransplantation. Curr Opin Organ Transplant 13(2):159–164
55. Buhler L, Sgroi A, Sykes M, Noel L (2006) International inventory of xenotransplantation practices in humans. Xenotransplantation 13 (6):482

INDEX

Cristina Costa (ed.), *Xenotransplantation: Methods and Protocols*, Methods in Molecular Biology, vol. 2110,
https://doi.org/10.1007/978-1-0716-0255-3, © Springer Science+Business Media, LLC, part of Springer Nature 2020

Made in United States
North Haven, CT
31 May 2022

19706516R00210